MARINE ECOSYSTEMS

Edited by **Antonio Cruzado**

Marine Ecosystems
Edited by Antonio Cruzado

Published by InTech
Janeza Trdine 9, 51000 Rijeka, Croatia

Publishing Process Manager Daria Nahtigal

Technical Editor Teodora Smiljanic

Cover Designer InTech Design Team

Additional hard copies can be obtained from orders@intechopen.com

Marine Ecosystems, Edited by Antonio Cruzado
p. cm.
ISBN 978-953-51-0176-5

For EU product safety concerns:
IN TECH d.o.o., Prolaz Marije Krucifikse Kozulić 3, 51000 Rijeka, Croatia,
info@intechopen.com or visit our website at intechopen.com.

Meet the editor

Antonio Cruzado PhD, was educated at the University of Barcelona (Spain) and at the University of Washington (USA). Trained as a chemist (MSc), biologist (PhD) and oceanographer, he is currently the Scientific Director of Oceans at Catalonia International (Blanes, Spain). He also holds academic appointments as Visiting Professor of Marine Biogeochemistry at the University of Barcelona, and Professor of Chemical Oceanography at the Polytechnical University of Catalonia. Antonio has authored more then 100 chapters and articles published on peer-review and other journals and was the supervisor of 12 PhD theses. For more than 40 years he was PI of numerous oceanographic projects at the Consejo Superior de Investigaciones Científicas (Spain), Senior Scientist in charge of UNEP's Mediterranean Pollution Monitoring and Research Programme (MedPol), as well as Consultant to UNESCO, IOC, WHO/PAHO, World Bank and other international and national organizations. Dr Cruzado is a member of The Oceanographic Society (Charter Life), and of the American Geophysical Union.

Contents

Preface

Marine ecosystems is as wide a topic as one can possibly think including so many different processes, groups of organisms and geographical peculiarities. A book like this, not a text book, forcedly has to be a collection of unrelated topics aiming at presenting research carried out by different groups of scientists in different parts of the world, the only common element being the search for excellence of the different authors that we encouraged to express in their contributions.

Chapter 1. *Threats to ultraoligotrophic marine ecosystems* summarily describes the extremely oligotrophic ecosystems existing in the subtropical gyres of the oceans as well as in the eastern Mediterranean and northern Red Sea. It also addresses some of the problems that these, like other marine ecosystems, experience by the pressure exerted by human activities. Chlorophyll a concentration is used to define the various classes of marine ecosystems. With regard to the threats endangering these systems, some of them are common to other marine ecosystems.

Chapter 2. *Modelling the pelagic ecosystem dynamics: the NW Mediterranean.* This chapter presents the progress made in coupling hydrodynamic and biogeochemical models and their application to the understanding of processes at the basin- meso- and small-scale, particularly applied to the NW Mediterranean Sea with special emphasis on the area around the Blanes submarine canyon, where an operational oceanographic observatory is maintained and were there are frequent observations both by automatic and discrete sampling/measurement.

Chapter 3. *The marine ecosystem of the Sub-antarctic, Prince Edward Islands.* The Prince Edward Islands are, with the Crozet Island Group and the Kerguelen, further to the east, the only emerging land existing in the Indian Ocean sector of the Southern Ocean. More than 2000 km from South Africa and almost half way to the Antarctic mainland, the two islands, Marion (the largest) and Prince Edward (only about 25 km^2), are volcanic outcrops seasonally visited by large populations of seabirds, penguins and seals, especially during the breeding season. This chapter is a good synthesis of the physical and ecological conditions generally found in the Prince Edward Islands.

Chapter 4. *Meiofauna as a tool for marine ecosystem biomonitoring* is a thorough review of the effects of different disturbances, mainly caused by anthropogenic pollution, on the well being of individuals and communities of two meiofaunal groups: Foraminifera and Nematoda. The possible effects range from changes in species abundance and diversity to malformations to lethal conditions when some taxa may be completely absent while other may survive to significantly altered environments.

Chapter 5. *Chemical interactions in Antarctic marine benthic ecosystems* is a succinct description of experiments carried out in the laboratory to address the issue of chemical biotic interactions among common benthic organisms from various sites of the Antarctic seas. Benthic organisms compete for space and food while avoiding predation by other organisms. Many of these organisms recur to chemical substances to hurt competitors for space occupation or to avoid being eaten. Behind the ecological interest lies the interest expressed by pharmacological research to find substances and/or active principles allowing their potential use as medicines for human diseases.

Chapter 6. *An Interdisciplinary Approach on Erosion Mitigation for Coral Reef Protection- A Case Study from the Eastern Caribbean* combines a study of the coastal biogeography of the south-east section of the Isla Culebra (eastern Puerto Rico archipelago) carried out from aerial imagery produced by NASA-NOS Biogeography Project and remote video surveys at a few sites of coral reef communities, with a model assessment of sediment loads produced by mostly unpaved roads inland. A cost-effectiveness study is included to assess the convenience of using road paving or other strategies to reduce the sediment loads in order to achieve Preservation, Prevention or Remediation of the coastal habitats.

Chapter 7. *A revisit to the evolution and ecophysiology of the Labyrinthulomycetes* is an extremely interesting presentation of a group of organisms, the *Labyrinthulomycetes*, half way between some photosynthetic microalgae and the fungi. Although they cannot photosynthesize, they keep in their morphology remnants of ancient chloroplasts. The chapter constitutes an excellent review of the taxonomy and filogenetics of this group of organisms.

Chapter 8. *Seabed mapping and marine spatial planning: a case-study from a Swedish marine protected area* is an excellent review of the various approaches to map the bottom fauna both sessile and free moving. The authors claim that the mapping technique may be useful to resolve the problem of conflicting uses such as the protection of a marine park with traditional fishing and other leisure-related activities. The bottom-up technique of *in situ* species inventorying and the top-down *remote* observation of communities are two basic techniques often employed to carry out the mapping.

Chapter 9. *Management strategies to limit the impact of bottom trawling on VMEs in the High Seas of the SW Atlantic* is a detailed account of the morphology, sedimentology, bottom fauna and pollution of the deeper shelf and slope off SW Argentina, corresponding to the Patagonian Sea. The main objective of the paper is to identify areas in which the bottom communities may be classified as vulnerable marine ecosystems (VME) in order to limit high seas fisheries activities that may endanger such communities.

Chapter 10. *Hydrocarbon contamination and the swimming behavior of the estuarine copepod Eurytemora affinis* presents a quantitative assessment of changes in copepod swimming behavior as a consequence of hydrocarbon contamination. Behavioral changes are used as important indicators for ecosystem health. While they are driven by biochemical processes, they also reflect the fitness of the individual organism as well as potential consequences at the population level, such as altered abundance of a species in an ecosystem.

Chapter 11. *Interactions between marine ecosystems and tourism on the Adriatic and Mediterranean* points out the situation of the touristic areas around the Mediterranean (with special focus on the Adriatic) and elsewhere (Canary Islands). It shows the deep knowledge the authors have of all the coastal and island areas and the great diversity of situations found in terms of urban development and environmental respect and/or protection with particular focus on effects on resources (mainly water) and quality of coastal areas.

The objective of this book is to present various topics of great importance for understanding the marine ecosystems, what they are, how they work and how we can model them in order to forecast their behaviour under changing conditions. The chapters presented have been thoroughly reviewed and accepted for their publication.

We would like to express our thanks to the contributing authors who are the key factor in this achievement. The Editor expresses his acknowledgement to InTech for providing the excellent technical setup that made the work very pleasant and efficient.

Prof. Antonio Cruzado, M.Sc, Ph.D.
Oceans Catalonia International SL, Blanes, Catalonia,
Spain

1

Threats to Ultraoligotrophic Marine Ecosystems

Demetris Kletou and Jason M. Hall-Spencer
University of Plymouth,
United Kingdom

Remote marine areas with low productivity are amongst the least explored and understood ecosystems of the biosphere

1. Introduction

Marine phytoplankton account for ~1% of the world's photosynthetic biomass but for nearly half of the world's primary production (Field et al., 1998; Bryant, 2003). Water bodies are often classified on the basis of surface chlorophyll a concentrations, the photosynthetic pigment that is present in all primary producers (Table 1).

Water body class	Chl. a (mg m^{-3})
Ultraoligotrophic	<0.06
Oligotrophic	0.06-0.1
Mesotrophic	0.1-0.3
Eutrophic	0.3-1
Hypertrophic	>1

Table 1. Classification scheme based on chlorophyll a concentrations proposed by Shushkina et al. (1997).

Data from the Sea-viewing Wide Field-of-view Sensor (SeaWiFS) show that ultraoligotrophic marine areas occur within subtropical gyres at mid-latitudes and cover about 16-28% of the Earth's surface (Fig. 1) (McClain et al., 2004). Despite their low productivity, subtropical gyres account for 30-50% of global oceanic primary productivity (Karl et al., 1996). The subtropical gyres of the North Pacific, North Atlantic, South Pacific, South Atlantic and South Indian Ocean are ultraoligotrophic year-round with the lowest productivity found in the South Pacific gyre near Easter Island (Morel et al., 2010). Periods of ultraoligotrophy also occur in the Eastern Mediterranean and the North Red Sea, particularly during summer (Labiosa et al., 2003; Siokou-Frangou et al., 2010). In this chapter we compare ultraoligotrophic areas and describe the main threats to these systems.

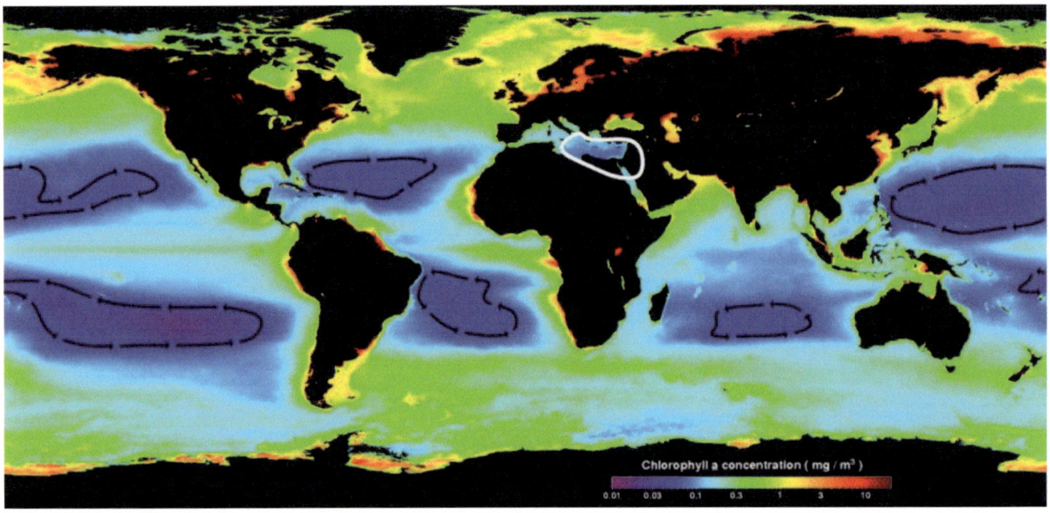

Fig. 1. World map of surface ocean chlorophyll-a concentration. Areas within black arrows and the white line indicate ultraoligotrophic (<0.06 mg m^{-3}) open ocean and enclosed systems, respectively. Image from SeaWiFS Project NASA/GSFC and GeoEYE, data 1997-2010.

2. Ultraoligotrophic ecosystems

Satellite data underestimate phytoplankton productivity in ultraoligotrophic waters since light penetrates deep into the highly transparent waters, with distinct phytoplankton communities found at different depths and a peak in chlorophyll concentrations as deep as 150 m (Morel et al., 2007; Malmstrom et al., 2010). Phytoplankton sizes range from picoplankton (0.2-2µm in diameter), through nanoplankton (2-20µm in diameter) to microplankton (>20-200µm). Nutrient rich conditions favour microplankton (e.g. diatoms and dinoflagellates), which are large enough to be eaten by copepods and krill, that in turn are consumed by zooplanktivorous fish. These short, simple food webs have efficient energy transfer to larger consumers (Sommer et al., 2002). In ultraoligotrophic waters, picoplankton (Fig. 2) seem better able to acquire nutrients than large phytoplankton as they have a higher surface area to volume ratio (Raven, 1998). Picoplankton are too small to be ingested by copepods and instead are eaten by microplanktonic protists which then feed mesozooplankton (Christaki et al., 2002; Calbet, 2008) or they form aggregates that can then be consumed by crustacean and gelatinous zooplankton (Lomas & Moran, 2011). The complex food webs that occur in ultraoligotrophic waters result in less efficient energy transfer to higher trophic levels. Ephemeral phytoplankton blooms can occur in ultraoligotrophic areas and during these events herbivorous plankton proliferate rapidly thanks to short generation times (Eden et al., 2009). During blooms myriads of vertically migrating grazers such as copepods, euphausiids and gelatinous zooplankton feed higher trophic groups such as squid, fish and other vertebrates (Seki & Polovina, 2001).

Low phosphorous (P) and nitrogen (N) concentrations normally limit primary production in ultraoligotrophic systems. A spring peak in Chl. a concentrations usually occurs when longer days allow phytoplankton to thrive due to the greater nutrient availability that follows winter mixing (Morel et al., 2010). Competition for P may have shaped the evolution of marine microbes; the dominance of picocyanobacteria genera *Prochlorococcus* and *Synechococcus* in low P environments is thought to be due in part to their ability to form lipid membranes that require less P than most other organisms (Van Mooy et al., 2006; Dyhrman et al., 2009). Picocyanobacteria and picoeukaryotes carry genes encoding for enzymes like alkaline phosphatase (AP) that hydrolyze dissolved organic phosphorous (DOP) and *PstS* genes which are related to the high-affinity uptake of phosphate (Moore et al., 2005; Martiny et al., 2009). Many plankton are able to fix N_2, although this ability can be limited by a lack of trace elements such as iron (Tyrrell, 1999; Kustka et al., 2003). The ability to fix N_2 should be ecologically advantageous in ultraoligotrophic environments where the most abundant

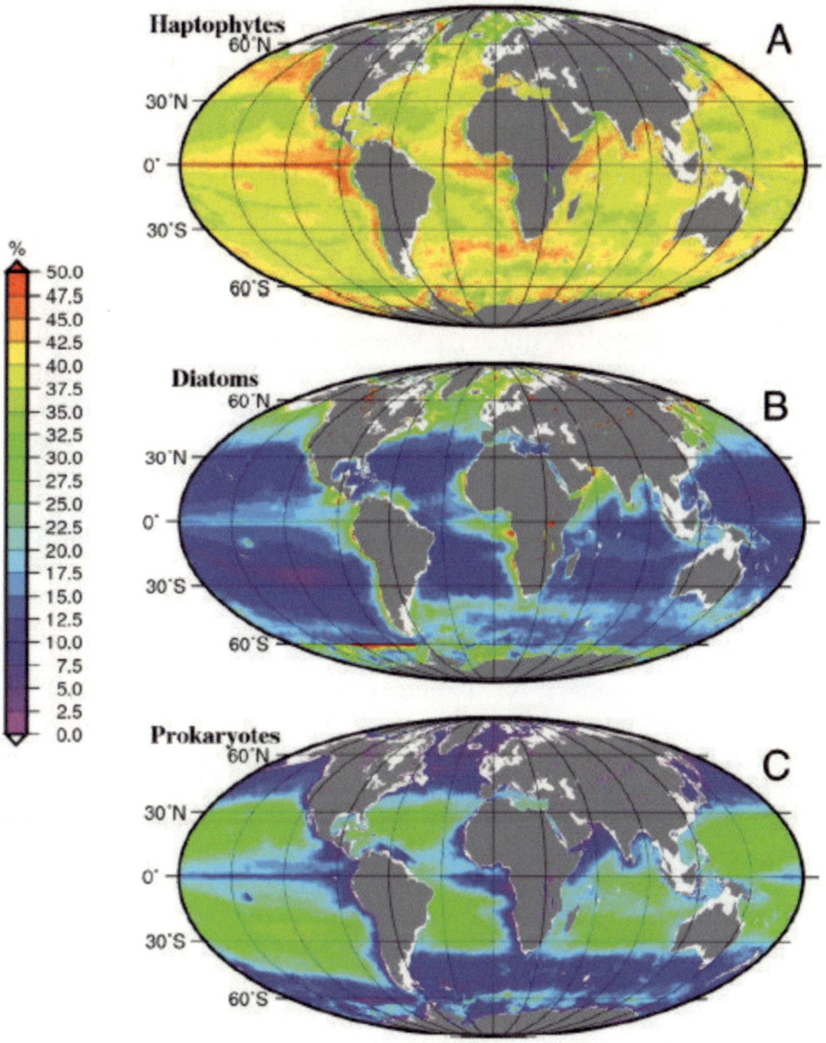

Fig. 2. Accessory pigments based on relative contribution of (A) haptophytes, (B) diatoms and (C) photosynthetic prokaryotes to total Chl. a biomass in the euphotic layer for the year 2000. Image from Liu et al. (2009).

forms of N are dissolved N_2 gas and dissolved organic nitrogen (DON). In ultraoligotrophic surface waters N_2 fixing bacteria typically have much lower abundances than non-N_2 fixing cyanobacteria and picoeukaryotes but N_2 fixation increases in importance with depth (Dekas et al., 2009). N_2 fixing cyanobacteria, such as *Trichodesmium* spp., occur in many warm, calm and oligotrophic waters (Capone et al., 1997) and are a seasonal and episodic phenomenon in ultraoligotrophic waters. So far, research efforts have focused on colonial *Trichodesmium* spp. but free trichomes, which seem more important in oligotrophic systems, have received little attention (Taboada et al., 2010). Primary production in ultraoligotrophic areas is usually dominated by unicellular N_2 fixing bacteria (e.g. *Crocosphaera* and UCYN clades), non- N_2 fixing picocyanobacteria (e.g. *Prochlorococcus* and *Synechococcus* spp.) and small eukaryotes (e.g. haptophytes) (Malmstrom et al., 2010; Moisander et al., 2010). Surface-ocean microbial growth is mostly supported by regenerated production, such as DON (e.g. urea) and ammonia oxidation by nitrification which occurs in bacteria and archaea (Zehr & Kudela, 2011).

Even though photosynthetic picoplankton are dominated numerically by *Prochlorococcus* and *Synechococcus*, much of the carbon is fixed by photosynthetic picoeukaryotes such as the exceptionally diverse haptophytes (Grob et al., 2011). Picohaptophytes are thought to contribute 30-50% of the total photosynthetic standing stock across the world ocean with their competitive success attributed to their mixed mode of nutrition as some are able to photosynthesize as well as engulf bacteria (Liu et al., 2009). Recent applications of molecular techniques demonstrate high diversity in the microbial parts of the food web (DeLong, 2009) and a previously unimagined diversity of eukaryotes (Massana & Pedrós-Alió, 2008). Even though the phytoplankton abundance is lowest in oligotrophic waters, the diversity of small-sized phytoplankton seems to peak in these areas (Cermeño & Figueiras, 2008; Kirkham et al., 2011). How such a diversity of plankton can coexist on limited resources is intriguing and was dubbed the 'paradox of plankton' by Hutchinson (1961). Explanations range from prolonged coexistence and niche segregation to mesoscale turbulence of the ocean (Roy & Chattopadhyay, 2007; Perruche et al., 2010).

2.1 Open ocean systems – Subtropical gyres

Data from monitoring stations off Bermuda and Hawaii are revolutionizing our understanding of mid-latitude gyre dynamics. Once thought of as homogeneous unchanging ocean desserts, we now know that these ultraoligotrophic ecosystems are both physically and biologically dynamic. The gyres circulate clockwise in the northern hemisphere and anticlockwise in the southern hemisphere due to the Coriolis effect. Ekman pumping (water moving to the right of the wind) and geostrophic flow cause downwelling of relatively warm surface waters at the subtropical convergence near 20° - 30° latitude (Pedlosky, 1998). The gyres have deep pycnoclines and even deeper nutriclines (e.g. nitrate, phosphate, and silicate) (McClain et al., 2004) and expand in area in summer. In most gyres Chl. a concentrations peak in spring following mixing in winter, while in the North Atlantic a secondary peak occurs at the end of September; in the North Pacific Chl. a concentration is higher during stratified conditions in the summer (Morel et al., 2010). Episodic blooms are also detected in all the gyres during stratified periods (Wilson & Qiu, 2008). The ultraoligotrophic gyres are each ecologically distinctive, as illustrated by the differences in their primary producers.

2.1.1 North Atlantic

The Sargasso Sea is probably the most studied open ocean system in the world (Steinberg et al., 2001). This subtropical gyre receives iron-rich Saharan dust (Marañón et al., 2010) but has extremely low P concentrations, possibly as a result of iron enhanced N_2 fixation (Wu et al., 2000). In January-April waves deepen the mixed layer and bring nutrients into the euphotic zone. Subsequent stratification retains nutrients in the surface waters, promoting N_2 fixation, primary production and blooms of phytoplankton such as *Trichodesmium* spp. (Taboada et al., 2010). As summer progresses the uptake of P by prokaryotes causes P limitation, although DOP is also utilised and can support ~25-30% of annual primary production (Mather et al., 2008; Lomas et al., 2010). In summer a distinct shallow-water microbial community develops in the region of lowest nutrients, with a deep chlorophyll maximum community and an upper mesopelagic community (Treusch et al., 2009). Bacteria seem to be more concentrated in the surface waters while Archaea (e.g. *Crenarchaeota*) seem better adapted in the mesopelagic layer (Schattenhofer et al., 2009). Picoplankton (*Prochlorococcus* and *Synechococcus* spp. and picoeukaryotes) dominate carbon fixation in the subsurface chlorophyll maximum, while in surface waters the nanoplankton (e.g. some haptophytes, pelagophytes, small diatoms and dinoflagellates) make significant contributions to productivity (Poulton et al., 2006). *Prochlorococcus* is twice as abundant in the deep chlorophyll zone than at the surface, but is almost absent below 200 m (Schattenhofer et al., 2009; Riemann et al., 2011). *Prochlorococcus* clades have a succession of blooms as each responds differently to seasonal changes in light, temperature and mixing. *Prochlorococcus* peak in abundance during late summer and autumn whilst *Synechococcus* is scarce then but can occasionally become more abundant than *Prochlorococcus* during winter when the Sargasso Sea is more deeply mixed (Malmstrom et al., 2010).

Even though photosynthetic picoeukaryotes are less abundant than picocyanobacteria, they cause the observed variations in Chl. a and peak in abundance during winter/spring (Riemann et al., 2011). They are extremely diverse and dominated by haptophytes and chrysophytes, neither of which was traditionally considered to be important in carbon fixation (Kirkham et al., 2011). Rates of carbon fixation are comparable to those in the South Atlantic subtropical gyre and peak during the spring blooms (Poulton et al., 2006). The North Atlantic gyre appears to be net heterotrophic in autumn and balanced in spring (Gist et al., 2009) although it may be net autotrophic annually (Kähler et al., 2010). Despite being ultraoligotrophic, the Sargasso Sea is the spawning site of Atlantic eels. We now know that the picoplankton and nanoplankton make significant contributions to carbon export into deeper zones via settling of aggregates and/or consumption of those aggregates by mesozooplankton (Lomas & Moran, 2011). In turn, mesozooplankton (e.g. heterotrophic athecate dinoflagellates and ciliates) feed copepods which may in turn be available to organisms at higher trophic levels, such as the larvae of Atlantic eels (Andersen et al., 2011).

2.1.2 South Atlantic

Much of our knowledge for the South Atlantic low nutrient gyre comes from the Atlantic Meridional Transect programme which has been undertaken semi-annually since 1995 along a 13,500km transect between 50°N and 52°S (Robinson et al., 2009). Nutrient concentrations are lower than in the North Atlantic gyre, yet the southern system appears to be more autotrophic (Gist et al., 2009). NO_3^- concentrations are below detection limits, and iron concentrations are also very low, but soluble reactive P is almost an order of magnitude higher than in the North Atlantic gyre. As P is a more bioavailable nutrient source than DOP reduced APA is detected which results in accumulation of DOP (Mather et al., 2008). The microbes seem to be adapted to higher organic loading and utilize organic inputs more efficiently than heterotrophic bacteria of the North Atlantic gyre (Martinez-Garcia et al., 2010). Unlike the North Atlantic gyre, N_2 fixation is very low and is possibly limited by iron (Moore et al., 2009).

Prochlorococcus is more abundant than in the North Atlantic gyre although its contribution in the mesopelagic zone is minimal (Schattenhofer et al., 2009). SAR11 heterotrophs occur at lower abundances than in the North Atlantic gyre, but still make up about 25% of all picoplankton cells (Mary et al., 2006). Larger picoprokaryotes are found in the South Atlantic gyre compared to the North Atlantic (Schattenhofer et al., 2009). Small photosynthetic picoeukaryotes of a size <3µm seem to play a crucial role in oceanic primary production. Phylogenetic analyses using both plastid and nuclear rRNA genes reveal a high diversity especially in the members of haptophytes and chrysophytes, with the latter dominating the South Atlantic gyre and associated with higher light intensities (Kirkham et al., 2011). Chrysophytes were previously only known from freshwater systems and their significance in marine oligotrophic areas is poorly understood.

2.1.3 North Pacific

Stratification of surface waters usually inhibits marine primary productivity as nutrients become depleted in the euphotic zone. However, at ALOHA monitoring station stratification and productivity are not strongly correlated (Dave & Lozier, 2010). Presumably allochthonous nutrients maintain new production during stratified periods but it is not well understood how these nutrients are supplied. Unicellular diazotrophs frequently dominate N_2 fixation in late winter and early spring, while filamentous diazotrophs (heterocyst-forming cyanobacteria and *Trichodesmium* spp.) fluctuate episodically during the summer (Church et al., 2009). The picocyanobacteria seem well adapted to P starvation by exhibiting significant increases in APA (Moore et al., 2005). In the past, a shift from eukaryotic to prokaryotic dominance transformed the North Pacific gyre from a N-limited to a P-limited system (Karl et al., 2001). There is now molecular evidence for an increase in N-limited strains of *Prochlorococcus* which may indicate that the gyre is returning to a N-limited phase (Van Mooy & Devol, 2008).

Picophytoplankton are dominant contributors (averaging 91%) to euphotic zone Chl. a concentrations (Li et al., 2011). Cyanobacteria such as *Prochlorococcus* spp. and heterotrophic bacteria, though incapable of N_2 fixation, represent the vast majority of the total cell abundance throughout the euphotic layer (Duhamel et al., 2011). *Proclorococcus* spp. are numerically dominant year-round. Here plankton communities can be distinguished as epipelagic, mesopelagic and bathypelagic (Eiler et al., 2011) with distinct *Prochlorococcus* clades at different depths (Malmstrom et al., 2010).

In summer Chl. a concentrations peak, the phytoplankton is supported by N_2 fixation and dominated by a few genera of large diatoms and the cyanobacterium *Trichodesmium* (Dore et al., 2008). Filamentous organisms, specifically heterocyst-forming cyanobacteria and *Trichodesmium* spp. fluctuate episodically during the summer resulting in highly variable fixation rates, possibly triggered by mesoscale physical processes (e.g. eddies, and wind-generated waves) that input nutrient rich waters in the euphotic zone and can cause blooms in the microbial communities (Fong et al., 2008).

There is uncertainty as to whether the system is a C sink or source. Net community production is calculated to be closely balanced or slightly negative (net heterotrophic) due to tightly coupled respiration and gross community production (Viviani et al., 2011), but high oxygen concentrations below the mixed layer may be consistent with an ecosystem that is a net producer of fixed C (net autotrophic) throughout the year (Riser & Johnson, 2008).

2.1.4 South Pacific

The South Pacific gyre is the largest oceanic gyre and has the clearest waters ever described with a chlorophyll maximum as deep as 180m (Ras et al., 2008). Far from continental sources it receives the lowest atmospheric iron flux in the world (Wagener et al., 2008). Both phytoplankton and heterotrophic bacteria are limited by N within the centre gyre, but not by iron which only limits primary production at the border of the gyre (Bonnet et al., 2008). In the surface waters (<180m), NO_3^- is undetected and only trace quantities of regenerated N are found. Despite N limitation, no evidence of N_2 fixation exists and *nifH* gene abundances are extremely low compared to North Pacific gyre (Bonnet et al., 2008). This suggests that the autotrophic communities are adapted to living at low iron levels, and that the common photoautotrophic N_2 fixing organisms are not favoured due to their elevated iron quotas. In spite of strong N depletion leading to low chlorophyll biomass, the South Pacific gyre with its characteristic reduced vertical mixing can accumulate organic matter (Raimbault et al., 2008) that can sustain active regeneration processes during stratification (Raimbault & Garcia, 2007).

In the clear waters of the gyre centre autotrophic eukaryotes shift to smaller cells (<2 µm) compared to more eutrophic conditions (Masquelier & Vaulot, 2008). Flow cytometry sorting carried out in the most oligotrophic areas of the gyre revealed several novel lineages of photosynthetic picoeukaryotes such as a clade of prasinophytes. Pelagophytes, chrysophytes and haptophytes are the dominant picophytoplankton (Shi et al., 2011). Coccolithophores are an important group of unicellular calcifying haptophytes, even though at low abundances they grow down to 300m deep with maximum cell concentrations recorded between the depths of 150–200m (Beaufort et al., 2008). In addition, high taxonomic diversity is also detected in the microzooplankton tintinnids that is inversely related to chlorophyll concentration and positively to the depth of the maximum chlorophyll layer (Dolan et al., 2007). Furthermore, larger microplankton (e.g. diatoms) can adapt to the ultraoligotrophic conditions of this region by forming symbiotic relationships with other species (Gómez, 2007).

There is now growing evidence that this oceanic expanse, once thought to be net heterotrophic may be net autotrophic. The deep layers, below the euphotic zones, may be significant contributors to C fixation fuelling heterotrophic processes in the upper layer (Claustre et al., 2008). However this remains a debate; as some studies show that net community production is closely balanced or slightly net heterotrophic (Viviani et al., 2011) while data from oxygen sensors deployed on profiling floats suggest that the system is net autotrophic throughout the year (Riser & Johnson, 2008).

2.1.5 South Indian

The Indian subtropical gyre is probably the least studied gyre. Research has so far focused in the Arabian Sea (north-western Indian Ocean) and extensive regions of the oceanic gyre remain unknown. In late winter (austral summer) warm and salty subtropical water is separated from deeper water (Tsubouchi et al., 2009). This pronounced vertical stratification impedes nutrient transport into the euphotic zone leading to low seasonal NO_3^- and $Si(OH)_4$ concentrations that limit primary production by microplankton so that nanoplankton and picoplankton dominate productivity. The turnover rate of nanoplankton and picoplankton seems to be closely coupled to microzooplankton grazing and low nutrient concentrations (Thomalla et al., 2010).

About 90% of Chl. a observed at the surface and at the deep chlorophyll maximum (up to 120m depth) is attributed to the picophytoplankton fraction, while picoeukaryotes account for up to 50% of the Chl. a measured (Not et al., 2008). Prokaryotic *Prochlorococcus* and eukaryotic prochlorophytes, haptophytes and pelagophytes seem to dominate the oligotrophic waters of the Indian Ocean, though a large fraction of the eukaryotic genomes sampled and a significant flagellate (small phototrophic protist) remain unidentified (Not et al., 2008; Schlüter et al., 2011). Greater variation in the picoeukaryotic assemblages has been observed vertically in the upper 200m of the water column than horizontally across the entire southern Indian oceanic expanse.

2.2 Enclosed systems

The low primary production observed in open-ocean subtropical gyres relates to their isolation from freshwater and airborne nutrient sources. Few coastal regions are ultraoligotrophic, although the Eastern Mediterranean and the Northern Red Sea become ultraoligotrophic during the warmer parts of the year (Labiosa et al., 2003; Siokou-Frangou et al., 2010).

2.2.1 Eastern Mediterranean

The Mediterranean connects through the Strait of Gibraltar to the Atlantic Ocean in the west, the Bosporus Strait to the smaller enclosed Black Sea in the northeast, and the Suez Canal to the Red Sea and Indian Ocean in the southeast. Evaporation exceeds precipitation and river run off (the main rivers are the Ebro, Rhone, Po, Danube and Nile) with surface waters increasing in salinity from west to east. Atlantic surface water enters through the Strait of Gibraltar and moves eastwards, sinking to 200-500m depth in the Eastern Mediterranean before circulating back west and exiting through the Strait after about 80-100 years and with nearly 10% more salt content (Bas, 2009).

Nutrients mainly enter the system through the Straits of Gibraltar and Bosporus, from wind-driven Saharan dust deposits and from river discharges mainly in the north. The Eastern Mediterranean has the lowest nutrient content. Here dams have resulted in drastic reductions in freshwater flow; the Aswan dam on the Nile, for example, restricts the amount of silica entering the Mediterranean (Turley, 1999). In the Eastern Mediterranean aeolian inputs can account for 60-100% of the bioavailable N and 30-50% of the bioavailable P (Krom et al., 2010). The unusually high ratio of N to P (~28:1) observed in the Eastern Mediterranean (it can sometimes reach 105:1) is due to high N inputs from rivers and atmospheric deposition (Krom et al., 2010; Markaki et al., 2010).

In the western Mediterranean, winter mixing of surface waters with nutrient-rich deeper waters causes a winter-spring phytoplankton bloom composed mostly of diatoms with some flagellates and coccolithophorids (Goffart et al., 2002). The bloom is less-pronounced in the Eastern Mediterranean (D'Ortenzio & Ribera d'Alcalà, 2009), Chl. a concentration is <0.1 mg m^{-3} on average, with the maxima occurring in late winter - early spring and minima in late summer (Siokou-Frangou et al., 2010). In summer a sharp thermocline at 10-20m results in nutrient depletion in the surface mixed layer. During this stratified period, primary production in the Eastern Mediterranean is both N and P limited, and during the winter mixing it becomes P limited (Thingstad et al., 2005; Tanaka et al., 2011). As in other ultraoligotrophic systems, the microbial loop is in a dynamic equilibrium in which grazing pressure, competition and nutrient concentrations can shift the limiting nutrient.

The importance of N_2 fixation in the Eastern Mediterranean is under investigation. There are low concentrations of diazotrophic cyanobacteria, possibly due to P and iron limitation. The N_2 fixation rates decrease from west-east but may sustain up to 35% of the primary production in the eastern basin and can be stimulated occasionally by Saharan dust events (Bonnet et al., 2011; Ridame et al., 2011) The diazotrophic community is dominated by unicellular picocyanobacteria, although N_2 fixation has also been detected within picoeukaryotes (Le Moal et al., 2011).

Picoplankton dominate the most nutrient limited areas of the Mediterranean (Tanaka et al., 2007). Larger diatoms and dinoflagellates become abundant after intermittent nutrient pulses associated with upwelling, fronts and gyres (Siokou-Frangou et al., 2010). Over 85% of Chl. a in the eastern basin is found in ultraplankton (<10μm), that comprises cyanobacteria (*Synechococcus* spp. are dominant), chlorophytes, prasinophytes and haptophytes (Denis et al., 2009). Coccolithophores are more abundant and diverse in the eastern basin (Ignatiades et al., 2009). In summer, dinoflagellates dominate the larger plankton fraction in offshore areas of the Eastern Mediterranean whereas diatoms are more prevalent during winter mixing and in inshore waters where anthropogenic eutrophication is evident (Aktan, 2011).

Most studies describing phytoplankton biomass dynamics in the Mediterranean Sea stress that low nutrients cause low primary production (bottom-up control). However, the planktonic food webs are very efficient at minimizing C export to deeper waters, benefiting predators that control the plankton biomass (top-down control) (Siokou-Frangou et al., 2010). A P addition experiment in the Eastern Mediterranean had an unexpected outcome because Chl. a concentrations decreased while egg-carrying copepods numbers increased (Krom et al., 2005; Thingstad et al., 2005). Efficient top-down control helps explain why Mediterranean fisheries are richer than anticipated based on Chl. a and nutrient concentrations. In addition to efficient C export to pelagic top predators, benthic primary producers also play an important role in sustaining Eastern Mediterranean food webs. Highly productive benthic primary producers, such as the seagrass *Posidonia oceanica* which grow at 0 – 50m in depth (Duarte, 1991) the coralline algal habitats (e.g. maerl) which grow in low light conditions (Ballesteros, 2006) and macroalgal assemblages (e.g. *Cystoseira* forests) in the shallows form diverse and complex habitats. The Mediterranean basin ranks among 25 'biodiversity hotspots' containing about 7% of the world's marine biodiversity (Bianchi & Morri, 2000; Myers et al., 2000). Even though it covers <0.8% of the world ocean surface and is <0.3% of its volume, it is home to 4-18% of the world's recorded species, depending on the phylum considered. Approximately 17 000 marine species occur in the Mediterranean Sea and this inventory is expanding rapidly, especially for microbes and deep sea species (Coll et al., 2010). An unusually high level of endemism is observed and the region hosts a number of species of conservation interest, such as 71 species of sharks, rays and chimaeras (Cavanagh & Gibson, 2007), sea turtles (*Dermochelys coriacea, Chelonia mydas, Caretta caretta*), nine permanent resident species of cetaceans (Reeves & Notarbartolo, 2006) and the critically endangered Mediterranean monk seal (*Monachus monachus*). The southeastern Mediterranean Sea has, on paper, the lowest species richness but this is influenced by the fact that there have been relatively sparse research efforts in this part of the Mediterranean.

2.2.2 North Red Sea

The Red Sea is thought to owe its name to intense phytoplankton blooms but they are very rare in this oligotrophic system. It connects to the Mediterranean Sea through the narrow and shallow (~8m) Suez Canal in the north and exchanges water with the Indian Ocean through the Bab el Mandeb strait (130m deep) and the Gulf of Aden in the south. There are no permanent rivers and scant rainfall so seawater entering through the Bab el Mandeb strait gets saltier as it progresses northwards. Like the Mediterranean Sea, the North Red Sea is heavily influenced by seasonal changes in physical and chemical characteristics of the water column. Oligotrophic to ultraoligotrophic conditions prevail in the northern region during the summer and autumn stratified period, while in the winter, conditions become eutrophic (Lindell & Post, 1995; Labiosa et al., 2003). The Gulf of Aqaba, at the northeast tip of the Red Sea is about 165km long, very deep (~1800m) but very narrow (max width <25km), bounded by desert and separated from the Red Sea by the shallow (240m) Strait of Tiran. Here, phytoplankton populations have a large spring bloom (with Chl. a peak at around 3 mg m^{-3}) and smaller autumn bloom but in the summer levels average ~0.2 mg m^{-3} (Labiosa et al., 2003).

Although N:P ratios are lower in the summer many phytoplankton species appear to be P limited and even though P is below detection limits, APA is consistently low in the picophytoplankton fraction indicating the absence of P limitation, while larger phytoplankton express increased APA especially during the stratified period indicating P limitation (Mackey et al., 2007). N_2 fixation rates are consistently low and are higher during the deep mixing season. Diazotrophic populations are dominated by the smaller N_2 fixing organisms (Foster et al., 2009). Small unicellular cyanobacteria (e.g. *Cyanothece* spp.), are the most abundant N_2 fixing organisms, while larger filamentous *Trichodesmium* occur in surface waters especially in the winter when soluble reactive P is more abundant (Mackey et al., 2007). Inputs of aerosol NO_3^- to surface waters represents an important source of 'new ' N in this region (Aberle et al., 2010).

The planktonic communities are characterized by low abundances and the dominance (95%) of ultraplankton (0.2-8μm) (Berninger & Wickham, 2005; Al-Najjar et al., 2007). During the summer and autumn, stratified surface waters become nutrient depleted and picophytoplankton dominate. In winter, nutrient concentrations increase and larger phytoplankton become more abundant. This pronounced seasonal succession of major taxonomic groups is observed with *Prochlorococcus* dominating during the stratified summer period but being almost absent during the winter and chlorophytes with cryptophytes dominating during the winter mixing but being almost absent during the summer (Al-Najjar et al., 2007). Larger cells (>8 μm) are dominated by dinoflagellates and ciliates (Berninger & Wickham, 2005). The ciliates prey on the dominant picoautotrophs so that this primary production then becomes available to metazoan grazers (Claessens et al., 2008). Stable isotope analyses revealed a complex and diverse planktonic community that included herbivores and a large variety of omnivores (e.g. non-calanoid copepods) (Aberle et al., 2010). It appears top-down and bottom-up controls operate simultaneously in the North Red Sea with small cells being controlled by grazing while larger cells (e.g. diatoms) are limited by nutrient availability (Berninger & Wickham, 2005).

Despite periods of ultraoligotrophic conditions, the Red Sea is a biodiversity and endemism hotspot (Roberts et al., 2002). The Gulf of Aqaba is characterized by very high levels of endemism, especially in the mollusc and echinoderm taxa and there are exceptionally diverse fringing reefs, steeply sloping to depths of up to 150m (Fricke & Schuhmacher, 1983).

3. Threats

The human population now exceeds 7 billion compared to around 800 million in the year 1750 and an estimated 9.4 billion by 2050 (Raleigh, 1999; United States Census Bureau, 2011). This rapid population increase has been matched with environmental degradation and global biodiversity loss. Marine litter is now ubiquitous, and resources are being exhausted at alarming rates. The major stressors of anthropogenic climate change on the world's marine ecosystems are warming, acidification and deoxygenation (Gruber, 2011) with impacts that range from decreased ocean productivity, altered food web dynamics, reduced abundance of habitat-forming species, shifting species distributions, and a greater incidence of disease (Hoegh-Guldberg & Bruno, 2010). In this section we consider how ultraoligotrophic marine ecosystems are being altered by Man, and to what extent these systems may be vulnerable to the multiple stressors that are present.

3.1 Open ocean systems

Subtropical open ocean ecosystems are far removed from human civilization yet despite this remoteness rapid changes are underway, such as ocean acidification and the accumulation of marine debris.

Climate change

Remotely-sensed ocean colour data show that ultraoligotrophic marine regions have expanded by about 15% in the past decade (Polovina et al., 2008) and that the growth of these provinces may be accelerating as they get larger (Irwin & Oliver, 2009). Significant decreases in Chl. a concentrations have also been recorded in most subtropical gyres (Signorini & McClain, 2011). Polovina et al. (2011) predict that ocean warming will expand the area of the subtropical biome by ~30% by 2100 due to increased water stratification and restricted supplies of nutrients to the upper water column. In such areas, large and efficient C fixing eukaryotic species are outcompeted by smaller eukaryotic and prokaryotic plankton causing productivity to fall.

During the past 100 years, rising atmospheric greenhouse gas concentrations have increased global surface ocean temperatures by ~0.7°C (Trenberth et al., 2007). The deep ocean remains relatively cool, so a density gradient is developed which increases upper ocean stratification which can lower the oxygen and nutrient contents of the water. Ocean warming and increased stratification of the upper ocean may lead to 1-7% declines in dissolved oxygen in the ocean interior with implications for ocean productivity and nutrient cycling (Keeling et al., 2010). Large expansions of the oxygen minimum zones have occurred horizontally and vertically in all tropical and subtropical oceans and it is estimated that since 1960 deoxygenated areas have increased by 4.5 million km^2 (Stramma et al., 2010). The implications of ocean warming and deoxygenation on the functioning of ultraoligotrophic systems are poorly known yet alterations in food webs can be expected since warming will favour some microbes and plankton over others (Marinov et al., 2010; Sarmento et al., 2010).

Ocean acidification results from the uptake of anthropogenic carbon dioxide (CO_2) of which around one third is absorbed by the oceans (Sabine et al., 2004) where it reacts with water to form carbonic acid (H_2CO_3) which further dissociates into hydrogen ions (H^+) and carbonate ions (CO_3^{2-}). Increased H^+ ions lower the pH of the water. Surface waters of the oceans have been acidified by an average of 0.1 pH units compared with pre-industrial levels (Doney, 2010). Model simulations predict that ocean pH will decrease by 0.2 to 0.3 pH units by the end of the twenty first century (Orr et al., 2005). The ecological effects of ocean acidification remain uncertain yet there are widespread concerns over the effects on calcified organisms since uptake of atmospheric CO_2 leads to a decrease in carbonate concentrations and increases $CaCO_3$ dissolution (Riebesell et al., 2009; Rodolfo-Metalpa et al., 2011). The calcifying plankton that occur in ultraoligotrophic systems (e.g. coccolithophores, foraminiferans, and pteropods) may have a reduced ability to construct their $CaCO_3$ shells. Beaufort et al. (2011) for example found a significant decrease in coccolith mass at sites all over the world as pCO_2 concentrations increase, although there were exceptions with a heavily calcified coccolith morphotype found in some low pH areas. Biogeochemical disruptions are also possible due to ocean acidification, although the ecological effects of these remain unknown. For example experimental decreases in pH lower microbial nitrification (oxidation of ammonia into nitrite) rates (Beman et al., 2011). When stimulated by pCO_2, N_2 fixation rates appear to increase in filamentous non-heterocystous *Trichodesmium* spp. (Barcelos e Ramos et al., 2007) and the unicellular *Crocosphaera watsonii* (Fu et al., 2008), but decrease in heterocystous diazotrophs (Czerny et al., 2009). Changes in nitrification and N_2 fixation rates have the potential to cause fundamental alterations to the marine environment. Elevated pCO_2 in cultured organisms and in a few mesocosms reveal contradicting results with some prokaryotic species and communities exhibiting increased production when others are adversely impacted (Liu et al., 2010). It is clear that our understanding of the potential impacts of acidification on the overall biogeochemistry of marine waters is limited by the lack of *in situ* experiments (except in some coastal areas with CO_2 vents) and the inconsistency or lack of data for several taxa. Predicting changes to marine ecosystems is also problematic since decreasing pH/increasing CO_2 is occurring in combination with other changes such as deoxygenation and warming (Denman et al., 2011).

Marine debris

During the last 60 years, the global production of plastic has increased from 1.5 million tonnes to 265 million tonnes (Plastics Europe, 2011). The light plastic particles (e.g.

polyethylene and polypropylene) that enter water bodies then float and drift with the currents and can be transported over large distances. The subtropical gyres trap floating debris in the central slower moving water masses. Accumulating plastic was discovered in 1972 in the Sargasso Sea, with increasing amounts recorded with time, such as in the North Pacific gyre where up to 334,271 pieces per km^2 and a startling 6:1 biomass ratio of zooplankton to plastic were recorded (Moore et al., 2001). Similar observations have been made in the North Atlantic gyre (Law et al., 2010). Models and observations show that all five subtropical gyres are litter aggregation hotspots (Maximenko et al., 2011).

Plastic can degrade to microscopic pieces (Thompson et al., 2004) that adsorb persistent organic pollutants such as PCBs, PAHs, DDTs, PBDEs, alkylphenols, and bisphenol A (Rios et al., 2010). Planktonic plastic loaded in organic pollutants can easily be mistaken for prey and upon ingestion the pollutants bioaccumulate (Harwani et al., 2011), while the plastic remains undigested and can sometimes clog the digestive tract of the organism leading to starvation and subsequent death. Top predators have been consistently reported victims of this plastic menace; 34% of 408 dissected leatherback turtles (Mrosovsky et al., 2009), 28% of 106 dolphins incidentally captured in artisanal fisheries (Denuncio et al., 2011) and 9.2% of 141 mesopelagic fishes from 27 species in the North Pacific subtropical gyre (Davison & Asch, 2011) had plastic in their stomachs. Every albatross chick egested bolus examined from the North Pacific colonies contained plastic (Young et al., 2009). 134 different types of nets causing stomach rupturing and emaciation were found inside two stranded male sperm whales in Argentina (Jacobsen et al., 2010), and the list goes on. It is now recognized that the environmental impacts of plastic debris are wide-ranging and include among others entanglement of marine fauna, ingestion by consumers from all trophic levels including the small heterotrophic plankton, dispersal of invasive species to non-native waters, and bioaccumulation of organic contaminants (Gregory, 2009).

How the biocommunities inhabiting the deoxygenated, acidified, warm waters of the ultraoligotrophic subtropical gyres will respond to changes brought about by the 'Marine Debris Era' remains to be seen.

3.2 Enclosed systems

Due to the proximity of humans, enclosed ultraoligotrophic systems are exposed to multiple anthropogenic stressors. The benefits supplied by marine biodiversity to human health are enormous and include: i) seafood (high-quality protein, minerals and vitamin D and omega-3 fatty acids) with antioxidant properties and cardio and cancer protective effects, ii) marine organisms such as sharks, algae and sponges supply a large variety of bioactive metabolites some of which are used to treat human diseases and, iii) maritime leisure activities such as recreational provide physical and psychological effects to users such as recreational fisheries, diving, snorkelling, and whale watching (Lloret, 2010). To sustain such benefits improvements are required in the ways that we manage ultraoligotrophic seas.

The North Red Sea is a biodiversity hotspot with high levels of endemism and stunning fringing reefs that can extend to depths of 150m. Protecting the threatened coral reefs of the enclosed North Red Sea is a real challenge as there are multiple stressors already in effect. Ocean warming slows coral growth and increases bleaching events (Cantin et al., 2010). Future acidification is a significant threat that is expected to increase bioerosion and decrease the net calcification rates (aragonite formation) of stony corals (Silverman et al., 2009; Rodolfo-Metalpa et al., 2011). Furthermore, the coral reefs of the North Red Sea attract thousands of visitors that can contribute to impacts on coral reefs (Hasler & Ott, 2008). Submerged marine litter in coral reefs of the North Red Sea with an overall mean density of 2.8 items/m^2 and overall mean weight of 0.31 kg/m^2 is another major concern (Abu-Hilal & Al-Najjar, 2009). Bioaccumulation of toxic contaminants in North Red Sea corals is high (Ali et al., 2010). Moreover, coastal development has resulted in increasing demand for freshwater. Seawater desalination plants are being constructed that discharge high salinity water often contaminated with other chemicals (Hoepner & Lattemann, 2003).

The Mediterranean coasts support a high density of inhabitants, distributed in 21 countries with a population of about 450 million (cf. 246 million in 1960), of which 132 million live on the coast (26,000 km in length). In addition, 200 million tourists per year visit Mediterranean coastal countries. During the past one hundred years, the Eastern Mediterranean has been subjected to the effects of two important events, the opening of the Suez Canal in 1869 (discussed below) and the construction of the Aswan High Dam in 1964. Before the construction of the High Dam, nutrient enrichment extended along the Egyptian coast and was detected off the Israeli coast and sometimes off southern Turkey. It provided for dense blooms of phytoplankton off the Nile Delta (Nile bloom) which in turn provided nourishment to sardines, other pelagic fishes and crustaceans. Huge declines have been observed in these fisheries in the years following the High Dam construction. Since the late 1980s the recovery of total fish landings in the region reveal that the pelagic ecosystem is adjusting but the mismatch between extremely low primary productivity and relatively high levels of fish production remains a puzzle 'the Levantine Basin Paradox' to scientists (Dasgupta & Chattopadhyay, 2004). Whether this recent increase in fisheries is due to increased fishing efforts, recovery of fish stocks or nutrient enrichments by anthropogenic activities is not yet clear.

Human activities have been reducing biodiversity of the Mediterranean Sea at all levels. The major stressors in the Eastern Mediterranean appear to be: climate change, alien species invasions, pollution, fishing impacts, eutrophication and aquaculture, and habitat loss (Claudet & Fraschetti, 2010; Coll et al., 2010; Durrieu de Madron et al., 2011). Often these stressors act synergistically and have cumulative negative impacts on a great number of taxonomic groups. The Mediterranean Sea is perhaps the most investigated marine environment in the world, however research efforts have been concentrated in the northwestern Mediterranean, so much less is known about human-environmental interaction in the ultraoligotrophic waters of southeastern Mediterranean.

Climate change

The effects of global climate change are likely to affect chemical and physical properties of the water and act synergistically with other anthropogenic stressors (Gambaiani et al., 2009). Climate change impacts in the Mediterranean may provide useful insights for potential impacts elsewhere as the region is well monitored. As in many other regions; sea temperatures are rising, acidification is underway, extreme climatic events and related disease outbreaks are becoming more frequent, native species are being displaced and invasive species are spreading (Lejeusne et al., 2010).

Increased warming across the Mediterranean increases stratification of the water column further restricting nutrient availability in ultraoligotrophic zones and is related to increased mortality of the endemic seagrass *Posidonia oceanica* (Diaz-Almela et al., 2009). Higher temperatures may disrupt juvenile life histories stages of numerous organisms (Hawkes et al., 2007; Byrne, 2011) and cause mass mortalities of adults (Garrabou et al., 2009). In addition increasing temperatures may also contribute to higher frequencies of disease outbreaks as tropical microbial pathogens are expected to spread (Danovaro et al., 2009). Rising water temperatures are altering biogeographic boundaries and leading to a progressive homogenization of Mediterranean marine biota. Changes include an increase in abundance of eurythermal species and a decrease in cold stenothermal species as well as northward species shifts and mass mortalities during unusually hot summers (Coll et al., 2010). Warm-water fish like *Thalassoma pavo, Sphyraena* spp., *Epinephelus* spp., *Sparisoma cretense* and, *Sardinella aurita* have spread northwestwards (Sara et al., 2005). Certain cold water species have been replaced, for example the distribution of the cave-dwelling crustacean *Hemimysis speluncola* has contracted and been replaced by *H. margalefi,* a warm water species that was previously unknown in the region (Chevaldonné & Lejeusne, 2003). Non-indigenous warm water species of algae, invertebrates and fish are enlarging their geographical ranges (Bianchi, 2007). Invasive tropical fauna and flora are most evident in the southern Mediterranean where they now form a significant portion of the biota and some outcompete native species (Lasram & Mouillot, 2009). Predicted levels of warming for the end of this century lie beyond the thermotolerance levels of the developmental stages of many metazoa (Byrne, 2011).

Ocean acidification may also alter the ecology of the Mediterranean, although the evidence to date is sparse. Israel and Hophy (2002), found that acidifying seawater to pH 7.8 with CO_2 did not adversely affect growth and photosynthesis in a wide range of Mediterranean chlorophyte, rhodophyte and phaeophyte algae whereas Invers et al. (1997) found that this level of acidification enhanced photosynthesis in the Mediterranean seagrasses *Posidonia oceanica* and *Cymodocea nodosa.* Martin and Gattuso (2009) found that the Mediterranean encrusting coralline alga *Lithophyllum cabiochae* decreased calcification when elevated pCO_2 conditions were combined with high temperatures (pH 7.8; seasonal temperature +3°C). Investigations into the effects of acidification at a natural volcanic CO_2 vent off Ischia in Italy show that seagrasses and certain seaweeds were able to benefit from the elevated CO_2 levels (Martin et al., 2008; Porzio et al., 2011) but that around 30% of the coastal biodiversity was lost at mean pH levels predicted for 2100 (Hall-Spencer et al., 2008). This is partly because ocean acidification disrupts recruitment of organisms from the plankton (Cigliano et al., 2010), and partly because peak summer temperatures increase the susceptibility of some organisms to shell and skeleton dissolution (Rodolfo-Metalpa et al., 2011). Calcareous systems such as vermetid reefs and, mussel beds, as well as deep and shallow coral communities, appear to be especially vulnerable in ultraoligotrophic regions where organisms lack food and are therefore less able to allocate resources for coping with multiple stressors. In contrast, carbon limited organisms, like seagrasses, may make use of the extra dissolved CO_2 and if their habitats are protected they may thrive due to higher photosynthetic rates.

Alien species

Warm-water species are found in the Mediterranean due to Atlantic influx, Lessepsian migration, introductions by humans and present-day sea warming (Bianchi, 2007). Most of the 955 alien species so far recorded occur in the oligotrophic Eastern Mediterranean (Zenetos et al., 2010). About 20% of Mediterranean alien species were accidentally introduced from biofouling on ship hulls or in ballast tanks (Galil, 2009). However most (about 67%) Mediterranean alien species came from the Red Sea since the Suez Canal was opened in 1869. More than 600 tropical Indo-Pacific species have been reported entering the Mediterranean where they have established reproducing populations in the Levantine basin and beyond (Coll et al., 2010; Costello et al., 2010). The rate of invasion of species from the Red Sea into the low nutrient waters of the eastern Mediterranean is increasing due to warming. Now nearly half of the trawl catches along the Levantine coast consist of Erythrean fish, but whilst some are now targeted commercially, others are detrimental to fisheries. In Cyprus, for example, the invasive puffer fish *Lagocephalus sceleratus* is out-competing native fishes and exhausting invertebrates such as the *Octopus vulgaris* and squid; in this region several other invasive species have caused substantial shifts in coastal ecosystems (Katsanevakis et al., 2009).

Pollution

Like all coastal systems the Mediterranean Sea is affected by numerous anthropogenic contaminants, but due to its enclosed and oligotrophic nature their impacts can be exacerbated. Marine litter is a major problem in the region, causing obstruction of digestive tracts and contaminant bioaccumulation in many marine animals. Persistent organic pollutants tend to bioaccumulate and come from maritime sources, aerosol deposits, urban/industrial activity, river discharges and accumulate in harbour sediments (Gómez-Gutiérrez et al., 2007; Thébault et al., 2008).

Riverine inputs and air masses from northern and central Europe carry persistent organic pollutants that can reach the Eastern Mediterranean basin (Mandalakis & Stephanou, 2002). Large commercial harbours are situated mostly in the northwest Mediterranean and maritime traffic causes noise pollution that adversely affects cetaceans (Dolman et al., 2011). Submarine drilling for oil and gas takes place in the south with exploration now underway in the eastern Mediterranean. About 300 000 tonnes of crude oil are released into the Mediterranean every year (Danovaro & Pusceddu, 2007) and can cause environmental damage, especially when chemical dispersants are used in clean-up procedures. An oil spill in Valencia in 1990 was followed by hundreds of dead dolphins being washed up along the Spanish, French, Italian and North African shores and a year later on the beaches of southern Italy and Greece, thought to be due to disease triggered by immunosuppressants in the oil spill (Zenetos et al., 2002).

Overexploitation of resources

Industrialized fishing has severe impacts on species, habitats and ecosystems (Tudela, 2004). Several fish resources are highly exploited or overexploited (Palomera et al., 2007; MacKenzie et al., 2009). A number of other organisms are also affected by exploitation and include unwanted by-catch (accidental capture in fishing gear). Bottom-trawling is a non-selective fishing method and causes a large mortality of discarded benthic invertebrates which can induce severe biodiversity and biogeochemical changes (Pusceddu et al., 2005). Severe population declines have occurred for all top predators during the last 50 years with the Mediterranean Sea described as the most dangerous sea in the world for cartilaginous fishes (Cavanagh & Gibson, 2007). See turtles face entangling, pollution and loss of habitat.

Population declines have also been recorded among marine mammals (such as sperm whales, short-beaked common dolphins, common bottlenose dolphins, striped dolphins and monk seals) that face prey depletion, direct killing and fishery by-catch (Reeves & Notarbartolo, 2006). The Mediterranean monk seal is the most endangered seal in the world with less than 600 individuals currently surviving. Remnant populations are fragmented and declining. The species faces a number of threats (i.e. accidental entanglement, exploitation, persecution and tourism) that caused severe declines in abundance (Karamanlidis et al., 2008).

There are clearly multiple threats acting synergistically on species of the Mediterranean Sea. For example, in December 2009, a pod of seven male sperm whales stranded along the coasts of Southern Italy. It appears the cause of death was prolonged starvation not from plastic obstruction (even though plastic was found in all dissected individuals) but due to a lack of prey. High concentrations of pollutants in the tissues of the stranded animals led researchers to conclude that prolonged starvation stimulated the mobilization of highly concentrated lipophilic contaminants from their adipose tissue which entered the blood circulation and may have impaired immune and nervous functions (Mazzariol et al., 2011).

Eutrophication and aquaculture

Eutrophication in the ultraoligotrophic Eastern Mediterranean is disrupting habitats and causing community shifts. Eutrophic conditions favour opportunistic species that may increase productivity and fishery catches but may out compete the highly diverse communities of ultraoligotrophic systems. Eutrophication sources from agriculture, urbanization, river run-offs, and aquaculture. Considering the exponential human population growth and the fact that fisheries are in global decline, aquaculture efforts are predicted to increase to meet growing demand (Duarte et al., 2009).

Fin-fish farming can have a number of environmental effects on the surrounding and downstream ecosystems (Holmer et al., 2008). Dissolved wastes increase the nutrient loading of the area and particulate wastes increase sediment deposition. In the benthos sedimentation and organic loading can cause biochemical changes affecting the composition and function of benthic communities (Karakassis et al., 2000), stimulating the growth of undesirable species that produce toxic metabolic waste that can kill species of conservation significance. Large-scale *Posidonia oceanica* losses adjacent to fish farm cages have been reported across the Mediterranean (Pergent-Martini et al., 2006) including the Eastern Mediterranean (Holmer et al., 2008; Apostolaki et al., 2009).

Improved fish farm management may increase their sustainability although culturing carnivorous fish is still likely to come at environmental costs. Integrated multi-trophic aquaculture (culturing organisms from different trophic levels, mimicking natural ecosystem interactions and producing less waste than monoculture systems) may be key to environmental sustainability of aquaculture practices in ultraoligotrophic waters (Chopin, 2006; Angel & Freeman, 2009).

Habitat loss

Coastal habitats such as seagrass meadows, mollusc (oyster, vermetid and mussel) reefs, coralligenous maerl formations, and macroalgal assemblages on shallow reefs are examples of complex and highly productive ecosystems. They supply food resources, nurseries and shelter for a large array of species that are protected by international conventions, directives and action plans. A meta-analysis of 158 experiments in the Mediterranean revealed that human activity caused adverse impacts on all habitat types. Fisheries, species invasion, aquaculture, sedimentation increase, water degradation, and urbanization can all have negative impacts on Mediterranean habitats and associated species assemblages (Claudet & Fraschetti, 2010).

Habitat destruction is considered one of the most pervasive threats to the diversity, structure and functioning of marine coastal ecosystems. The loss of habitat structure generally leads to lower abundances and species richness that usually allows opportunistic species to prosper (Airoldi et al., 2008). Habitat destruction can also impair the integrity, connectivity and functioning of large-scale processes decreasing population stability and isolating communities (Thrush et al., 2006). Continued losses of habitats to coastal development has triggered several international protective measures such as the development of Marine Protected Areas (MPAs), but their efficacy is much questioned (García-Charton et al., 2008; Montefalcone et al., 2009) as habitat loss continues apace.

Oligotrophic coastal habitats are dominated by slow growing species and intricate food webs. Habitat losses can be considered irreversible, as it would take centuries following the cessation of disturbances for ecosystems to return to their climax state.

4. Conclusions

Ultraoligotrophic marine ecosystems cover almost a third of the earth's surface and contribute significantly to global productivity and biogeochemistry. They are, however, amongst the least understood systems on this planet. Once considered to be monotonous oceanic desserts, they are now known to have highly dynamic physical and biological properties with extremely diverse and vertically-distinct planktonic communities.

There is increasing evidence that these systems may be net autotrophic. The water column is dominated by the smallest eukaryotic and prokaryotic picoplankton, which seem well adapted for surviving in oligotrophic conditions. Adaptations range from niche segregation through prolonged coexistence, symbiotic associations, mixed modes of nutrition, lower cellular nutrient requirements, genes encoding for enzymes that regenerate nutrients from allochthonous sources, genes involved in high affinity uptake of nutrients and efficient nutrient uptake due to large surface: volume ratios. Unicellular cyanobacteria and extremely diverse picoeukaryotes dominate primary production in the deep euphotic zones of ultraoligotrophic waters. This production is channelled through the microbial food web (e.g. small ciliates and nanoflagellates) to vertically-migrating gelatinous and crustacean zooplankton and then to higher trophic levels. Phytoplankton blooms mainly occur after winter mixing events but sporadic blooms can occur during the stratified periods. Such blooms can favour larger planktonic species that in turn may sustain large predators (e.g. leatherback turtles, elasmobranchs, cetaceans, tunas and billfishes).

Environmental metagenomics has revealed the high biodiversity observed in ultraoligotrophic marine systems, although the causes for this high biodiversity remain puzzling (Roy & Chattopadhyay, 2007). In the Eastern Mediterranean and North Red Sea biogenic engineers such as corals, seagrasses, and macroalgae form habitats that are biodiversity hotspots of international commercial significance. Exponential growth in the human population has resulted in multiple stressors that act synergistically in the marine environment reducing biodiversity. We believe that in ultraoligotrophic environments, where resources are scarce, organisms are particularly vulnerable to multiple stressors. Climate change is underway and its impacts may continue for many millennia after cessation of anthropogenic CO_2 emissions (Tyrrell, 2011). Warming increases stratification that keeps nutrients below the thermocline. Deoxygenated regions are expanding and acidification may impair ecological functioning (Byrne, 2011). Predictions for 2100 include substantial changes in biogeochemical processes and the extinction of many tropical coral reefs (Silverman et al., 2009). In addition to climate change, marine litter continues to accumulate in ultraoligotrophic subtropical gyres where it is physically degraded to microscopic pieces adsorbing persistent organic contaminants from the surrounding water. Plastic has been found in many consumer species ranging from copepods to large mammals. It may cause starvation, contaminant bioaccumulation, alien species transportation and entanglement. Enclosed ultraoligotrophic systems face additional threats due to their close proximity to Man. Toxic pollutants bioaccumulate and impair the normal physiological functions of organisms causing for example, cetacean strandings. Invasive alien species are spreading and are competing, predating and infecting indigenous species and altering ancient food webs. Marine fish stocks are overexploited with most top predators in decline. Eutrophication decreases water quality which can add pressure on coastal systems subjected to habitat loss and degradation. It is clear that past methods have failed to ensure environmental sustainability yet there are several reasons to be optimistic.

It is now realized that marine ecosystem degradation is a global concern. International efforts to reduce rates of biodiversity loss have led to numerous agreements, conventions or other legal instruments that are coming into force. Such international agreements form the basis of long-term collaboration that is necessary for improved environmental management. For example, the Kyoto Protocol came into force on 2005 and commits the 191 member states to tackle the issue of global warming by reducing greenhouse gas emissions. Annex 1 countries pledged to reduce their emissions by 5.2% from 1990 levels by the end of 2012. The United Nations Convention on the Law of the Sea (UNCLOS) signed by 161 countries helps control pollution and set guidelines for the protection of the environment and the management of marine natural resources in the world's oceans. Inter-governmental organizations, like the International Commission for the Conservation of Atlantic Tunas (ICCAT), are charged with the conservation of stocks of highly migratory species. In Europe, the Marine Strategy Framework Directive aims to achieve healthy waters by 2020 with an unprecedented level of cooperation between countries in developing a network of MPAs. Monitoring of environmental quality, biodiversity and long-term changes in community structure through an international coordinated network of MPAs is an approaching reality. Cautious use of Integrated Coastal Zone Management and Environmental Impact Assessments can help slow the rate of coastal environmental degradation. International partnerships like the Global Ocean Biodiversity Initiative (GOBI) are promising and the identification of Ecologically or Biologically Significant Areas (EBSAs) in the open oceans and deep seas is well underway. It is clear that these international efforts are required to slow the rates of marine environmental degradation.

There are now ample examples where interventions have had positive environmental outcomes. A primary goal among nations should be to raise awareness of effective marine environmental protection. For example, the most viable option to reduce litter is to reduce its production in the first place and then to improve reuse and recycling through enhanced environmental awareness (Thiel et al., 2011). There is now scientific clarity that ocean warming, acidification and deoxygenation are underway due to CO_2 emissions so the primary mitigation strategy is to reduce these emissions (Gruber, 2011). There are reasons to be optimistic about improved management of ultraoligotrophic systems as a growing awareness of their value is being accompanied by shifts towards more sustainable ways of obtaining resources (e.g. marine renewables) and dealing with wastes (e.g. carbon capture and storage).

Change is underway…

5. Acknowledgment

This review is a contribution to the EU Framework 7 Program funded by MedSeA grant 265103 (Mediterranean acidification under a changing climate) and KnowSeas grant 226675 (Knowledge-based Sustainable Management for Europe's Seas). The European Mediterranean Sea Acidification in a changing climate (MedSeA) http://medsea-project.eu/.

6. References

Aberle, N.; Hansen, T.; Boettger-Schnack, R.; Burmeister, A.; Post, A.&Sommer, U., (2010). Differential routing of "new" nitrogen toward higher trophic levels within the marine food web of the Gulf of Aqaba, Northern Red Sea. *Marine Biology* Vol. 157, No. 1, pp. (157-169), 1432-1793

Abu-Hilal, A.&Al-Najjar, T., (2009). Marine litter in coral reef areas along the Jordan Gulf of Aqaba, Red Sea. *Journal of Environmental Management* Vol. 90, No. 2, pp. (1043-1049), 0301-4797

Airoldi, L.; Balata, D.&Beck, M.W., (2008). The Gray Zone: Relationships between habitat loss and marine diversity and their applications in conservation. *Journal of Experimental Marine Biology and Ecology* Vol. 366, No. 1-2, pp. (8-15), 0022-0981

Aktan, Y., (2011). Large-scale patterns in summer surface water phytoplankton (except picophytoplankton) in the Eastern Mediterranean. *Estuarine, Coastal and Shelf Science* Vol. 91, No. 4, pp. (551-558), 0272-7714

Al-Najjar, T.; Badran, M.; Richter, C.; Meyerhoefer, M.&Sommer, U., (2007). Seasonal dynamics of phytoplankton in the Gulf of Aqaba, Red Sea. *Hydrobiologia* Vol. 579, No. 1, pp. (69-83), 1573-5117

Ali, A.-h.; Hamed, M.&Abd El-Azim, H., (2010). Heavy metals distribution in the coral reef ecosystems of the Northern Red Sea. *Helgoland Marine Research* Vol. 65, No. 1, pp. (67-80), 1438-3888

Andersen, N.G.; Nielsen, T.G.; Jakobsen, H.H.; Munk, P.&Riemann, L., (2011). Distribution and production of plankton communities in the subtropical convergence zone of the Sargasso Sea. II. Protozooplankton and copepods. *Marine Ecology Progress Series* Vol. 426, No. pp. (71-86), 1616-1599

Angel, D.&Freeman, S., (2009). Integrated aquaculture (INTAQ) as a tool for an ecosystem approach in the Mediterranean Sea. *FAO Fisheries and Aquaculture Technical Paper* Vol. 529, No. pp. (133-183), 2070-7010

Apostolaki, E.T.; MarbГ, N.; Holmer, M.&Karakassis, I., (2009). Fish farming enhances biomass and nutrient loss in *Posidonia oceanica* (L.) Delile. *Estuarine, Coastal and Shelf Science* Vol. 81, No. 3, pp. (390-400), 0272-7714

Ballesteros, E., (2006). Mediterranean coralligenous assemblages : A synthesis of present knowledge. *Anglais* Vol. 44, No. (0078-3218), pp. (123-196), 0078-3218

Barcelos e Ramos, J.; Biswas, H.; Schulz, K.G.; LaRoche, J.&Riebesell, U., (2007). Effect of rising atmospheric carbon dioxide on the marine nitrogen fixer Trichodesmium. *Global Biogeochemical Cycles* Vol. 21, No. 2, pp. (GB2028), 0886-6236

Bas, C., (2009). The Mediterranean: a synoptic overview *Contributions to Science* Vol. 5, No. 1, pp. (25-39), 1575-6343

Beaufort, L.; Couapel, M.; Buchet, N.; Claustre, H.&Goyet, C., (2008). Calcite production by coccolithophores in the south east Pacific Ocean. *Biogeosciences* Vol. 5, No. 4, pp. (1101-1117), 1726-4189

Beaufort, L.; Probert, I.; de Garidel-Thoron, T.; Bendif, E.M.; Ruiz-Pino, D.; Metzl, N., et al., (2011). Sensitivity of coccolithophores to carbonate chemistry and ocean acidification. *Nature* Vol. 476, No. 7358, pp. (80-83), 0028-0836

Beman, J.M.; Chow, C.-E.; King, A.L.; Feng, Y.; Fuhrman, J.A.; Andersson, A., et al., (2011). Global declines in oceanic nitrification rates as a consequence of ocean acidification. *Proceedings - National Academy Of Sciences USA* Vol. 108, No. 1, pp. (208-213), 1091-6490

Berninger, U.-G.&Wickham, S.A., (2005). Response of the microbial food web to manipulation of nutrients and grazers in the oligotrophic Gulf of Aqaba and northern Red Sea. *Marine Biology* Vol. 147, No. 4, pp. (1017-1032), 1432-1793

Bianchi, C.N., (2007). Biodiversity issues for the forthcoming tropical Mediterranean Sea. *Hydrobiologia* Vol. 580, No. 1, pp. (7-21), 0018-8158

Bianchi, C.N.&Morri, C., (2000). Marine Biodiversity of the Mediterranean Sea: Situation, Problems and Prospects for Future Research. *Marine Pollution Bulletin* Vol. 40, No. 5, pp. (367-376), 0025-326X

Bonnet, S.; Grosso, O.&Moutin, T., (2011). Planktonic dinitrogen fixation in the Mediterranean Sea: a major biogeochemical process during the stratified period? *Biogeosciences Discuss* Vol. 8, No. 1, pp. (1197-1225), 1810-6285

Bonnet, S.; Guieu, C.; Bruyant, F.; Prášil, O.; Van Wambeke, F.; Raimbault, P., et al., (2008). Nutrient limitation of primary productivity in the Southeast Pacific (BIOSOPE cruise). *Biogeosciences* Vol. 5, No. 1, pp. (215-225), 1726-4189

Bryant, D.A., (2003). The beauty in small things revealed. *Proceedings - National Academy Of Sciences USA* Vol. 100, No. 17, pp. (9647-9649), 0027-8424

Byrne, M., (2011). Impact of ocean warming and ocean acidification on marine invertebrate life history stages: Vulnerabilities and potential for persistence in a changing ocean, in: *Oceanography and Marine Biology: An Annual Review*, R. N. Gibson, R. J. A. Atkinson, J. D. M. Gordon, I. P. Smith and D. J. Hughes (Editors), pp. (1-42), CRC Press, 978-1-4398536-4-1,

Calbet, A., (2008). The trophic roles of microzooplankton in marine systems. *ICES Journal of Marine Science: Journal du Conseil* Vol. 65, No. 3, pp. (325-331), 1054-3139

Cantin, N.E.; Cohen, A.L.; Karnauskas, K.B.; Tarrant, A.M.&McCorkle, D.C., (2010). Ocean Warming Slows Coral Growth in the Central Red Sea. *Science* Vol. 329, No. 5989, pp. (322-325), 1095-9203

Capone, D.G.; Zehr, J.P.; Paerl, H.W.; Bergman, B.&Carpenter, E.J., (1997). *Trichodesmium*, a Globally Significant Marine Cyanobacterium. *Science* Vol. 276, No. 5316, pp. (1221-1229), 0036-8075

Cavanagh, R.D.&Gibson, C., (2007). *Overview of the conservation status of cartilaginous fishes (Chondrichthyans) in the Mediterranean Sea*, The World Conservation Union (IUCN), 978-2-8317-0997-0, Gland, Switzerland and Malaga, Spain.

Cermeño, P.&Figueiras, F.G., (2008). Species richness and cell-size distribution: size structure of phytoplankton communities. *Marine Ecology Progress Series* Vol. 357, No. pp. (79-85), 0171-8630

Chevaldonné, P.&Lejeusne, C., (2003). Regional warming-induced species shift in north-west Mediterranean marine caves. *Ecology Letters* Vol. 6, No. 4, pp. (371-379), 1461-0248

Chopin, T., (2006). Integrated Multi-Trophic Aquaculture. What it is and why you should care… and don't confuse it with polyculture. *Northern Aquaculture* Vol. 12, No. 4, pp. (4), 1183-2428

Christaki, U.; Courties, C.; Karayanni, H.; Giannakourou, A.; Maravelias, C.; Kormas, K.A., et al., (2002). Dynamic Characteristics of Prochlorococcus and Synechococcus Consumption by Bacterivorous Nanoflagellates. *Microbial Ecology* Vol. 43, No. 3, pp. (341-352), 1432-184X

Church, M.J.; Mahaffey, C.; Letelier, R.M.; Lukas, R.; Zehr, J.P.&Karl, D.M., (2009). Physical forcing of nitrogen fixation and diazotroph community structure in the North Pacific subtropical gyre. *Global Biogeochem. Cycles* Vol. 23, No. 2, pp. (GB2020), 0886-6236

Cigliano, M.; Gambi, M.; Rodolfo-Metalpa, R.; Patti, F.&Hall-Spencer, J., (2010). Effects of ocean acidification on invertebrate settlement at volcanic CO_2 vents. *Marine Biology* Vol. 157, No. 11, pp. (2489-2502), 1432-1793

Claessens, M.; Wickham, S.A.; Post, A.F.&Reuter, M., (2008). Ciliate community in the oligotrophic Gulf of Aqaba, Red Sea. *Aquatic Microbial Ecology* Vol. 53, No. 2, pp. (181-190), 1616-1564

Claudet, J.&Fraschetti, S., (2010). Human-driven impacts on marine habitats: A regional meta-analysis in the Mediterranean Sea. *Biological Conservation* Vol. 143, No. 9, pp. (2195-2206), 0006-3207

Claustre, H.; Huot, Y.; Obernosterer, I.; Gentili, B.; Tailliez, D.&Lewis, M., (2008). Gross community production and metabolic balance in the South Pacific Gyre, using a non intrusive bio-optical method. *Biogeosciences* Vol. 5, No. 2, pp. (463-474), 1726-4189

Coll, M.; Piroddi, C.; Steenbeek, J.; Kaschner, K.; Ben Rais Lasram, F.; Aguzzi, J., et al., (2010). The biodiversity of the mediterranean sea: estimates, patterns, and threats. *PLoS One* Vol. 5, No. 8, (eng), pp. (e11842-e11842), 1932-6203

Costello, M.J.; Coll, M.; Danovaro, R.; Halpin, P.; Ojaveer, H.&Miloslavich, P., (2010). A Census of Marine Biodiversity Knowledge, Resources, and Future Challenges. *PLoS ONE* Vol. 5, No. 8, pp. (e12110), 1932-6203

Czerny, J.; Barcelos e Ramos, J.&Riebesell, U., (2009). Influence of elevated CO_2 concentrations on cell division and nitrogen fixation rates in the bloom-forming cyanobacterium *Nodularia spumigena*. *Biogeosciences Discussions* Vol. 6, No. 2, pp. (4279-4304), 1810-6285

D'Ortenzio, F.&Ribera d'Alcalà, M., (2009). On the trophic regimes of the Mediterranean Sea: a satellite analysis. *Biogeosciences* Vol. 6, No. 2, pp. (139-148), 1726-4189

Danovaro, R.; Fonda Umani, S.&Pusceddu, A., (2009). Climate Change and the Potential Spreading of Marine Mucilage and Microbial Pathogens in the Mediterranean Sea. *PLoS ONE* Vol. 4, No. 9, pp. (e7006), 1932-6203

Danovaro, R.&Pusceddu, A., (2007). Ecomanagement of biodiversity and ecosystem functioning in the Mediterranean Sea: concerns and strategies. *Chemistry and Ecology* Vol. 23, No. 5, pp. (347-360), 1029-0370

Dasgupta, T.&Chattopadhyay, R.N., (2004). Ecological Contradictions Through Ages: Growth and Decay of the Indus and Nile Valley Civilizations. *Journal of Human Ecology* Vol. 16, No. 3, pp. (197-201), 0970-9274

Dave, A.C.&Lozier, M.S., (2010). Local stratification control of marine productivity in the subtropical North Pacific. *J. Geophys. Res.* Vol. 115, No. C12, pp. (C12032), 0148-0227

Davison, P.&Asch, R.G., (2011). Plastic ingestion by mesopelagic fishes in the North Pacific Subtropical Gyre. *Marine Ecology Progress Series* Vol. 432, No. pp. (173-180), 1616-1599

Dekas, A.E.; Poretsky, R.S.&Orphan, V.J., (2009). Deep-Sea Archaea Fix and Share Nitrogen in Methane-Consuming Microbial Consortia. *Science* Vol. 326, No. 5951, pp. (422-426), 1095-9203

DeLong, E.F., (2009). The microbial ocean from genomes to biomes. *Nature* Vol. 459, No. 7244, pp. (200-206), 0028-0836

Denis, M.; Thyssen, M.; Martin, V.; Manca, B.&Vidussi, F., (2009). Ultraphytoplankton distribution and upper ocean dynamics in the eastern Mediterranean during winter. *Biogeosciences Discuss* Vol. 6, No. 4, pp. (6839-6887), 1810-6285

Denman, K.; Christian, J.R.; Steiner, N.; Pörtner, H.-O.&Nojiri, Y., (2011). Potential impacts of future ocean acidification on marine ecosystems and fisheries: current knowledge and recommendations for future research. *ICES Journal of Marine Science: Journal du Conseil* Vol. 68, No. 6, pp. (1019-1029), 1095-9289

Denuncio, P.; Bastida, R.; Dassis, M.; Giardino, G.; Gerpe, M.&Rodríguez, D., (2011). Plastic ingestion in Franciscana dolphins, Pontoporia blainvillei (Gervais and d'Orbigny, 1844), from Argentina. *Marine Pollution Bulletin* Vol. 62, No. 8, pp. (1836-1941), 0025-326X

Diaz-Almela, E.; Marba, N.; Martinez, R.; Santiago, R.&Duarte, C., M., (2009). Seasonal dynamics of Posidonia oceanica in Magalluf Bay (Mallorca, Spain): Temperature effects on seagrass mortality. *Limnology and Oceanography* Vol. 54, No. 6, (0024-3590), pp. (2170-2182), 1939-5590

Dolan, J.R.; Ritchie, M.E.&Ras, J., (2007). The "neutral" community structure of planktonic herbivores, tintinnid ciliates of the microzooplankton, across the SE Tropical Pacific Ocean. *Biogeosciences* Vol. 4, No. 3, pp. (297-310), 1726-4189

Dolman, S.J.; Evans, P.G.H.; Notarbartolo-di-Sciara, G.&Frisch, H., (2011). Active sonar, beaked whales and European regional policy. *Marine Pollution Bulletin* Vol. 63, No. 1-4, pp. (27-34), 0025-326X

Doney, S.C., (2010). The Growing Human Footprint on Coastal and Open-Ocean Biogeochemistry. *Science* Vol. 328, No. 5985, pp. (1512-1516), 1095-9203

Dore, J.E.; Letelier, R.M.; Church, M.J.; Lukas, R.&Karl, D.M., (2008). Summer phytoplankton blooms in the oligotrophic North Pacific Subtropical Gyre:

Historical perspective and recent observations. *Progress In Oceanography* Vol. 76, No. 1, pp. (2-38), 0079-6611

Duarte, C.; Holmer, M.; Olsen, Y.; Soto, D.; Marbà, N.; Guiu, J., et al., (2009). Will the oceans help feed humanity? *BioScience* Vol. 59, No. 11, pp. (967-976), 0006-3568

Duarte, C.M., (1991). Seagrass depth limits. *Aquatic Botany* Vol. 40, No. 4, pp. (363-377), 0304-3770

Duhamel, S.; Dyhrman, S.T.&Karl, D.M., (2011). Alkaline phosphatase activity and regulation in the North Pacific Subtropical Gyre. *Limnology and Oceanography* Vol. 55, No. 3, (0024-3590), pp. (1414-1425), 1939-5590

Durrieu de Madron, X.; Guieu, C.; Sempéré, R.; Conan, P.; Cossa, D.; Dβ€™Ortenzio, F., et al., (2011). Marine ecosystems' responses to climatic and anthropogenic forcings in the Mediterranean. *Progress In Oceanography* Vol. 91, No. 2, pp. (97-166), 0079-6611

Dyhrman, S.T.; Ammerman, J.W.&Van Mooy, B.A.S., (2009). Microbes and the Marine Phosphorus Cycle. *Oceanography* Vol. 20, No. 2, pp. (110-116), 1042-8275

Eden, B.R.; Steinberg, D.K.; Goldthwait, S.A.&Mcgillicuddy, D.J., (2009). Zooplankton community structure in a cyclonic and mode-water eddy in the Sargasso Sea. *Deep Sea Research Part I: Oceanographic Research Papers* Vol. 56, No. 10, pp. (1757-1776), 0967-0637

Eiler, A.; Hayakawa, D.H.&Rapp?, M.S., (2011). Non-random assembly of bacterioplankton communities in the subtropical North Pacific Ocean. *Frontiers in Microbiology* Vol. 2, No. 140, (English), pp. (1-12), 1664-302X

Field, C.B.; Behrenfeld, M.J.; Randerson, J.T.&Falkowski, P., (1998). Primary Production of the Biosphere: Integrating Terrestrial and Oceanic Components. *Science - New York then Washington* Vol. 281, No. 5374, pp. (237-240), 0036-8075

Fong, A.A.; Karl, D.M.; Lukas, R.; Letelier, R.M.; Zehr, J.P.&Church, M.J., (2008). Nitrogen fixation in an anticyclonic eddy in the oligotrophic North Pacific Ocean. *ISME J* Vol. 2, No. 6, pp. (663-676), 1751-7370

Foster, R.A.; Paytan, A.&Zehr, J.P., (2009). Seasonality of N_2 fixation and nifH gene diversity in the Gulf of Aqaba (Red Sea). *Limnology and Oceanography* Vol. 54, No. 1, pp. (219-233), 1939-5590

Fricke, H.W.&Schuhmacher, H., (1983). The Depth Limits of Red Sea Stony Corals: An Ecophysiological Problem (A Deep Diving Survey by Submersible). *Marine Ecology* Vol. 4, No. 2, pp. (163-194), 1616-1599

Fu, F.-X.; Mulholland, M.R.; Garcia, N.S.; Beck, A.; Bernhardt, P.W.; Warner, M.E., et al., (2008). Interactions between changing pCO sub(2), N sub(2) fixation, and Fe limitation in the marine unicellular cyanobacterium Crocosphaera. *Limnology and Oceanography* Vol. 53, No. 6, pp. (2472-2484), 1939-5590

Galil, B., (2009). Taking stock: inventory of alien species in the Mediterranean sea. *Biological Invasions* Vol. 11, No. 2, pp. (359-372), 1573-1464

Gambaiani, D.D.; Mayol, P.; Isaac, S.J.&Simmonds, M.P., (2009). Potential impacts of climate change and greenhouse gas emissions on Mediterranean marine ecosystems and cetaceans. *Journal of the Marine Biological Association of the United Kingdom* Vol. 89, No. 01, pp. (179-201), 1469-7769

García-Charton, J.A.; Pérez-Ruzafa, A.; Marcos, C.; Claudet, J.; Badalamenti, F.; Benedetti-Cecchi, L., et al., (2008). Effectiveness of European Atlanto-Mediterranean MPAs:

Do they accomplish the expected effects on populations, communities and ecosystems? *Journal for Nature Conservation* Vol. 16, No. 4, pp. (193-221), 1617-1381

Garrabou, J.; Coma, R.; Bensoussan, N.; Bally, M.; ChevaldonnÉ, P.; Cigliano, M., et al., (2009). Mass mortality in Northwestern Mediterranean rocky benthic communities: effects of the 2003 heat wave. *Global Change Biology* Vol. 15, No. 5, pp. (1090-1103), 1365-2486

Gist, N.; Serret, P.; Woodward, E.M.S.; Chamberlain, K.&Robinson, C., (2009). Seasonal and spatial variability in plankton production and respiration in the Subtropical Gyres of the Atlantic Ocean. *Deep Sea Research Part II: Topical Studies in Oceanography* Vol. 56, No. 15, pp. (931-940), 0967-0645

Goffart, A.; Hecq, J.H.&Legendre, L., (2002). Changes in the development of the winterspring phytoplankton bloom in the Bay of Calvi (NW Mediterranean) over the last two decades: a response to changing climate? *Marine Ecology Progress Series* Vol. 236, No. 45-60, pp. 0171-8630

Gómez-Gutiérrez, A.; Garnacho, E.; Bayona, J.M.&Albaigés, J., (2007). Assessment of the Mediterranean sediments contamination by persistent organic pollutants. *Environmental Pollution* Vol. 148, No. 2, pp. (396-408), 0269-7491

Gómez, F., (2007). On the consortium of the tintinnid Eutintinnus and the diatom Chaetoceros in the Pacific Ocean. *Marine Biology* Vol. 151, No. 5, pp. (1899-1906), 0025-3162

Gregory, M.R., (2009). Environmental implications of plastic debris in marine settings - entanglement, ingestion, smothering, hangers-on, hitch-hiking and alien invasions. *Philosophical Transactions of the Royal Society B: Biological Sciences* Vol. 364, No. 1526, pp. (2013-2025), 1471-2970

Grob, C.; Hartmann, M.; Zubkov, M.V.&Scanlan, D.J., (2011). Invariable biomass-specific primary production of taxonomically discrete picoeukaryote groups across the Atlantic Ocean. *Environmental Microbiology* Vol., No. pp. 1462-2920

Gruber, N., (2011). Warming up, turning sour, losing breath: ocean biogeochemistry under global change. *Phil. Trans. R. Soc. A.* Vol. 369, No. 1943, pp. (1980-1996), 1471-2962

Hall-Spencer, J.M.; Rodolfo-Metalpa, R.; Martin, S.; Ransome, E.; Fine, M.; Turner, S.M., et al., (2008). Volcanic carbon dioxide vents show ecosystem effects of ocean acidification. *Nature* Vol. 454, No. 7200, pp. (96-99), 1476-4687

Harwani, S.; Henry, R.W.; Rhee, A.; Kappes, M.A.; Croll, D.A.; Petreas, M., et al., (2011). Legacy and contemporary persistent organic pollutants in North Pacific albatross. *Environmental Toxicology and Chemistry* Vol. 30, No. 11, pp. (2562-2569), 1552-8618

Hasler, H.&Ott, J.A., (2008). Diving down the reefs? Intensive diving tourism threatens the reefs of the northern Red Sea. *Marine Pollution Bulletin* Vol. 56, No. 10, pp. (1788-1794), 0025-326X

Hawkes, L.A.; Broderick, A.C.; Godfrey, M.H.&Godley, B.J., (2007). Investigating the potential impacts of climate change on a marine turtle population. *Global Change Biology* Vol. 13, No. 5, pp. (923-932), 1365-2486

Hoegh-Guldberg, O.&Bruno, J.F., (2010). The Impact of Climate Change on the World's Marine Ecosystems. *Science* Vol. 328, No. 5985, pp. (1523-1528), 1095-9203

Hoepner, T.&Lattemann, S., (2003). Chemical impacts from seawater desalination plants – a case study of the northern Red Sea. *Desalination* Vol. 152, No. 1-3, pp. (133-140), 0011-9164

Holmer, M.; Argyrou, M.; Dalsgaard, T.; Danovaro, R.; Diaz-Almela, E.; Duarte, C.M., et al., (2008). Effects of fish farm waste on *Posidonia oceanica* meadows: Synthesis and provision of monitoring and management tools. *Marine Pollution Bulletin* Vol. 56, No. 9, pp. (1618-1629), 0025-326X

Hutchinson, G.E., (1961). The Paradox of the Plankton. *The American Naturalist* Vol. 95, No. 882, pp. (137-145), 00030147

Ignatiades, L.; Gotsis-Skretas, O.; Pagou, K.&Krasakopoulou, E., (2009). Diversification of phytoplankton community structure and related parameters along a large-scale longitudinal eastβ€"west transect of the Mediterranean Sea. *Journal of Plankton Research* Vol. 31, No. 4, pp. (411-428), 1464-3774

Invers, O.; Romero, J.&Pérez, M., (1997). Effects of pH on seagrass photosynthesis: a laboratory and field assessment. *Aquatic Botany* Vol. 59, No. 3-4, pp. (185-194), 0304-3770

Irwin, A.J.&Oliver, M.J., (2009). Are ocean deserts getting larger? *Geophysical Research Letters* Vol. 36, No. 18, pp. (L18609), 0094–8276

Israel, A.&Hophy, M., (2002). Growth, photosynthetic properties and Rubisco activities and amounts of marine macroalgae grown under current and elevated seawater CO2 concentrations. *Global Change Biology* Vol. 8, No. 9, pp. (831-840), 1365-2486

Jacobsen, J.K.; Massey, L.&Gulland, F., (2010). Fatal ingestion of floating net debris by two sperm whales (*Physeter macrocephalus*). *Marine Pollution Bulletin* Vol. 60, No. 5, pp. (765-767), 0025-326X

Kähler, P.; Oschlies, A.; Dietze, H.&Koeve, W., (2010). Oxygen, carbon, and nutrients in the oligotrophic eastern subtropical North Atlantic. *Biogeosciences* Vol. 7, No. 3, pp. (1143-1156), 1726-4189

Karakassis, I.; Tsapakis, M.; Hatziyanni, E.; Papadopoulou, K.N.&Plaiti, W., (2000). Impact of cage farming of fish on the seabed in three Mediterranean coastal areas. *ICES Journal of Marine Science* Vol. 57, No. 5, pp. (1462-1471), 1095-9289

Karamanlidis, A.A.; Androukaki, E.; Adamantopoulou, S.; Chatzispyrou, A.; Johnson, W.M.; Kotomatas, S., et al., (2008). Assessing accidental entanglement as a threat to the Mediterranean monk seal *Monachus monachus Endangered Species Research* Vol. 5, No. 2-3, pp. (205-213), 1613-4796

Karl, D.M.; BjoKrkman, K.M.; Dore, J.E.; Fujieki, L.; Hebel, D.V.; Houlihan, T., et al., (2001). Ecological nitrogen-to-phosphorus stoichiometry at station ALOHA. *Deep Sea Research Part II: Topical Studies in Oceanography* Vol. 48, No. 8-9, pp. (1529-1566), 0967-0645

Karl, D.M.; Christian, J.R.; Dore, J.E.; Hebel, D.V.; Letelier, R.M.; Tupas, L.M., et al., (1996). Seasonal and interannual variability in primary production and particle flux at Station ALOHA. *Deep Sea Research Part II: Topical Studies in Oceanography* Vol. 43, No. 2-3, pp. (539-568), 0967-0645

Katsanevakis, S.; Tsiamis, K.; Ioannou, G.; Michailidis, N.&Zenetos, A., (2009). Inventory of alien marine species of Cyprus *Mediterranean Marine Science* Vol. 10, No. 2, pp. (109-133), 1791-6763

Keeling, R.F.; Körtzinger, A.&Gruber, N., (2010). Ocean Deoxygenation in a Warming World. *Annual Review of Marine Science* Vol. 2, No. 1, pp. (199-229), 1941-0611

Kirkham, A.R.; Jardillier, L.E.; Tiganescu, A.; Pearman, J.; Zubkov, M.V.&Scanlan, D.J., (2011). Basin-scale distribution patterns of photosynthetic picoeukaryotes along an

Atlantic Meridional Transect. *Environmental Microbiology* Vol. 13, No. 4, pp. (975-990), 1462-2920

Krom, M.D.; Emeis, K.C.&Van Cappellen, P., (2010). Why is the Eastern Mediterranean phosphorus limited? *Progress In Oceanography* Vol. 85, No. 3-4, pp. (236-244), 0079-6611

Krom, M.D.; Thingstad, T.F.; Brenner, S.; Carbo, P.; Drakopoulos, P.; Fileman, T.W., et al., (2005). Summary and overview of the CYCLOPS P addition Lagrangian experiment in the Eastern Mediterranean. *Deep Sea Research Part II: Topical Studies in Oceanography* Vol. 52, No. 22-23, pp. (3090-3108), 0967-0645

Kustka, A.; Sañudo-Wilhelmy, S.; Carpenter, E.J.; Capone, D.G.&Raven, J.A., (2003). A revised estimate of the iron use efficiency of nitrogen fixation, with special reference to the marine cyanobacterium *Trichodesmium* spp. (Cyanophyta). *Journal of Phycology* Vol. 39, No. 1, pp. (12-25), 0022-3646

Labiosa, R.G.; Arrigo, K.R.; Genin, A.; Monismith, S.G.&Van Dijken, G., (2003). The interplay between upwelling and deep convective mixing in determining the seasonal phytoplankton dynamics in the Gulf of Aqaba: Evidence from SeaWiFS and MODIS. *Limnology and Oceanography* Vol. 48, No. 6, pp. (2355-2368), 0024-3590

Lasram, F.B.R.&Mouillot, D., (2009). Increasing southern invasion enhances congruence between endemic and exotic Mediterranean fish fauna. *Biological Invasions* Vol. 11, No. 3, pp. (697-711), 1573-1464

Law, K.L.; Morét-Ferguson, S.; Maximenko, N.A.; Proskurowski, G.; Peacock, E.E.; Hafner, J., et al., (2010). Plastic Accumulation in the North Atlantic Subtropical Gyre. *Science* Vol. 329, No. 5996, pp. (1185-1188), 1095-9203

Le Moal, M.; Collin, H.&Biegala, I.C., (2011). Intriguing diversity among diazotrophic picoplankton along a Mediterranean transect: a dominance of rhizobia. *Biogeosciences* Vol. 8, No. 3, pp. (827-840), 1726-4189

Lejeusne, C.; Chevaldonné, P.; Pergent-Martini, C.; Boudouresque, C.F.&Pérez, T., (2010). Climate change effects on a miniature ocean: the highly diverse, highly impacted Mediterranean Sea. *Trends in ecology & evolution (Personal edition)* Vol. 25, No. 4, pp. (250-260), 0169-5347

Li, B.; Karl, D.M.; Letelier, R.M.&Church, M.J., (2011). Size-dependent photosynthetic variability in the North Pacific Subtropical Gyre. *Marine Ecology Progress Series* Vol. 440, No. pp. (27-40), 1616-1599

Lindell, D.&Post, A.F., (1995). Ultraphytoplankton succession is triggered by deep winter mixing in the Gulf of Aqaba (Eilat), Red Sea. *Limnology and Oceanography* Vol. 40, No. 6, (0024-3590), pp. (1130-1141), 0024-3590

Liu, H.; Probert, I.; Uitz, J.; Claustre, H.; Aris-Brosou, S.; Frada, M., et al., (2009). Extreme diversity in noncalcifying haptophytes explains a major pigment paradox in open oceans. *Proceedings - National Academy Of Sciences USA* Vol. 106, No. 31, pp. (12803-12808), 0027-8424

Liu, J.; Weinbauer, M.G.; Maier, C.; Dai, M.&Gattuso, J.-P., (2010). Effect of ocean acidification on microbial diversity and on microbe-driven biogeochemistry and ecosystem functioning *Aquatic Microbial Ecology* Vol. 61, No. 3, pp. (291-305), 1616-1564

Lloret, J., (2010). Human health benefits supplied by Mediterranean marine biodiversity. *Marine Pollution Bulletin* Vol. 60, No. 10, pp. (1640-1646), 0025-326X

Lomas, M.W.; Burke, A.L.; Lomas, D.A.; Bell, D.W.; Shen, C.; Dyhrman, S.T., et al., (2010). Sargasso Sea phosphorus biogeochemistry: an important role for dissolved organic phosphorus (DOP). *Biogeosciences* Vol. 7, No. 2, pp. (695-710), 1726-4189

Lomas, M.W.&Moran, S.B., (2011). Evidence for aggregation and export of cyanobacteria and nano-eukaryotes from the Sargasso Sea Euphotic zone. *Biogeosciences* Vol. 8, No. 1, pp. (203–216), 1726-4189

MacKenzie, B.R.; Mosegaard, H.&Rosenberg, A.A., (2009). Impending collapse of bluefin tuna in the northeast Atlantic and Mediterranean. *Conservation Letters* Vol. 2, No. 1, pp. (26-35), 1755-263X

Mackey, K.R.M.; Labiosa, R.G.; Calhoun, M.; Street, J.H.; Post, A.F.&Paytan, A., (2007). Phosphorus availability, phytoplankton community dynamics, and taxon-specific phosphorus status in the Gulf of Aqaba, Red Sea. *Limnology and Oceanography* Vol. 52, No. 2, pp. (873-885), 0024-3590

Malmstrom, R.R.; Coe, A.; Kettler, G.C.; Martiny, A.C.; Frias-Lopez, J.; Zinser, E.R., et al., (2010). Temporal dynamics of *Prochlorococcus* ecotypes in the Atlantic and Pacific oceans. *ISME Journal* Vol. 4, No. 10, pp. (1252-1264), 1751-7362

Mandalakis, M.&Stephanou, E.G., (2002). Study of atmospheric PCB concentrations over the eastern Mediterranean Sea. *Journal of Geophysical Research* Vol. 107, No. D23, pp. (4716), 2156-2202

Marañón, E.; Fernández, A.; Mouriño-Carballido, B.; Martínez-García, S.; Teira, E.; Cermeño, P., et al., (2010). Degree of oligotrophy controls the response of microbial plankton to Saharan dust *Limnology and Oceanography* Vol. 55, No. 6, pp. (2339-2352), 0024-3590

Marinov, I.; Doney, S.C.&Lima, I.D., (2010). Response of ocean phytoplankton community structure to climate change over the 21st century: partitioning the effects of nutrients, temperature and light. *Biogeosciences Discuss* Vol. 7, No. 3, pp. (4565-4606), 1810-6285

Markaki, Z.; Loÿe-Pilot, M.D.; Violaki, K.; Benyahya, L.&Mihalopoulos, N., (2010). Variability of atmospheric deposition of dissolved nitrogen and phosphorus in the Mediterranean and possible link to the anomalous seawater N/P ratio. *Marine Chemistry* Vol. 120, No. 1-4, pp. (187-194), 0304-4203

Martin, S.&Gattuso, J.-P., (2009). Response of Mediterranean coralline algae to ocean acidification and elevated temperature. *Global Change Biology* Vol. 15, No. 8, pp. (2089-2100), 1365-2486

Martin, S.; Rodolfo-Metalpa, R.; Ransome, E.; S, R.; Buia, M.C.; Gattuso, J.P., et al., (2008). Effects of naturally acidified seawater on seagrass calcareous epibionts. *Biology Letters* Vol. 4, No. 6, pp. (689-692), 1744-957X

Martinez-Garcia, S.; Fernandez, E.; Calvo-Diaz, A.; Maranon, E.; Moran, X.A.G.&Teira, E., (2010). Response of heterotrophic and autotrophic microbial plankton to inorganic and organic inputs along a latitudinal transect in the Atlantic Ocean. *Biogeosciences* Vol. 7, No. 5, pp. (1701-1713), 1726-4189

Martiny, A.C.; Huang, Y.&Li, W., (2009). Occurrence of phosphate acquisition genes in Prochlorococcus cells from different ocean regions. *Environmental Microbiology* Vol. 11, No. 6, pp. (1340-1347), 1462-2920

Mary, I.; Heywood, J.L.; Fuchs, B.M.; Amann, R.; Tarran, G.A.; Burkill, P.H., et al., (2006). SAR11 dominance among metabolically active low nucleic acid bacterioplankton in surface waters along an Atlantic meridional transect. *Aquatic Microbial Ecology* Vol. 45, No. 2, pp. (107-113), 0948-3055

Masquelier, S.&Vaulot, D., (2008). Distribution of micro-organisms along a transect in the South-East Pacific Ocean (BIOSOPE cruise) using epifluorescence microscopy. *Biogeosciences* Vol. 5, No. 2, pp. (311-321), 1726-4189

Massana, R.&Pedrós-Alió, C., (2008). Unveiling new microbial eukaryotes in the surface ocean. *Current Opinion in Microbiology* Vol. 11, No. 3, pp. (213-218), 1369-5274

Mather, R.L.; Reynolds, S.E.; Wolff, G.A.; Williams, R.G.; Torres-Valdes, S.; Woodward, E.M.S., et al., (2008). Phosphorus cycling in the North and South Atlantic Ocean subtropical gyres. *Nature Geoscience* Vol. 1, No. 7, pp. (439-443), 1752-0894

Maximenko, N.; Hafner, J.&Niiler, P., (2011). Pathways of marine debris derived from trajectories of Lagrangian drifters. *Marine Pollution Bulletin* Vol., No. pp. 0025-326X

Mazzariol, S.; Di Guardo, G.; Petrella, A.; Marsili, L.; Fossi, C.M.; Leonzio, C., et al., (2011). Sometimes Sperm Whales (*Physeter macrocephalus*) Cannot Find Their Way Back to the High Seas: A Multidisciplinary Study on a Mass Stranding. *PLoS ONE* Vol. 6, No. 5, pp. (e19417), 1932-6203

McClain, C.R.; Signorini, J.R.&Christian, S.R., (2004). Subtropical gyre variability observed by ocean-color satellites. *Deep Sea Research Part II* Vol. 51, No. 1-3, pp. (281-301), 09670645

Moisander, P.H.; Beinart, R.A.; Hewson, I.; White, A.E.; Johnson, K.S.; Carlson, C.A., et al., (2010). Unicellular Cyanobacterial Distributions Broaden the Oceanic N_2 Fixation Domain. *Science* Vol. 327, No. 5972, pp. (1512-1514), 1095-9203

Montefalcone, M.; Albertelli, G.; Morri, C.; Parravicini, V.&Bianchi, C.N., (2009). Legal protection is not enough: *Posidonia oceanica* meadows in marine protected areas are not healthier than those in unprotected areas of the northwest Mediterranean Sea. *Marine Pollution Bulletin* Vol. 58, No. 4, pp. (515-519), 0025-326X

Moore, C.J.; Moore, S.L.; Leecaster, M.K.&Weisberg, S.B., (2001). A Comparison of Plastic and Plankton in the North Pacific Central Gyre. *Marine Pollution Bulletin* Vol. 42, No. 12, pp. (1297-1300), 0025-326X

Moore, L.R.; Ostrowski, M.; Scanlan, D.J.; Feren, K.&Sweetsir, T., (2005). Ecotypic variation in phosphorus-acquisition mechanisms within marine picocyanobacteria. *Aquatic Microbial Ecology* Vol. 39, No. 3, pp. (257-269), 1616-1564

Moore, M.C.; Mills, M.M.; Achterberg, E.P.; Geider, R.J.; LaRoche, J.; Lucas, M.I., et al., (2009). Large-scale distribution of Atlantic nitrogen fixation controlled by iron availability. *Nature Geoscience* Vol. 2, No. 12, pp. (867-871), 1752-0894

Morel, A.; Claustre, H.&Gentili, B., (2010). The most oligotrophic subtropical zones of the global ocean: similarities and differences in terms of chlorophyll and yellow substance. *Biogeosciences* Vol. 7, No. 10, pp. (3139-3151), 1726-4189

Morel, A.; Gentili, B.; Claustre, H.; Babin, M.; Bricaud, A.; Ras, J., et al., (2007). Optical properties of the "clearest" natural waters. *Limnology and Oceanography* Vol. 52, No. 1, pp. (217-229), 0024-3590

Mrosovsky, N.; Ryan, G.D.&James, M.C., (2009). Leatherback turtles: The menace of plastic. *Marine Pollution Bulletin* Vol. 58, No. 2, pp. (287-289), 0025-326X

Myers, N.; Mittermeier, R.A.; Mittermeier, C.G.; da Fonseca, G.A.B.&Kent, J., (2000). Biodiversity hotspots for conservation priorities. *Nature* Vol. 403, No. 6772, pp. (853-858), 1476-4687

Not, F.; Latasa, M.; Scharek, R.; Viprey, M.; Karleskind, P.; Balagué, V., et al., (2008). Protistan assemblages across the Indian Ocean, with a specific emphasis on the picoeukaryotes. *Deep Sea Research Part I: Oceanographic Research Papers* Vol. 55, No. 11, pp. (1456-1473), 0967-0637

Orr, J.C.; Fabry, V.J.; Aumont, O.; Bopp, L.; Doney, S.C.; Feely, R.A., et al., (2005). Anthropogenic ocean acidification over the twenty-first century and its impact on calcifying organisms. *Nature* Vol. 437, No. 7059, pp. (681-686), 1476-4687

Palomera, I.; Olivar, M.P.; Salat, J.; SabatΓ©s, A.; Coll, M.; GarcΓa, A., et al., (2007). Small pelagic fish in the NW Mediterranean Sea: An ecological review. *Progress In Oceanography* Vol. 74, No. 2-3, pp. (377-396), 0079-6611

Pedlosky, J., (1998). *Ocean Circulation Theory* (2 ed.), Springer-Verlag Berlin Heidelberg, 3-540-60489-8, New York.

Pergent-Martini, C.; Boudouresque, C.-F.; Pasqualini, V.&Pergent, G., (2006). Impact of fish farming facilities on Posidonia oceanica meadows: a review. *Marine Ecology* Vol. 27, No. 4, pp. (310-319), 1439-0485

Perruche, C.; RiviΓ˙re, P.; Pondaven, P.&Carton, X., (2010). Phytoplankton competition and coexistence: Intrinsic ecosystem dynamics and impact of vertical mixing. *Journal of Marine Systems* Vol. 81, No. 1-2, pp. (99-111), 0924-7963

PlasticsEurope, 2011. Plastics - the Facts 2011 - An analysis of European plastics production, demand and recovery for 2010. http://www.plasticseurope.org/.

Polovina, J.J.; Dunne, J.P.; Woodworth, P.A.&Howell, E.A., (2011). Projected expansion of the subtropical biome and contraction of the temperate and equatorial upwelling biomes in the North Pacific under global warming. *ICES Journal of Marine Science: Journal du Conseil* Vol. 68, No. 6, pp. (986-995), 1095-9289

Polovina, J.J.; Howell, E.A.&Abecassis, M., (2008). Ocean's least productive waters are expanding. *Geophys. Res. Lett.* Vol. 35, No. 3, pp. (L03618), 0094-8276

Porzio, L.; Buia, M.C.&Hall-Spencer, J.M., (2011). Effects of ocean acidification on macroalgal communities. *Journal of Experimental Marine Biology and Ecology* Vol. 400, No. 1-2, pp. (278-287), 0022-0981

Poulton, A.J.; Holligan, P.M.; Hickman, A.; Kim, Y.-N.; Adey, T.R.; Stinchcombe, M.C., et al., (2006). Phytoplankton carbon fixation, chlorophyll-biomass and diagnostic pigments in the Atlantic Ocean. *Deep Sea Research Part II: Topical Studies in Oceanography* Vol. 53, No. 14-16, pp. (1593-1610), 0967-0645

Pusceddu, A.; Fiordelmondo, C.; Polymenakou, P.; Polychronaki, T.; Tselepides, A.&Danovaro, R., (2005). Effects of bottom trawling on the quantity and biochemical composition of organic matter in coastal marine sediments (Thermaikos Gulf, northwestern Aegean Sea). *Continental Shelf Research* Vol. 25, No. 19-20, pp. (2491-2505), 0278-4343

Raimbault, P.&Garcia, N., (2007). Carbon and nitrogen uptake in the South Pacific Ocean: evidence for efficient dinitrogen fixation and regenerated production leading to large accumulation of dissolved organic matter in nitrogen-depleted waters. *Biogeosciences Discuss* Vol. 4, No. 5, pp. (3531-3579), 1810-6285

Raimbault, P.; Garcia, N.&Cerutti, F., (2008). Distribution of inorganic and organic nutrients in the South Pacific Ocean – evidence for long-term accumulation of organic matter in nitrogen-depleted waters. *Biogeosciences* Vol. 5, No. 2, pp. (281-298), 1810-6285

Raleigh, V.S., (1999). Trends in world population: how will the millenium compare with the past? *Human Reproduction Update* Vol. 5, No. 5, pp. (500-505), 1355-4786

Ras, J.; Claustre, H.&Uitz, J., (2008). Spatial variability of phytoplankton pigment distributions in the Subtropical South Pacific Ocean: comparison between in situ and predicted data. *Biogeosciences* Vol. 5, No. 2, pp. (353-369), 1726-4189

Raven, J.A., (1998). The twelfth Tansley lecture. Small is beautiful: the picophytoplankton. *Functional ecology*. Vol. 12, No. 4, pp. (505-513), 1365-2435

Reeves, R.&Notarbartolo, d.S.G., (2006). *The status and distribution of cetaceans in the Black Sea and Mediterranean Sea*, The World Conservation Union (IUCN), 137 pp., Malaga, Spain.

Ridame, C.; Le Moal, M.; Guieu, C.; Ternon, E.; Biegala, I.C.; L'Helguen, S., et al., (2011). Nutrient control of N_2 fixation in the oligotrophic Mediterranean Sea and the impact of Saharan dust events. *Biogeosciences* Vol. 8, No. 9, pp. (2773-2783), 1726-4189

Riebesell, U.; Körtzinger, A.&Oschlies, A., (2009). Sensitivities of marine carbon fluxes to ocean change. *Proceedings of the National Academy of Sciences* Vol. 106, No. 49, pp. (20602-20609), 1091-6490

Riemann, L.; Nielsen, T.G.; Kragh, T.; Richardson, K.; Parner, H.; Jakobsen, H.H., et al., (2011). Distribution and production of plankton communities in the subtropical convergence zone of the Sargasso Sea. I. Phytoplankton and bacterioplankton. *Marine Ecology Progress Series* Vol. 426, No. pp. (57-70), 1616-1599

Rios, L.M.; Jones, P.R.; Moore, C.&Narayan, U.V., (2010). Quantitation of persistent organic pollutants adsorbed on plastic debris from the Northern Pacific Gyre's "eastern garbage patch". *Journal of Environmental Monitoring* Vol. 12, No. 12, pp. (2226-2236), 1464-0333

Riser, S.C.&Johnson, K.S., (2008). Net production of oxygen in the subtropical ocean. *Nature* Vol. 451, No. 7176, pp. (323-325), 1476-4687

Roberts, C.M.; McClean, C.J.; Veron, J.E.N.; Hawkins, J.P.; Allen, G.R.; McAllister, D.E., et al., (2002). Marine Biodiversity Hotspots and Conservation Priorities for Tropical Reefs. *Science* Vol. 295, No. 5558, pp. (1280-1284), 0036-8075

Robinson, C.; Holligan, P.; Jickells, T.&Lavender, S., (2009). The Atlantic Meridional Transect Programme (1995-2012). *Deep Sea Research Part II: Topical Studies in Oceanography* Vol. 56, No. 15, pp. (895-898), 0967-0645

Rodolfo-Metalpa, R.; Houlbreque, F.; Tambutte, E.; Boisson, F.; Baggini, C.; Patti, F.P., et al., (2011). Coral and mollusc resistance to ocean acidification adversely affected by warming. *Nature Climate Change* Vol. 1, No. 6, pp. (308-312), 1758-6798

Roy, S.&Chattopadhyay, J., (2007). Towards a resolution of 'the paradox of the plankton': A brief overview of the proposed mechanisms. *Ecological Complexity* Vol. 4, No. 1-2, pp. (26-33), 1476-945X

Sabine, C.L.; Feely, R.A.; Gruber, N.; Key, R.M.; Lee, K.; Bullister, J.L., et al., (2004). The Oceanic Sink for Anthropogenic CO_2. *Science* Vol. 305, No. 5682, pp. (367-371), 0036-8075

Sara, G.; Bianchi, C.N.&Morri, C., (2005). Mating behaviour of the newly-established ornate wrasse Thalassoma pavo (Osteichthyes: Labridae) in the Ligurian Sea (north-western Mediterranean). *Journal of the Marine Biological Association of the United Kingdom* Vol. 85, No. 01, pp. (191-196), 1469-7769

Sarmento, H.; Montoya, J.M.; Vázquez-Domínguez, E.; Vaqué, D.&Gasol, J.M., (2010). Warming effects on marine microbial food web processes: how far can we go when it comes to predictions? *Philosophical Transactions of the Royal Society B: Biological Sciences* Vol. 365, No. 1549, pp. (2137-2149), 1471-2970

Schattenhofer, M.; Fuchs, B.M.; Amann, R.; Zubkov, M.V.; Tarran, G.A.&Pernthaler, J., (2009). Latitudinal distribution of prokaryotic picoplankton populations in the Atlantic Ocean. *Environmental Microbiology* Vol. 11, No. 8, pp. (2078-2093), 1462-2920

Schlüter, L.; Henriksen, P.; Nielsen, T.G.&Jakobsen, H.H., (2011). Phytoplankton composition and biomass across the southern Indian Ocean. *Deep Sea Research Part I: Oceanographic Research Papers* Vol. 58, No. 5, pp. (546-556), 0967-0637

Seki, M.P.&Polovina, J.J., (2001). Ocean gyre ecosystems, in: *The Encyclopedia of Ocean Sciences*, Steele, J. H. et al (eds.), pp. (1959-1964), Academic Press, Ca.3399p, San Diego

Shi, X.L.; Lepère, C.; Scanlan, D.J.&Vaulot, D., (2011). Plastid 16S rRNA Gene Diversity among Eukaryotic Picophytoplankton Sorted by Flow Cytometry from the South Pacific Ocean. *PLoS ONE* Vol. 6, No. 4, pp. (e18979), 1932-6203

Shushkina, E.A.; Vinogradov, M.E.; Lebedeva, L.P.&Anokhina, L.L., (1997). Productivity Characteristics of Epipelagic Communities of the World's Oceans. *Oceanology* Vol. 37, No. 3, pp. (346-353), 0001-4370

Signorini, S.R.&McClain, C.R., (2011). Subtropical gyre variability as seen from satellites. *Remote Sensing Letters* Vol. 3, No. 6, pp. (471-479), 2150-704X

Silverman, J.; Lazar, B.; Cao, L.; Caldeira, K.&Erez, J., (2009). Coral reefs may start dissolving when atmospheric CO_2 doubles. *Geophysical Research Letters* Vol. 36, No. 5, pp. (L05606), 0094-8276

Siokou-Frangou, I.; Christaki, M.; Mazzocchi, G.; Montresor, M.; Ribera D'Alcalá, M.&Vaqué D, Z.A., (2010). Plankton in the open Mediterranean Sea: a review. *Biogeosciences* Vol. 7, No. 5, pp. (1543-1586), 1726-4189

Sommer, U.; Stibor, H.; Katechakis, A.; Sommer, F.&Hansen, T., (2002). Pelagic food web configurations at different levels of nutrient richness and their implications for the ratio fish production:primary production. *Hydrobiologia* Vol. 484, No. 1, pp. (11-20), 1573-5117

Steinberg, D.K.; Carlson, C.A.; Bates, N.R.; Johnson, R.J.; Michaels, A.F.&Knap, A.H., (2001). Overview of the US JGOFS Bermuda Atlantic Time-series Study (BATS): a decade-scale look at ocean biology and biogeochemistry. *Deep Sea Research Part II: Topical Studies in Oceanography* Vol. 48, No. 8-9, pp. (1405-1447), 0967-0645

Stramma, L.; Schmidtko, S.; Levin, L.A.&Johnson, G.C., (2010). Ocean oxygen minima expansions and their biological impacts. *Deep Sea Research Part I: Oceanographic Research Papers* Vol. 57, No. 4, pp. (587-595), 0967-0637

Taboada, G., Fernando; González Gil, R.; Höfer, J.; González, S.&Anadón, R., (2010). Trichodesmium spp. population structure in the eastern North Atlantic subtropical gyre. *Deep Sea Research Part I: Oceanographic Research Papers* Vol. 57, No. 1, pp. (65-77), 0967-0637

Tanaka, T.; Thingstad, T.F.; Christaki, U.; Colombet, J.; Cornet-Barthaux, V.; Courties, C., et al., (2011). Lack of P-limitation of phytoplankton and heterotrophic prokaryotes in surface waters of three anticyclonic eddies in the stratified Mediterranean Sea. *Biogeosciences* Vol. 8, No. 2, pp. (525-538), 1726-4189

Tanaka, T.; Zohary, T.; Krom, M.D.; Law, C.S.; Pitta, P.; Psarra, S., et al., (2007). Microbial community structure and function in the Levantine Basin of the eastern Mediterranean. *Deep Sea Research Part I: Oceanographic Research Papers* Vol. 54, No. 10, pp. (1721-1743), 0967-0637

Thébault, H.; Rodriguez y Baena, A.M.; Andral, B.; Barisic, D.; Albaladejo, J.B.; Bologa, A.S., et al., (2008). 137Cs baseline levels in the Mediterranean and Black Sea: A cross-basin survey of the CIESM Mediterranean Mussel Watch programme. *Marine Pollution Bulletin* Vol. 57, No. 6-12, pp. (801-806), 0025-326X

Thiel, M.; Bravo, M.; Hinojosa, I.A.; Luna, G.; Miranda, L.; Núñez, P., et al., (2011). Anthropogenic litter in the SE Pacific: an overview of the problem and possible solutions. *Journal of Integrated Coastal Zone Management* Vol. 11, No. 1, pp. (115-134), 1477-7835

Thingstad, T.F.; Krom, M.D.; Mantoura, R.F.C.; Flaten, G.A.F.; Groom, S.; Herut, B., et al., (2005). Nature of Phosphorus Limitation in the Ultraoligotrophic Eastern Mediterranean. *Science* Vol. 309, No. 5737, pp. (1068-1071), 1095-9203

Thomalla, S.J.; Waldron, H.N.; Lucas, M.I.; Read, J.F.; Ansorge, I.J.&Pakhomov, E., (2010). Phytoplankton distribution and nitrogen dynamics in the Southwest Indian subtropical gyre and Southern Ocean Waters. *Ocean Science Discussions* Vol. 7, No. 4, pp. (1347-1403), 1812-0822

Thompson, R.C.; Olsen, Y.; Mitchell, R.P.; Davis, A.; Rowland, S.J.; John, A.W.G., et al., (2004). Lost at Sea: Where Is All the Plastic? *Science* Vol. 304, No. 5672, pp. (838), 1095-9203

Thrush, S.F.; Gray, J.S.; Hewitt, J.E.&Ugland, K.I., (2006). Predicting the effects of habitat homogenization on marine biodiversity. *Ecological Applications* Vol. 16, No. 5, pp. (1636-1642), 1051-0761

Trenberth, K.E.; Jones, P.D.; Ambenje, P.; Bojariu, R.; Easterling, D.; Klein, T.A., et al., (2007). Series. *Observations: Surface and Atmospheric Climate Change. In: Climate Change 2007: The Physical Science Basis. Contribution of Working Group I to the Fourth Assessment Report of the Intergovernmental Panel on Climate Change,* (Solomon, S., Qin, D., Manning, M., Chen, Z., Marquis, M., Averyt, K.B., Tignor, M., Miller, H.L., (eds.)), Cambridge University Press, Cambridge and New York.

Treusch, A.H.; Vergin, K.L.; Finlay, L.A.; Donatz, M.G.; Burton, R.M.; Carlson, C.A., et al., (2009). Seasonality and vertical structure of microbial communities in an ocean gyre. *ISME Journal* Vol. 3, No. 10, pp. (1148-1163), 1751-7370

Tsubouchi, T.; Suga, T.&Hanawa, K., (2009). Indian Ocean subtropical mode water: its water characteristics and spatial distribution. *Ocean Science Discussions* Vol. 6, No. 1, pp. (723-739), 1812-0822

Tudela, S., (2004). Ecosystem effects of fishing in the Mediterranean: an analysis of the major threats of fishing gear and practices to biodiversity and marine habitats *Studies and reviews. General Fisheries Commission for the Mediterranean* Vol., No. 74, pp. (1-44), 1020-7236

Turley, C.M., (1999). The changing Mediterranean Sea – a sensitive ecosystem? *Progress In Oceanography* Vol. 44, No. 1-3, pp. (387-400), 0079-6611

Tyrrell, T., (1999). The relative influences of nitrogen and phosphorus on oceanic primary production. *Nature* Vol. 400, No. 6744, pp. (525-531), 0028-0836

Tyrrell, T., (2011). Anthropogenic modification of the oceans. *Phil. Trans. R. Soc. A.* Vol. 369, No. 1938, pp. (887-908), 1471-2962

UnitedStatesCensusBureau, 2011. World Population, http://www.census.gov/main/www/popclock.html.

Van Mooy, B.A.S.&Devol, A.H., (2008). Assessing nutrient limitation of Prochlorococcus in the North Pacific subtropical gyre by using an RNA capture method. *Limnology and Oceanography* Vol. 53, No. 1, pp. (78-88), 0024-3590

Van Mooy, B.A.S.; Rocap, G.; Fredricks, H.F.; Evans, C.T.&Devol, A.H., (2006). Sulfolipids dramatically decrease phosphorus demand by picocyanobacteria in oligotrophic marine environments. *Proceedings - National Academy Of Sciences USA* Vol. 103, No. 23, pp. (8607-8612), 0027-8424

Viviani, D.A.; Björkman, K.M.; Karl, D.M.&Church, M.J., (2011). Plankton metabolism in surface waters of the tropical and subtropical Pacific Ocean. *Aquatic Microbial Ecology* Vol. 62, No. 1, pp. (1-12), 1616-1564

Wagener, T.; Guieu, C.; Losno, R.; Bonnet, S.&Mahowald, N., (2008). Revisiting atmospheric dust export to the Southern Hemisphere ocean: Biogeochemical implications. *Global Biogeochem. Cycles* Vol. 22, No. 2, pp. (GB2006), 0886-6236

Wilson, C.&Qiu, X., (2008). Global distribution of summer chlorophyll blooms in the oligotrophic gyres. *Progress In Oceanography* Vol. 78, No. 2, pp. (107-134), 0079-6611

Wu, J.; Sunda, W.; Boyle, E.A.&Karl, D.M., (2000). Phosphate Depletion in the Western North Atlantic Ocean. *Science* Vol. 289, No. 5480, pp. (759-762), 0036-8075

Young, L.C.; Vanderlip, C.; Duffy, D.C.; Afanasyev, V.&Shaffer, S.A., (2009). Bringing Home the Trash: Do Colony-Based Differences in Foraging Distribution Lead to Increased Plastic Ingestion in Laysan Albatrosses? *PLoS ONE* Vol. 4, No. 10, pp. (e7623), 1932-6203

Zehr, J.P.&Kudela, R.M., (2011). Nitrogen cycle of the open ocean: from genes to ecosystems. *Annual Review of Marine Science* Vol. 3, No. pp. (197-225), 1941-1405

Zenetos, A.; Gofas, S.; Verlaque, M.; Cinar, M.E.; García, R.E.; Azzurro, E., et al., (2010). Alien species in the Mediterranean by 2010. A contribution to the application of European Union's Marine Strategy Framework Directive (MSFD). Part I. Spatial distribution. *Mediterranean Marine Science* Vol. 11, No. 2, pp. (481-493), 1791-6763

Zenetos, A.; Siokou-Frangou, I.&Gotsis Skretas, O., (2002). *The Mediterranean Sea - blue oxygen-rich, nutrient-poor waters*, Technical Report. European Environment Agency, Copenhagen, Denmark.

2

Modelling the Pelagic Ecosystem Dynamics: The NW Mediterranean

Antonio Cruzado[1,*], Raffaele Bernardello[2], Miguel Ángel Ahumada-Sempoal[3] and Nixon Bahamon[4]
[1]Oceans Catalonia International SL,
[2]Earth and Environmental Science, University of Pennsylvania,
[3]Resources Institute, University del Mar,
[4]Center for Advanced Studies of Blanes (CEAB-CSIC),
[1,4]Spain
[2]USA
[3]México

1. Introduction

The word *pelagic* comes from the Greek πέλαγος meaning *open sea* and refers to the marine and oceanic domain away from the shore line and from surface to bottom (Wikipedia). The *pelagic ecosystem* includes the ever-moving and continuously changing waters, the *habitat*, and the diverse and inter-related groups of organisms or *communities*. Hydrodynamics, forced by external, mostly atmospheric processes set the very special physical conditions that, to a great extent, control the functioning of the biological processes. Currents, waves, mixing, turbulence, air/sea exchanges or fertilization are all mechanisms allowing planktonic communities, the most important in terms of biomass and fluxes of matter and energy, to develop and sustain other communities higher up in the trophic chain.

Since the initial times of Oceanography, modelling the marine system has been mainly applied to the behaviour of its physical properties: temperature, salinity, density, circulation. Forces driving the dynamics of the ocean are heat and water fluxes and wind stress at the free surface, friction between water layers and between water and the solid boundaries and inertia related to the rotation of the Earth. The Navier-Stokes momentum equation

$$\frac{Dv}{Dt} = -\frac{1}{\rho}\nabla p - 2\Omega \times v + g + F_r$$

relates the rate of change of velocity (v) to the field of pressure (p), including Coriolis inertia force (2Ω), gravitation (g), and friction (F_r). This equation has been widely used by physical modellers together with the equation of state of seawater relating density (ρ) to temperature

---

* Corresponding Author

and salinity (heat and water balance). Since density determines the field of pressure, horizontal gradients of pressure are a key variable in dynamic models for the computation of horizontal and vertical velocities. On the other hand, the continuity condition requires that no gains or losses of water or heat take place other than at the open boundaries (evaporation, precipitation).

In the second half of the 20th century, theoretical ecologists began developing the rationale to simulate the behaviour of both freshwater and marine ecosystems. Much of the work carried out by theoretical ecologists has ended up with highly complex pictorial models (**Figure 1**) not always amenable to numerical simulations. However, as early as in 1949, Gordon A. Riley, a biological oceanographer, and Henry Stommel and Dean Bumpus, two physical oceanographers, published a seminal paper on *Quantitative ecology of the plankton of the western North Atlantic* (Riley at al., 1949). On the other hand, Howard T. Odum developed a theory of ecosystems based on electrical analogs (Odum, 1960), at a time when digital computers were still unavailable, carried out simulations by means of electric circuits. Since then, numerous efforts have been carried out in coupling physical and biological models.

Numerical modelling of the ocean ecosystems was initiated, with some degree of success, in the late 1960s with application to relatively closed coastal lagoons and estuaries (Kremer and Nixon, 1978). However, the complexity of natural biological communities could hardly be modelled by means of "deterministic" causal equations and many of the models made use of stochastic relations to obviate the great variability associated to the genetic and ecological diversity (Margalef, 1972).

Fig. 1. Pictorial model of an ecosystem model (from **wordpress**)

One group of ecological models (today called biogeochemical models) were developed to cope with the need to understand the variability of the planktonic system. Based on the principle that biomass is the material basis of the ecosystem and that organisms are made up of carbon requiring, for their growth and development, the availability of nutrients (nitrogen, orthophosphate, orthosilicic acid, etc.) and light to form such biomass (organic matter) of one or more virtual groups of organisms (photosynthetic microalgae, Bacteria, etc.).

Transfer of biomass through the trophic chain by grazing and predation processes was represented by various forms of equations simulating prey-predator (Lotka-Volterra) dynamics (Lotka, 1925; Volterra, 1926). Part of the carbon and nutrients taken up by primary producers are recycled within the system or exported via faecal detritus or dead organisms settling out of the system. In models of the pelagic system, all or some of the state-variables are subject to physical processes, namely advection and diffusion that, in addition, control the fertility of the system (Steele, 1970).

Biogeochemical models may be as simple as the NPZD (nutrient, phytoplankton, zooplankton, detritus) in which the nutrient (usually nitrogen) is not only the rate-controlling factor, together with light, for photosynthetic growth but also the building block of the entire ecosystem, the assumption being that all other variables are controlled by the Redfield ratio (N:P:C) (Redfield et al., 1963). In these models, availability of light and/or nutrient modifies the maximum theoretical growth rate of the phytoplankton population (nutrient uptake) and both of them modify the density of this population and that of herbivorous, thus controlling the transfer of nutrients from one trophic level to the next. Part of the biomass transferred to the grazing zooplankton is used for the growth of its population while part is excreted back as nutrient to the water or goes into the detritus thus closing the system (Cruzado, 1982; Wroblewski et al., 1988; Wroblewski, 1989; Fasham et al., 1990; Varela et al., 1992, 1994; Bahamon and Cruzado, 2003).

All-inclusive biogeochemical models consider carbon as the building block, with light and various forms of nitrogen (nitrate, ammonia), orthophosphate, silicic acid, iron, etc. controlling primary production (Varela et al., 1995). Dissolved oxygen may or may not be included in the processes of photosynthesis and respiration. However, there seems to be a trend away from modelling entire ecosystems, e.g. large food-web models (Baretta et al., 1995).

In future, ecosystem models will continue to be developed which may be used in support of decision-making. New databases and new measurement (observing) tools will contribute this trend. Nevertheless, important questions still remain open and may be answered with a new generation of ecosystem models which will explicitly contain general principles of ecosystem behaviour. Such models shall aim at process reduction and will be derived from an analysis of present knowledge (Ebenhoh, 2000).

Variability in the physical environment at proper time scales increases the biological activities in the model while the estimate of the primary production is a direct consequence of the dynamic environment. Future operational model applications should include the highest possible resolution of the surface forcing fields and, as the next step, the collection of biological observations necessary at the same scales will be considered to verify the model's capabilities (Vichi, 2000).

Ecosystem modelling has made great advances in the last decades during which it progressed from naive mechanistic and process-oriented modelling to data-driven approaches and individual-based models. Large projects that tried to model whole ecosystems have proven to be of limited use and the trend today appears to go towards more modest and perhaps more successful models of limited aspects of the ecosystems (e.g. special events like toxic algal blooms, etc.)

The coupling of hydrodynamic and biogeochemical models in a three-dimensional framework is a key step toward understanding marine systems and management of marine resources. Ecological processes are strongly influenced by the high heterogeneity in both vertical and horizontal hydrodynamic processes that can only be approached by means of high resolution 3-D models. The purpose of this paper is to highlight the main features of several complementary models applied to parts of the Mediterranean Sea resulting in the best available tools for linking observations (in situ and remote) with theory (both hydrodynamics and ecosystems).

2. The NW Mediterranean

The Mediterranean Sea area, roughly located north of 39°N and limited by the mainland to the north, the Balearic islands to the southwest and the islands of Corsica and Sardinia to the east, is usually called NW Mediterranean sea (NW Med) although different names may be given to subareas such as Gulf of Valencia to the west, Catalan Sea and Gulf of Lions, in the central part, and Ligurian Sea to the east (**Figure 2**).

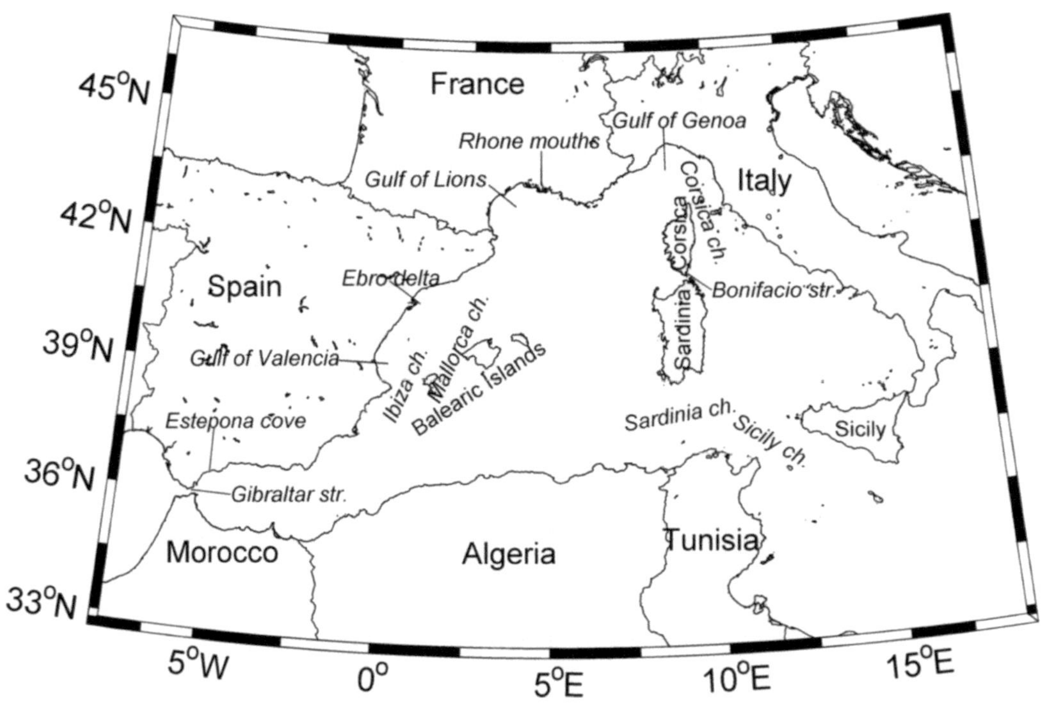

Fig. 2. Map of the Western Mediterranean Sea showing locations cited in this paper.

The NW Med is essentially considered a three layer system: 1) an upper layer between the surface and ~100/200m; 2) an intermediate layer between ~100/200m and ~600/800m; and 3) a deep layer down to the bottom. Three water masses defined mostly on the basis of salinity occupy these layers (**Figure 3**): Modified Atlantic Water, MAW; Levantine Intermediate Water, LIW; and Western Mediterranean Deep Water, WMDW.

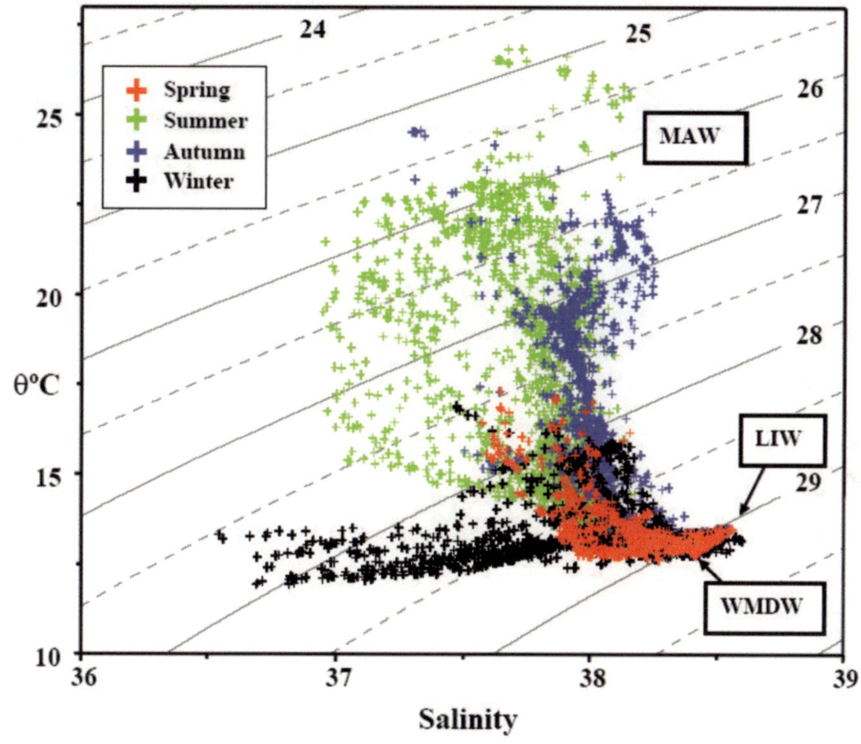

Fig. 3. T-S diagram for the major water masses (MAW, LIW and WMDW) in the NW Mediterranean Sea (Modified from Velasquez, 1997).

Early works pointed out that the major large-scale hydrodynamic feature in the NW Med is a well-defined cyclonic circulation, involving both the MAW surface layer and the LIW layer below (Béthoux et al., 1982). The main component of this circulation is the Northern Current (NC), which is fed by the Eastern and Western Corsican Currents. According to Astraldi et al. (1994), a marked frontal structure found nearly parallel to the coast, separates the Northern Current from the open sea, characterized by a doming of the internal hydrological structure extending from the Ligurian Sea to the Catalan Sea. A general view of the NW Med circulation (**Figure 4**), from large-scale to mesoscale, has been proposed by Millot (1999).

The Ligurian Sea plays an important role for understanding the dynamic characteristics of the whole NW Med (Astraldi and Gasparini, 1992). MAW and LIW enter the sea via northward flows along the east and west coasts of Corsica (Astraldi et al., 1990) converging just north of this island into the Ligurian Current (name given here to the NC). A large-scale cyclonic circulation characterizes the Ligurian Sea, affecting all water masses. The presence of a thermal front separating the warmer coastal water from those of the interior and a doming of the hydrological structure in the central part of the sea are common features associated to the cyclonic circulation (Crépon and Boukthir, 1987).

The general circulation in the Gulf of Lions, a very complex hydrodynamic area, goes along the continental slope (**Figure 4**). Formation of dense waters both on the shelf and offshore and a seasonal variation of the stratification compete simultaneously in the control of the mesoscale circulation (Millot, 1990).

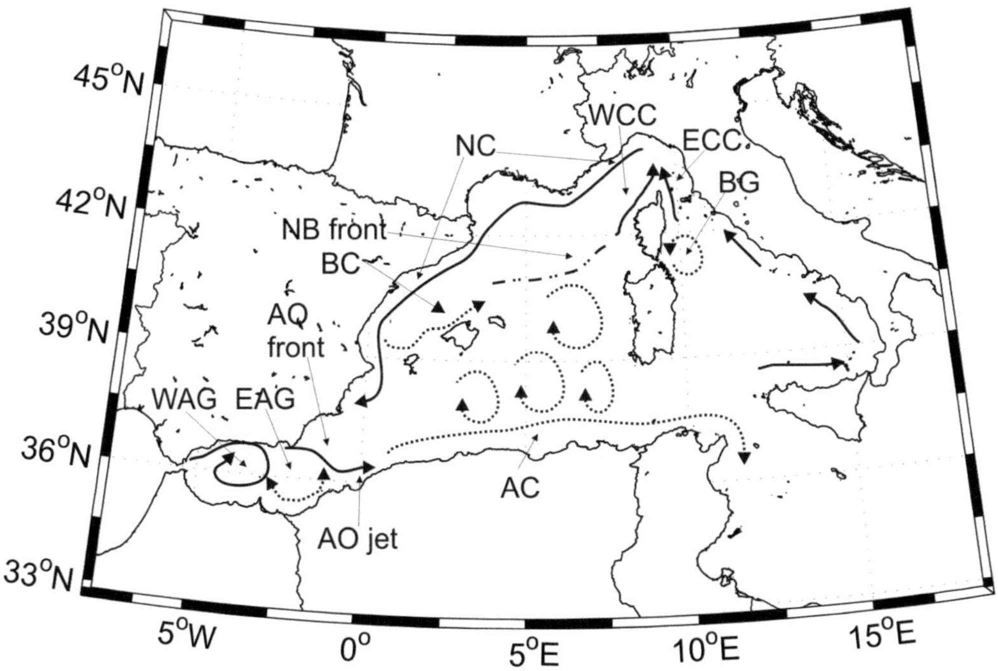

Fig. 4. Major features of the upper layer circulation in the NW Mediterranean Sea (redrawn from Millot, 1999). WAG: Western Alboran Gyre; EAG: Eastern Alboran Gyre; AO jet: Almeria-Oran jet; AC: Algerian Current; AO front: Almeria-Oran front; NB front: North Balearic Front; NC: Northern Current; WCC: Western Corsican Current; ECC: Eastern Corsican Current; BG: Bonifacio Gyre.

During the summer, the Provençal Current (name given here to the NC) flows along the continental slope of the Gulf into the Catalan Sea but, in winter, one part of it deviates from the continental slope off Cap Creus and then follows eastwards. A cyclonic gyre of ~100 km in diameter induces a horizontal divergence at the surface and a convergence at depth, with large vertical motions occurring in the centre (Gascard, 1978). Deep-water formation processes in winter are recognized as a major characteristic of this area (MEDOC Group, 1970; Schott et al., 1996), other significant features being meanders, eddies, "chimneys" and fronts that have also been reported in this part of the NW Med.

The Catalan Sea is a transition zone playing a key role in the general circulation of the NW Med since it connects the two different regimes that configure the Western Mediterranean dynamics (López-García et al., 1994). The Catalan Sea is characterized by the presence of two permanent density fronts along the continent (the Catalan Front) and along the north slope of the islands (the Balearic Front) and their associated currents. The Catalan Front is a shelf/slope front mainly produced by salinity gradients (Font et al., 1988), that separate the surface layer with denser water in the centre of the sea from the water transported by the Catalan Current (name given here to the NC). In contrast, the Balearic Front is mostly characterized by temperature gradients associated with the so-called Balearic Current (Font et al., 1988; Salat 1995). This front separates the "old" MAW located at the surface in the centre of the Catalan Sea from lighter (and warmer) "recent" MAW flowing northward from the Algerian basin through the Balearic channels. Consequently, the formation of the Balearic Front is determined both by the presence of denser "old" MAW in the centre of the sea and the bottom topography (López-García et al., 1994).

Regarding the biogeochemical properties, the NW Med is known as an oligotrophic environment characterised by high rates of ammonium-based primary production (regenerated production) with relatively scarce contribution from deep water based nitrate (new production) (Dugdale and Goering, 1967). The new to total production (new+regenerated production) ratio (*f*-ratio) in oligotrophic ecosystems is relatively low around 20-30% (*f*-ratio=0.20-0.30) (Eppley and Peterson, 1979). Nevertheless, in temperate zones such as the NW Med, these percentages can strongly vary with seasons. For the summer period with thermal stratification (**Figure 5**), the nutrient shortage in the surface makes the *f*-ratio to be about that expected for oligotrophic ecosystems. However, in winter time, with the breakdown of thermal stratification, nutrient-rich deep waters are brought up to the surface thus providing new nitrogen making the new to total production ratio to increase up to about 0.80 (80%) (Bahamon and Cruzado, 2003).

Phytoplankton (primary) production takes place in the euphotic zone that may show variable thickness with a generally accepted lower limit at the depth of about 3% to 1% the surface irradiance (e.g. Bricaud et al., 1992). At 41° N latitude, the lower limit of the euphotic zone in summer is located at about 50-60 m depth in summer and about 30 m depth in winter (**Figure 5**), as observed at the station OOCS in the Catalan Sea (Bahamon et al., 2011). In the Catalan Sea, the lower limit of the euphotic zone is characterised by a yearly upward nitrogen flux of 0.64 mol N m^{-2} with variations more dependent on the vertical nitrogen gradients than on water density gradients (Bahamon and Cruzado, 2003). This makes the difference with oligotrophic ecosystems at lower latitudes, such as the NE Atlantic Ocean, showing yearly upward nitrogen flux of 0.22 mol N m^{-2}.

The primary production in oligotrophic ecosystems is mostly controlled by nutrient fluxes, PAR availability and zooplankton grazing that has been tested with the use of relatively simple one-dimensional numerical models (e.g. Doney et al, 1996; Bahamon and Cruzado, 2003). In coastal areas of the NW Med, the main fertilization processes fuelling primary production, are the nutrient-rich river water discharges by the Ebro River in the coast of Catalonia and the Rhône River in the Gulf of Lions. In open sea areas, the vertical convection in winter time is the main responsible for fertilising the surface ocean. The central areas of the cyclonic gyres and hydrographic fronts produced by temporary eddies are also responsible for bringing nutrients to the surface from deeper water layers (Estrada, 1995). In the Catalan Sea, despite the spatial chlorophyll variability, coastal and oceanic regions are separated by a distinctive region showing chlorophyll concentration quite

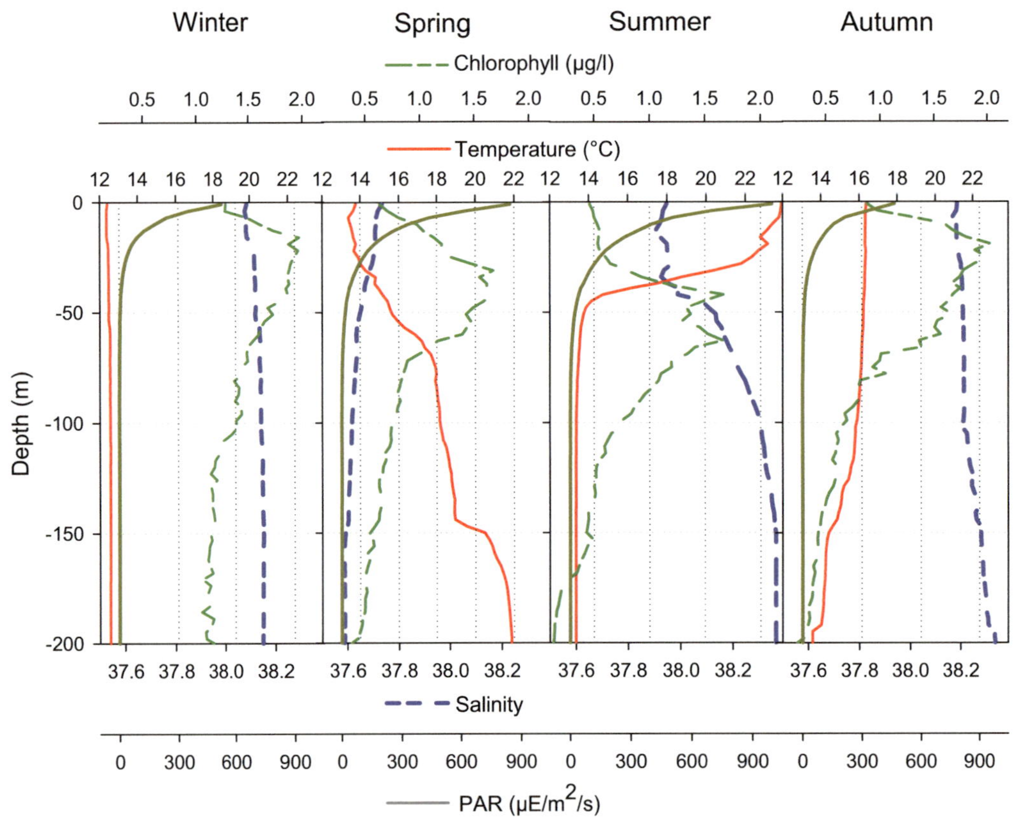

Fig. 5. Average temperature, salinity, PAR and chlorophyll profiles at the station OOCS for the middle of seasons (February, May, August, November) between March 2009 and Feb 2011.

constant throughout the year, following the course for the Northern Current (Gordoa et al., 2008). Although most of surface chlorophyll variability is generally inversely related to temperature this connection is unclear in areas with influence of freshwater discharges (Bahamon et al., 2010; Volpe et al., 2011).

The NW Med shows a subtropical climate with characteristics also found in other areas overseas such as California (USA), central Chile, southern Australia and South Africa (also called Mediterranean climate zones). However, the climate characteristics might be changing in the last decades as suggested by the increasing occurrence of tropical species (Bianchi, 2007) in line with indications of warming of Mediterranean waters (Vargas-Yáñez et al., 2010). In fact, the connection of upper water layers of the pelagic ecosystem to the lower atmosphere allows the strong atmospheric changes recorded in the last years (Vargas-Yáñez et al., 2010; Camuffo et al., 2010) to produce still uncertain effects on the pelagic ecosystem, thus making even more difficult to understand the ecosystem dynamics.

3. Hydrodynamic and biogeochemical modelling rationale

In order to understand and quantitatively describe the functioning of the oceans, physics, chemistry and biology need to be integrated. The fluxes of energy and matter are regulated by the interactions between the physical environment and the elements contained in the waters. The description of the oceanic biogeochemical cycles needs to start from a realistic representation of the oceanic circulation.

We present here results from two hydrodynamic models that represent the three-dimensional circulation on two high-resolution grids (~5 km and ~1.1 km). The first model (WMED) (Bernardello 2010) covers the whole Western Mediterranean Sea, from the Sicilian channel to the straits of Gibraltar. The second model (BLANCA) covers a smaller area in the Catalan Sea centred over the Blanes submarine canyon.

Submarine canyons are irregular steep interruptions of the shelf edge of many continental margins of the world oceans. Because of their interactions with the overlying circulation, submarine canyons are of high biological importance and productivity (Gili et al., 1999; Sardà et al., 2009; Vetter et al., 2010) and often play a significant role on the exchange of energy and matter between the coastal zone and the deep basin (Allen and Durrieu de Madron, 2009).

Since this work is primarily focused on the biogeochemical behaviour of the NW Med and its relationship with the hydrodynamic constraint, we will focus on the surface circulation at basin-scale analysing results from WMED. We will discuss then the circulation within and in the vicinity of the Blanes Canyon as well as the interaction between the incident along-slope flow and the canyon bottom topography using results from BLANCA.

4. The 3D hydrodynamic model

In this work, the codes sbPOM (Stony Brook Parallel Ocean Model) and POM2k are used to simulate hydrodynamics in the NW Med. The models are based on the Princeton Ocean Model (POM). The sbPOM works on a parallelized configuration while POM2k makes use of a single CPU. POM is a finite difference, primitive equations ocean circulation model whose principal attributes, according to Mellor (2004), are: 1) it contains a second order momentum turbulence closure sub-model to provide vertical mixing coefficients; 2) it is a sigma-coordinate model; 3) the horizontal grid uses curvilinear orthogonal coordinates and an "Arakawa C" differencing scheme; 4) the horizontal time differencing is explicit whereas the vertical differencing is implicit; and 5) it has a free surface and a split time step.

In POM, the surface elevation, temperature, salinity, and velocity fields are prognosticated assuming as fundamental hypotheses: 1) the seawater is incompressible; 2) the pressure at any point of the ocean is equal to the weight of the column of water and air above it (hydrostatic approximation); and 3) the density can be expressed in terms of a mean value and a small fluctuation (Boussinesq approximation).

a. Grid set-up, initial and lateral open boundary conditions

The sbPOM was configured for the whole Western Mediterranean (WMED) while POM2k was configured for the North-western Mediterranean (NMS) and for the Blanes Canyon area (BLANCA). The models WMED (resolution of 1/20° in the horizontal and 52 sigma layers in the vertical) and NMS (resolution of 1/20° in the horizontal and 32 sigma layers in the vertical) get initial and lateral open boundary conditions (salinity, temperature, and velocity) from an operational basin-scale model of the Mediterranean Sea (MFS1671) based on the Océan PArallélisé (OPA) code, which has been configured with a resolution of 1/16° in the horizontal and 72 unevenly distributed z-level layers in the vertical. All the models are run with atmospheric forcing from the European Centre for Medium-Range Weather Forecasts (ECMWF) and ocean (http://www.ecmwf.int/research/era/do/get/era-interim) data assimilation (for more details see Tonani et al., 2008, 2009; Pinardi and Coppini, 2010). The BLANCA model (resolution of 1/60° in the horizontal and 32 sigma layers in the vertical) in particular gets initial and lateral open boundary conditions from the NMS outputs. The WMED model was run for the period 2001-2008 and the NMS and BLANCA models were run for the year 2001.

The WMED model has two lateral open boundaries at the strait of Gibraltar and the Sicilian channel, the NMS model has one lateral open boundary at 40°N and the BLANCA model has two lateral open boundaries at 41° 18.6′N and 3° 6.0′E. The first two models are fed with salinity, temperature, and velocity fields from MFS outputs. Given that WMED and NMS have different horizontal resolution than MFS, a horizontal bilinear interpolation was necessary. Since the vertical coordinate system is also different, MFS outputs were transformed from z-levels to sigma layers using a linear interpolation. The BLANCA, on the other hand, is fed with salinity, temperature and velocity fields from NMS outputs.

Since BLANCA has different horizontal resolution than NMS, a horizontal bilinear interpolation was necessary while, in the vertical, the grid is not refined as both models have the same number (32) of vertical layers. Furthermore, in order to avoid inaccuracies in the interpolation, which can generate errors leading to distortions of the model solution at the lateral open boundary or to violation of mass conservation, a volume conservation constraint [Marchesiello et al., 2001; Korres and Lascaratos, 2003; Sorgente et al., 2003; Zavatarelli and Pinardi, 2003] was imposed on the interpolated normal velocity across the lateral open boundaries thus guaranteeing volume conservation between MFS and WMED, between MFS and NMS and between NMS and BLANCA. Finally, to avoid temporal discontinuities at the lateral open boundaries, salinity, temperature, and velocity were specified at each time step using a linear interpolation in time between consecutive fields.

b. Mean dynamic topography and geostrophic circulation in the Western Mediterranean Sea from WMED model outputs

The mean dynamic topography (MDT) is the sea elevation due to the mean oceanic circulation. The only MDT available for the Mediterranean Sea was reconstructed combining oceanic observations as altimetry and in-situ measurements and outputs from an ocean general circulation model with no data assimilation for the period 1993-1999 (Rio et al., 2007). This MDT (hereafter called RioMDT) and the associated geostrophic circulation are compared to those estimated by the WMED model (**Figure 6**).

The NW Med is characterized by a mean cyclonic circulation whose northern side corresponds to the Northern current (NC; Millot, 1999). In the RioMDT the NC is clearly visible along the northern coast from Italy up to the Ibiza channel while in the model MDT it appears somewhat more intense on the Italian coast and, then, starts decreasing along the Spanish coast where it appears weak and hardly noticeable at the level of the Ibiza channel.

The progression of the NC southwestwards is characterized by a weakened flow and increased variability caused by complex interactions with incoming southern waters near the Balearic Islands (Garcia-Ladona et al., 1994). From current measurements performed 35 Km off the Ebro delta, Font et al. (1995) reported a mean speed of the order of 5 cm s^{-1}. The RioMDT shows high speed values (~35 cm s^{-1}) even south of the Ebro delta while the model MDT shows low values and the mean flow is not well defined. The general pattern obtained by Font et al., (1988) from climatological studies, describes the bifurcation of the NC at the height of the Ibiza channel. One branch of the current would then cross the channel transporting water southwards in the Algerian Basin becoming older Mediterranean water while the other branch returns cyclonically to the northeast forming the Balearic current (BC).

Garcia-Ladona et al., (1996) described the BC as formed by an incoming flux of Mediterranean Atlantic Water (MAW) through the Ibiza channel and the recirculation of old MAW from the NC. This description seems to agree with both MDT. In the RioMDT the BC seems to be fed by both the deflection of the NC and new MAW entering through the Ibiza channel. In the model MDT, the latter factor seems to be determinant for the existence of the current up to the island of Mallorca, where the eastward deflection of the NC strengthens the flow towards the center of the Algero-provençal Basin (**Figure 6**).

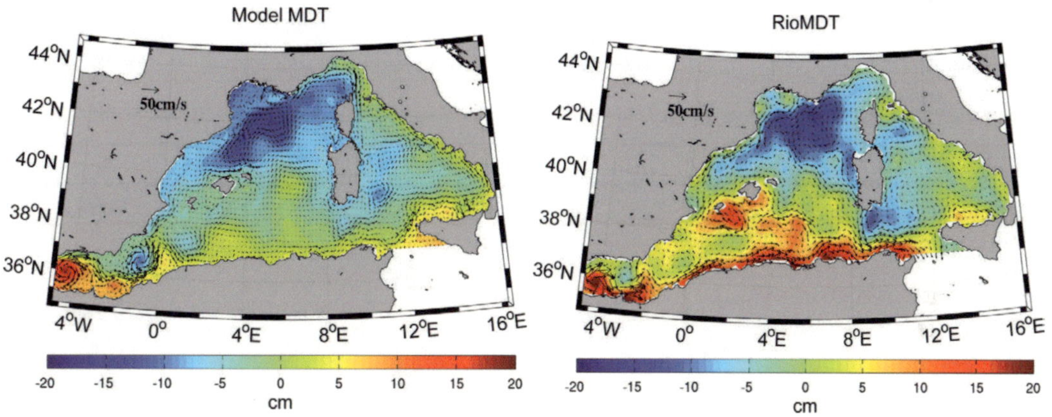

Fig. 6. Model MDT and RioMDT with associated geostrophic currents

Despite the uncertainty regarding the variability of the flux, specially in the Ibiza channel (Astraldi et al., 1999), the Balearic channels are commonly accepted to be the way through which MAW is transported to maintain the Balearic density front. This front is associated to the BC and has been described as a salinity front (Lopez-Garcia et al., 1994; Garcia-Ladona et al., 1996). The overall averages of model surface salinity and temperature are shown with the geostrophic currents superposed in **Figure 7**. The model reproduces the wavelike shape of the salinity front and the geostrophic current associated on the northern side of Mallorca, as described by La Violette et al. (1990).

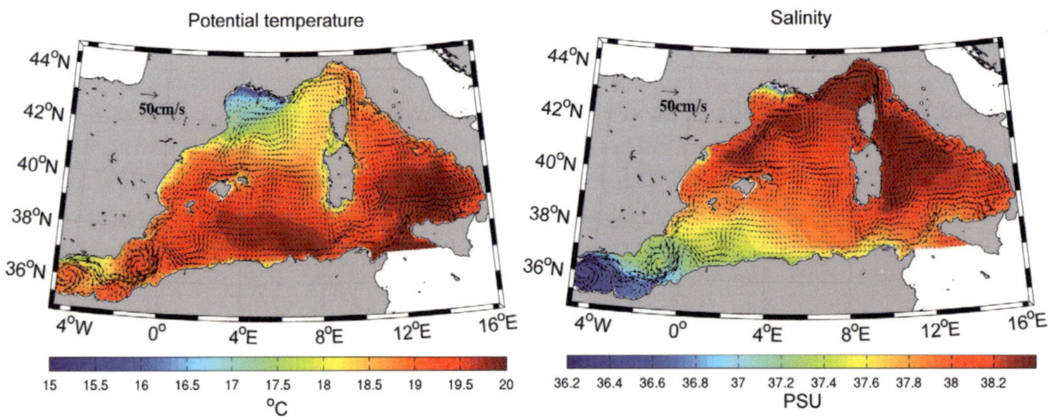

Fig. 7. Mean sea surface temperature and salinity for the whole simulation period. Geostrophic currents associated to model MDT are superposed

The NC is also associated to a permanent horizontal density gradient called the Catalan front. This front is maintained by a cool plume of water flowing along the Iberian Peninsula and originated in the Gulf of Lions (La Violette et al., 1990) or by low salinity due to the continental runoff (Font et al., 1988) or both (Garcia-Ladona et al., 1996). The model reproduces the Catalan front in both salinity and temperature, the latter being more pronounced (**Figure 7**).

c. The circulation within and around the Blanes submarine canyon from BLANCA outputs

Month	N (s^{-1})	R_d (km)	T_r (m)	Ro
January	0.0018	18.55	209.4	0.44
February	0.0015	15.46	253.7	0.47
March	0.0015	15.46	244.0	0.49
April	0.0018	18.55	206.3	0.44
May	0.0021	21.64	182.1	0.40
June	0.0026	26.80	144.3	0.44
July	0.0027	27.83	139.7	0.47
August	0.0028	28.86	135.0	0.41
September	0.0031	31.95	120.7	0.36
October	0.0032	32.98	119.7	0.40
November	0.0029	29.89	130.5	0.51
December	0.0022	22.68	169.2	0.50

N is the mean buoyancy frequency ($[-g/\rho_0(\Delta\rho/\Delta z)]^{1/2}$), $\mathbf{R_d}$ is the internal Rossby radius of deformation (NH/f), $\mathbf{T_r}$ is the vertical stratification scale (fL/N) and **Ro** is the Rossby number (U/fL). **H** is the canyon depth at the mouth= 1000m, $f= 2\Omega\sin\varphi$, $\Omega = 7.292\text{x}10^{-5}$ rad s^{-1}, φ=41.363°N, **f** = $9.7\text{x}10^{-5}$ s^{-1}, **W** is the canyon width at the mid-upper canyon= 8.0km, **L**=$W/2$= 4km, $\mathbf{H_s}$ is the shelf-break depth=150m (Flexas et al., 2008).

Table 1. Variability of hydrodynamic properties for the Blanes submarine canyon

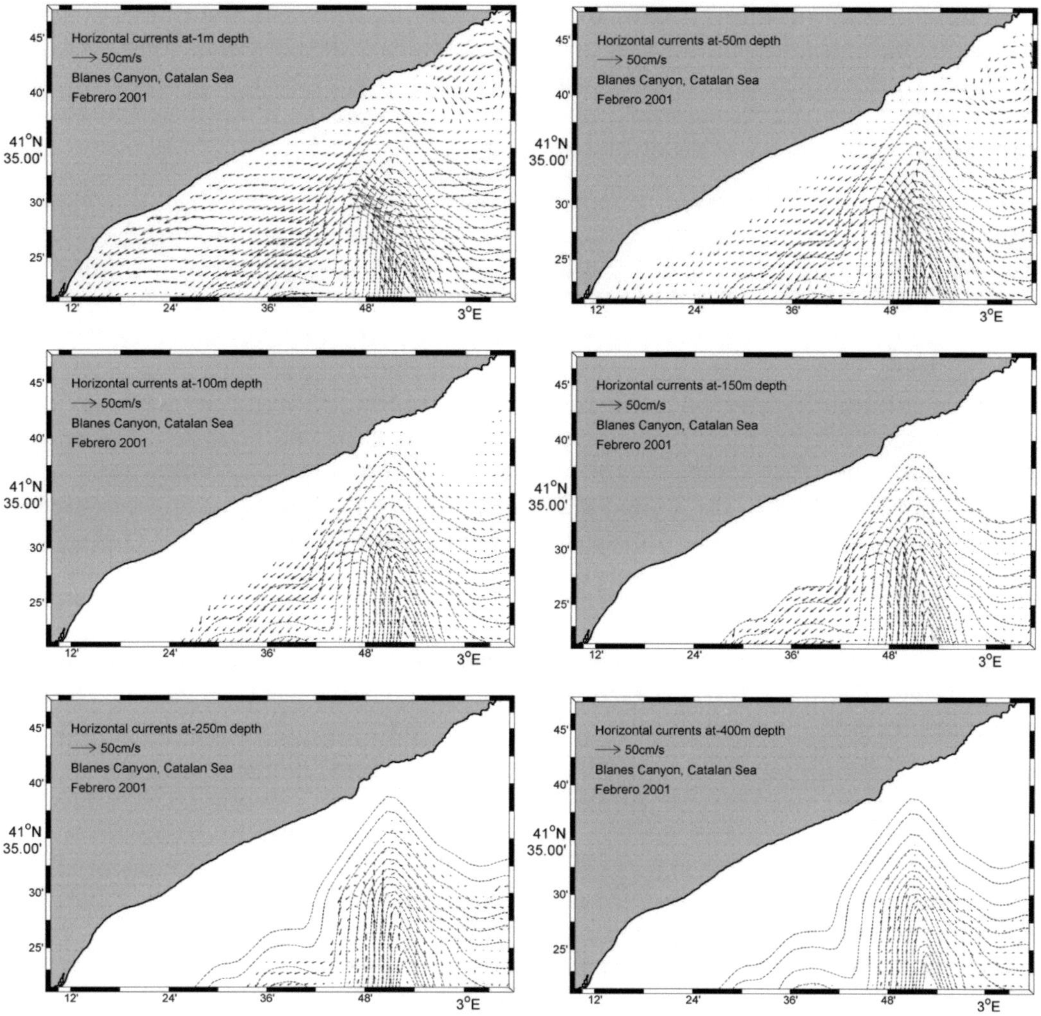

Fig. 8. Surface velocity at 1, 50, 100, 150, 250 and 400 m depth simulated with the BLANCA model

Since this section is mainly on the circulation within and in the vicinity of the Blanes submarine canyon, we will focus on analysing the BLANCA model results. In order to do this, the water column stratification and some dynamical considerations are presented (**Table 1**). On the other hand, horizontal velocity fields at different depths (1, 50, 100, 150, 250, and 400m) are used to provide an analysis of the circulation during a mid-winter month (February) characterized by moderate NW winds, net heat loss through the sea surface and weak stratification of the water column (**Figure 8**).

The seasonal variation of the water column stratification in the Blanes canyon is confirmed. Strongest stratification is displayed in September (N=0.0031s-1) and October (N=0.0032s-1) when the sea surface begins to lose heat and wind is still relatively weak. During November

(N=0.0029s-1), after strong wind bursts, stratification begins to be eroded. From December to March (N=0.0022-0.0015s-1), under the effects of strong wind and progressive cooling of the sea surface, stratification becomes increasingly weaker. Finally, from April to August (N=0.0018-0.0028s-1), as wind weakens and the sea surface gains heat, stratification becomes progressively stronger. In summary, stratification increases from April to October and decreases from November to March (Table 1, column 2).

Before examining the circulation within the Blanes canyon, it is important to review the relevant scales of motion. In this context, a key parameter for the resulting interactions between canyon topography and the incident flow is the ratio of the canyon width (W) to the local internal radius of deformation (Rd). A submarine canyon is considered narrow when it is narrower than half the smallest local internal radius of deformation (i.e. W<Rd/2) (Klinck, 1988; 1989). If a canyon is narrow, then cross-canyon geostrophic flow is inhibited, while along-canyon ageostrophic flow caused by the unbalanced pressure gradients would create substantial cross-shelf exchange. In contrast, if a canyon is wide, a geostrophic balance can be reached within the canyon, resulting in a cross-canyon flow, that is, the canyon merely steers the flow around the isobaths and so the horizontal pressure gradient remains balanced and no up- or downwelling occurs (Klinck, 1988; 1989; 1996; Hickey, 1995; Chen and Allen, 1996; Garcia-LaFuente et al., 1999; Jordi et al., 2005; Flexas et al., 2008; Allen and Durrieu de Madron, 2009).

According to the above-mentioned criterion, the Blanes canyon is moderately narrow, that is, the canyon width (~8 km) is slightly more than half the smallest local internal radius of deformation (15.46 km; Table 1, column 3). It is also deep, that is, the depth of the canyon below its rim is approximately three times the depth of the simulated incident flow (~1000 m versus ~300 m). On the other hand, the vertical stratification scale (Tr), that provides the distance above the canyon in which the flow will remain unaffected by the presence of the canyon (Craig, 2006; Flexas et al., 2008), exhibits values between ~119 m and ~253 m for strong and weak stratification conditions respectively (Table 1, column 4). Furthermore, an indicator of nonlinearity within a submarine canyon is the Rossby number (Ro) representing the relative importance of momentum advection (Klinck, 1996; Hickey, 1997; Skliris et al., 2001; Allen and Durrieu de Madron, 2009). For the simulated incident flow, which displays velocities between ~10 and 15 cm $\cdot s^{-1}$ at the shelf-break, Ro ranges from 0.36 to 0.51 (Table 1, column 5) indicating that momentum advection plays an important role on the flow pattern.

In mid-winter, the Blanes canyon topographic effects on the incident flow field are clearly observed from the sea surface to the bottom (**Figure 8**). Above the canyon rim (i.e. from the sea surface down to 100 m depth), the NC entering the canyon is deflected toward the coast at the same time that the increase of the depth gradient leads to an increase in the current velocity (maximum daily-averaged current speeds near the canyon axis of about 35, 30 and 20 cm $\cdot s^{-1}$ at 1 m, 50 m and 100 m depth respectively). While the onshore edge of the NC flows toward the west wall of the canyon head crossing the canyon isobaths, the offshore edge flows toward the upper canyon across them. Passing the canyon axis, the NC is again deflected and flows south-westward crossing the isobaths of the west wall of the canyon.

Over the shelf, there is also a south-westward flow with maximum daily-averaged near-surface (at 1 m depth) current speeds of about 25-30 cm s^{-1}. The overall pathway and the offshore intensification of this current suggest a possible link with the local dynamics of the NC resulting from its interaction with the canyon topography. Further north-east, over the north-easternmost inner shelf, there is a cyclonic gyre characterized by maximum daily-averaged near-surface current speeds of about 20-25 cm s^{-1}. At 150 m depth (i.e. below the canyon rim), the circulation patterns are very similar to those observed in the upper layers. The NC is also deflected toward the coast over the east wall of the canyon and then, passing the canyon axis, it is steered south-westward.

The main difference with the upper circulation patterns is a weak anticyclonic circulation (daily-averaged current speeds less than 15 cm s^{-1}) observed over the east wall of the upper canyon. These circulation patterns are rather different at 250 m. At this depth, a dipole-like structure with anticyclonic (cyclonic) circulation over the east (west) wall of the canyon is observed. In this structure, the divergence zone is located on the western side of the canyon axis. At greater depths (i.e. from 400 m depth down to the bottom), the flow steered by the canyon topography tends to follow along the canyon walls describing an anticyclonic path with a maximum daily-averaged current speed ranging 8 to 10 cm s^{-1}.

5. The 3D biogeochemical model

The appropriate basis for the study of marine biogeochemical cycles is the coupling of models that integrate physical, chemical and biological processes. In this section we present results from the biogeochemical component of WMED over the period 2001-2008.

The quantitative assessment of the predictive capability of a model is necessary before it is used with any degree of confidence for either scientific or operational purposes (Holt et al., 2005). Validating three-dimensional models is difficult because of the general paucity of observational data at the proper spatial and temporal scales (Lehmann et al., 2009). In this context, satellite imagery is a valuable data source because of its synopticity over wide areas.

Weekly composite images of chlorophyll (Chl) for the Western Mediterranean are detailed enough to fill most of the gaps caused by cloud cover. The resulting data set is robust and can be used for objective validation of model outputs. Quantitative metrics that measure the agreement between model predictions and observational data have recently received increasing attention (Allen et al., 2007). Remotely sensed chlorophyll was used for this purpose by Lacroix et al., (2007) in the North Sea and by Lehmann et al., (2009) in the western North Atlantic Ocean. Some of these metrics are used here to perform a quantitative validation of the model skill as a predictor of surface variability.

The biogeochemical model developed for this study is an aggregated-type model based on previous developments by Cruzado (1982), Fasham et al. (1990), Varela et al. (1992) and Bahamon and Cruzado (2003). It consists of different compartments representing nitrate, ammonium, phytoplankton, bacteria, zooplankton, detritic matter and dissolved organic matter and uses nitrogen as currency. Nitrogen fluxes among these compartments are parameterized in order to describe the main biogeochemical processes occurring at the lowest levels of the marine pelagic food-web (**Figure 9**).

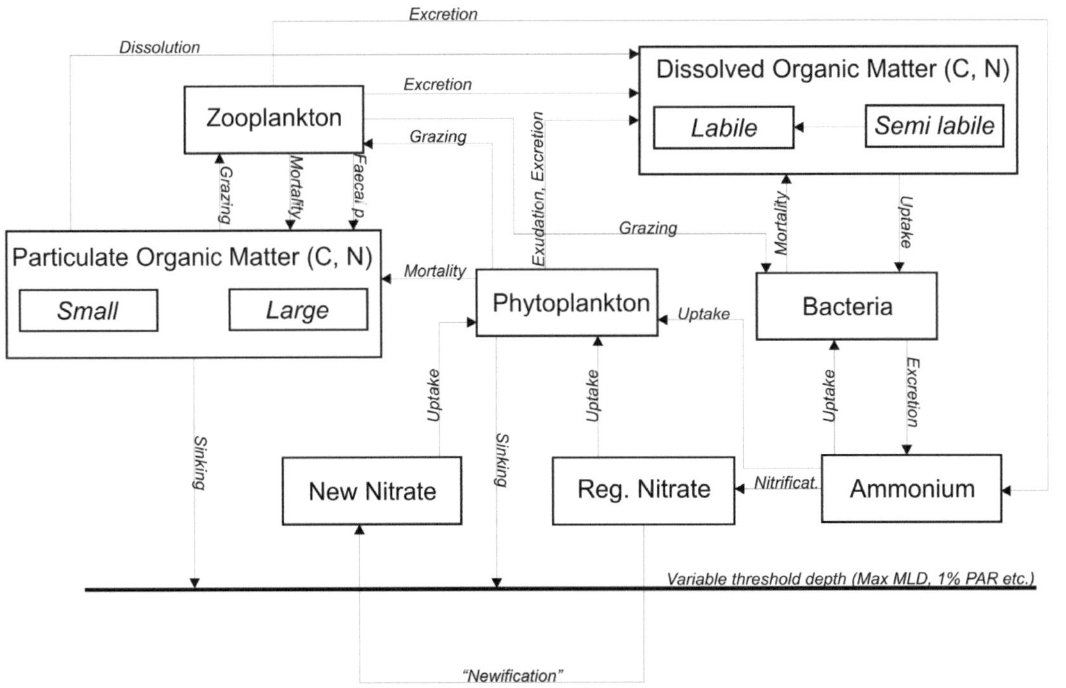

Fig. 9. Biogeochemical model description. Boxes show compartments and arrows correspond to fluxes.

Nitrogen is a key element for understanding the nutrient flow in the Mediterranean because its fractionation covers different aspects of processes occurring in the ecosystem (Dugdale and Wilkerson, 1988). Mediterranean waters show a general deficiency of phosphorus with respect to the world ocean with high N:P ratios (about 20-23). Several studies in the Western Mediterranean have pointed to phosphorus as an important limiting nutrient for phytoplankton growth (Thingstad et al., 1998; Diaz et al., 2001). Nevertheless, other scientists proved nitrogen to be the main limiting factor in the photic zone, phosphorus being potentially limiting only when other factors (light or nitrogen) are limiting at the same time (Bahamon and Cruzado, 2003; Lucea et al., 2003; Leblanc et al., 2003).

Compartments were introduced to better represent the vertical flux of particulate and dissolved organic matter. Particulate matter was split into small and large detritus and dissolved organic matter into labile and semi-labile fractions. Some degree of complexity was allowed by considering variable C:N ratios of decaying matter (both detritus and dissolved organic matter). The phytoplankton and zooplankton compartments include all the autotrophic and heterotrophic pelagic organisms without any functional or dimensional distinction. The number of state variables has been set to fourteen, ten for nitrogen and four for carbon. This configuration allows closing the cycle of nitrogen in a mass-conservative way while the cycle of carbon is only partially described and its total mass is not conserved. In fact, dissolved inorganic carbon is not considered as a state variable and this is equivalent to assume that it never limits primary production.

a. Statistics of model/observation fit

Level 3 weekly composite maps from sensor Aqua-MODIS were downloaded from NASA Ocean Colour Home Page. The images were interpolated to the model grid obtaining a data series that spans from June 2002 to December 2008. Four statistics frequently used to quantify agreement between model (*M*) and observations (*O*) are calculated for Chl: model bias (*Bias*), root mean square error (*RMSE*), model efficiency (*ME*) and correlation coefficient (*CRC*). In the following equations the total number of model/observation data pairs is indicated as *n* and summations are performed over the time dimension. Bi-dimensional maps are obtained for each index. The average over time allows visualizing spatially explicit error statistics that quantify the temporal agreement between model and observations at the spatial resolution of the model. This is useful in highlighting regional differences in the performance of the model.

The model bias represents the mean deviation between model estimates and observations. It provides a measure of whether the model is systematically underestimating (*Bias*<0) or overestimating (*Bias*>0) the observations. Note that *Bias* is able to reveal only a persistent error in magnitude of the modelled variable because negative and positive deviations tend to cancel each other in the summation.

$$Bias = \frac{1}{n}\sum(M-O)$$

Root mean square error (*RMSE*) measures the misfit between model and observations by neutralizing the sign of the deviation:

$$RMSE = \sqrt{\frac{\sum(M-O)^2}{n}}$$

Negative and positive contributions are added; then the square-root restores the unit to that of the variable considered.

Model efficiency (*ME*) is the proportion of the initial variance accounted for by the model:

$$ME = 1 - \frac{\sum(M-O)^2}{\sum(O-\overline{O})^2}$$

ME gives the deviation of the predicted values from the observed values in relation to the scattering of the latter. The maximum value for this indicator is 1 which corresponds to an explained variance of 100%. The over-bar denotes the average over *n*.

The correlation coefficient (*CRC*) indicates the quality and direction of a linear relationship between two variables:

$$CRC = \frac{\sum\left(O-\overline{O}\right)\left(M-\overline{M}\right)}{\sqrt{\sum(O-\overline{O})^2\sum(M-\overline{O})^2}}$$

b. Chlorophyll validation

The surface averages of model and MODIS chlorophyll for the whole time-series are shown in **Figure 10**. The overall agreement is reasonable as the model reproduces well the seasonal cycle with a good timing for the spring bloom and the summer oligotrophy. In late summer and autumn the model tends to overestimate the surface chlorophyll concentrations specially during 2003 and 2004 when the model simulates a secondary autumn peak not visible in the MODIS series. From autumn to winter the decrease in chlorophyll concentration in the MODIS series is hardly noticeable while in the model seems to be a recurrent feature causing the winter underestimation of chlorophyll. The interannual variability seems to be restricted to the duration and intensity of the spring bloom and is well captured by the model.

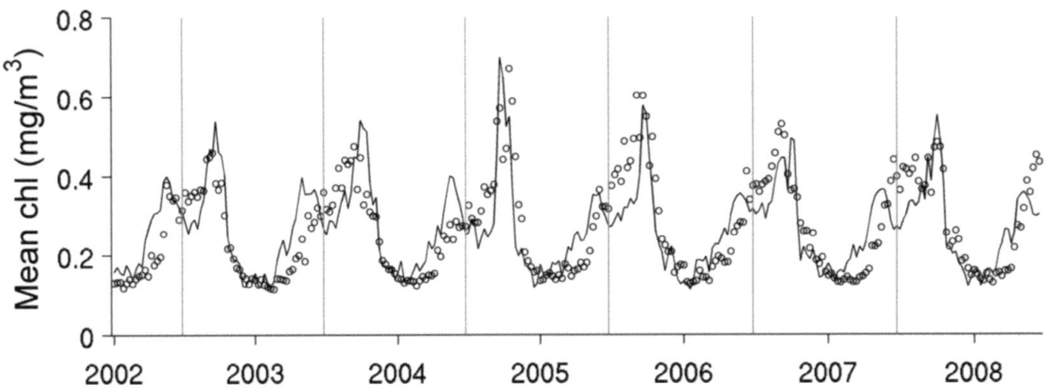

Fig. 10. Area-averaged chlorophyll for MODIS (circles) and model (continuous line). R=0.79, p< 0.01

The spatially resolved statistic indexes of the fit between model and MODIS chlorophyll are shown in **Figure 11**. The bias shows a general model underestimation along the coast in the Gulf of Lions, the Gulf of Valencia, the western Alboran Sea and along the Italian coast. It should be recalled that, in coastal areas, remotely sensed surface chlorophyll is likely to be overestimated due to the presence of inorganic sediments and coloured dissolved organic matter. This is certainly the case for the Gulf of Lions and the northern part of the gulf of Valencia where continental freshwater inputs are important. The Rhône and the Ebro rivers are the most important effluents in the NW Med and, though their influence is taken into account by the model as total nitrogen input, the inorganic sediment load can significantly complicate the interpretation of the colour signal. The same happens along the Italian coast where a certain number of smaller rivers (Liri-Garigliano, Tevere, Arno, Magra etc.) can bring considerable amount of sediments to the coastal waters.

In seasonal composite maps of remote sensing chlorophyll, the northern part of the western Mediterranean Sea is characterized by the presence of an area with concentrations higher than in the southern part. This is true during spring, summer and autumn while during winter the picture is opposite. This area interests the Gulf of Genova up to the Balearic Sea and is limited on the southern edge by the presence of islands (Corsica and Balearic) at its longitudinal extremes while in the centre, in front of the Gulf of Lions, the southern limit roughly coincides with the North-Balearic (NB) front. This area is enclosed by the general cyclonic circuit formed by the Northern Current (NC), the Balearic Current (BC), the NB front and the Western Corsican Current (WCC). Furthermore, it is an important site of deep water formation, specially in its central part.

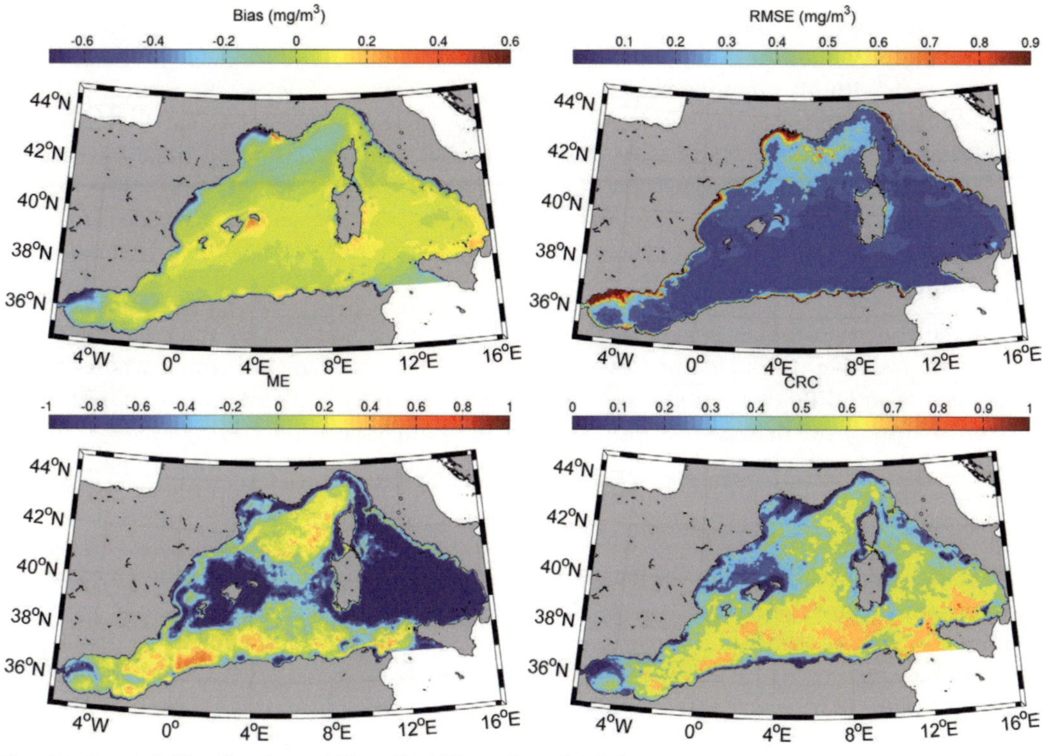

Fig. 11. Spatial distribution of Bias, RMSE, ME and CRC

This means that the vertical dynamics are particularly energetic during winter determining at the same time a dramatic nutrient replenishment of the euphotic zone and a reduction of light exposure for phytoplankton. This explains the so-called ``blue holes'' that correspond to reduced chlorophyll concentrations during winter (Barale et al., 2008).

With the progression of winter, the wind forcing decreases leading to the stabilization of the water column what results in the start of the spring bloom. Once the spring bloom has exhausted the surface nutrients, oligotrophic conditions prevail with the onset of summer stratification. During summer, the cyclonic-induced doming of the isopycnals and of the nutricline is still able to determine an enhancement of the nutrient flux into the euphotic zone. This flux is reduced by the thermal stratification but is sufficient to determine the presence of higher chlorophyll concentrations with respect to the surrounding areas. The area of higher chlorophyll is now narrower and reduced to its north-eastern part from the Gulf of Lions to the Gulf of Genova. The *Bias* for chlorophyll concentration shows model underestimation over an area that roughly coincides with the summer chlorophyll distribution. The shape and extension of the area is confirmed by *RMSE* while both *ME* and *CRC* point to a good performance of the model. This means that the underestimation of chlorophyll by the model is probably restricted to summer being the signal of the spring bloom determinant to obtain good results for *ME* and *CRC*.

6. Summary and conclusions

The three-dimensional modelling of hydrodynamic and biogeochemical processes taking place in the NW Med pelagic ecosystem (WMED and BLANCA models) presented here compare to in-situ and remote sensing observations. This is achieved because the models simulate key physical processes shaping the hydrodynamics over the Western Mediterranean basin and sub-basins (WMED model) and in a coastal area (mesoscale) strongly influenced by the submarine Blanes canyon bathymetry (BLANCA). The WMED model also allowed a satisfactory representation of the biogeochemical processes (e.g. heat transport, momentum and biogeochemical tracers) conducting the seasonal fluctuations of phytoplankton primary production.

Modelling the oligotrophic pelagic ecosystem dynamics is suitable for understanding and quantifying relatively complex processes related to energy and matter transfer constrained by advective and diffusive processes. Although not developed with all limiting nutrients for phytoplankton growth, e.g. the biogeochemical model assessed nitrogen-based matter fluxes, it allows explaining the primary production based on diffusivity and advection processes through the water column and based on the mixing produced by winter convection fuelling the phytoplankton bloom in late spring.

Both models are excellent tools for the study of past and future evolution of the physical and biogeochemical environment in an area that is subject to important anthropogenic pressures (navigation, fishing, tourism, industry, agriculture, coastal development, etc.). Robust tools such as these are crucial for assessing the impact exerted up to the present or expected to be made in future. Marine ecosystem analysis and operational oceanography are two scientific and technical fields in which combined disciplines are key for their success. On the other hand, observations by means of satellites or autonomous moored or drifting sensors are also crucial to validate models making them indispensable in the forecasting of future scenarios.

7. References

Allen, J.I., Somerfeld, P.J. and Gilbert, F.J. (2007). Quantifying uncertainty in high-resolution coupled hydrodynamic-ecosystem models. *Journal of Marine Systems* 64(1-4): 3 - 14.

Allen S.E. and Durrieu de Madron, X. (2009). A review of the role of submarine canyons in deep-ocean exchange with the shelf. *Ocean Science* 5: 607-620.

Astraldi, M., Gasparini, G.P. and Manzella, G.M.R. (1990). Temporal variability of currents in the Eastern Ligurian Sea. *Journal of Geophysical Research* 95(C2):1515-1522.

Astraldi M. and Gasparini, G.P. (1992). The seasonal characteristics of the circulation in the North Mediterranean Basin and their relationship with the Atmospheric-Climate Conditions. *Journal of Geophysical Research* 97(C6): 9531-9540.

Astraldi, M., Gasparini, G.P. and Sparnocchia, S. (1994). The seasonal and interannual variability in the Ligurian-Provencal Basin. *Coastal and Estuaries Studies* 46:93-113.

Astraldi, M., Balopoulos, S., Candela, J., Font, J., Gacic, M., Gasparini, G.P., Manca, B., Theocharis, A. and Tintore, J. (1999). The role of straits and channels in understanding the characteristics of Mediterranean circulation. *Progress in Oceanography* 44(1-3): 65-108. ISSN0079-6611.

Bahamon, N. and Cruzado, A. (2003). Modelling nitrogen uxes in oligotrophic environments: NW Mediterranean and NE Atlantic. *Ecological Modelling* 163(3): 223-244. ISSN 0304-3800

Bahamon, N., Cruzado, A., Velasquez, Z., Bernardello, R. and Donis, D. (2010). Patterns of phytoplankton chlorophyll variability in Mediterranean Seas. Rapp. *Comm. int. Mer Médit.* 39: 436.

Bahamon, N., Aguzzi, J., Bernardello, R., Ahumada-Sempoal, M-A., Puigdefabregas, J., Cateura, J., Muñoz, E., Velásquez, Z. and Cruzado, A. (2011). The new pelagic Operational Observatory of the Catalan Sea (OOCS) for the multisensor coordinated measurement of atmospheric and oceanographic conditions. *Sensors* 11: 11251-11272.

Barale, V., Jaquet, J.-M. and Ndiaye, M. (2008). Algal blooming patterns and anomalies in the Mediterranean Sea as derived from the SeaWiFS data set (1998-2003). *Remote Sensing of En- vironment*: 112(8), 3300 - 3313. ISSN 0034-4257.

Baretta, J.W., Ebenhöh, W. and Ruardij, P. (1995): The european regional seas ecosystem model, a complex marine ecosystem model. *Netherlands Journal of Sea Research* 33(3/4), 233-246.

Bethoux, J.P., Prieur, L. and Nyffeler, F. (1982). The water circulation in the North-western Mediterranean Sea, its relations with wind and atmospheric pressure. Hydrodynamics of semi-enclosed seas. Proceeding of the 13th International Liege Colloquium on Ocean Hydrodynamics. J.C.J., Nihoul (editor). *Elsevier Oceanography Series* 34:129-142.

Bianchi, C.N. (2007). Biodiversity issues for the forthcoming tropical Mediterranean Sea. *Hydrobiologia* 580:7–21.

Bricaud, A., Morel, A. and Tailliez, D. (1992). Mesures optiques. In: Neveux, J. (Ed.), Les maximums profonds de chl a en mer des Sargasses. Donnees physiques, chimiques et biologiques. Campagne Chlomax. Campagnes Oceanographiques.

Camuffo, D., Bertolin, C., Diodato, N., Barriendos, M., Dominguez-Castro, F., Cocheo, C., della Valle, A., Garnier, E., and Alcoforado, M. (2010). The western Mediterranean climate: How will it respond to global warming? *Climatic Change* 100:137-142.

Chen, X. and Allen, S.E. (1996). The influence of canyons on shelf currents: A theoretical study. *J. Geophysical Research* 101(C8): 18043-18059.

Craig, W. A. (2006). The Flinders Current and Upwelling in Submarine Canyons. *Master Thesis, UNSW*, Australia, 118 pp.

Crépon, M. and Boukthir, M. (1987). Effect of deep water formation on the circulation of the Ligurian Sea. *Annals of Geophysics* 5B (1):43-48.

Cruzado, A. (1982). Simulation model of primary production in coastal upwelling off western sahara. *J. Cons. perm. int. Explor Mer.* 180: 5-6.

Diaz, F., Raimbault, P., Boudjellal, B., Garcia, N. and Moutin, T. (2001). Early spring phosphorus limitation of primary productivity in a NW Mediterranean coastal zone (Gulf of Lions). *Marine Ecology Progress Series* 211: 51–62.

Doney, S.C., Glover, D.M. and Najjar, R.G. (1996). A new coupled, one-dimensional biological-physical model for the upper ocean: applications to the JGOFS Bermuda Atlantic Time-Series Study (BATS) site. *Deep-Sea Research II* 43: 591–62

Dugdale, R.C. and Goering, J.J. (1967). Uptake of new and regenerated forms of nitrogen in primary productivity. *Limnology and Oceanography* 12: 6: 206.

Dugdale, R.C. and Wilkerson, F. P. (1988). Nutrient sources and primary production in the Eastern Mediterranean. *Oceanologica Acta* 11: 179 – 184

Ebenhoh, W. (2000). Critical analysis of the *status quo* in marine ecosystem modeling. Biological Observations in Operational Oceanography. EuroGOOS Publication N° 15, EuroGOOS Office, Southampton.

Eppley, R. W. and Peterson, B. J. (1979). Particulate organic matter flux and planktonic new production in the deep ocean. *Nature* 282: 677-680.

Estrada, M. (1995). Primary production in NW Mediterranean. *Scientia Marina* 60(Supl.2): 55-64.

Fasham, M.J.R., Ducklow, H.W. and McKelvie, S.M. (1990). A nitrogen-based model of plankton dynamics in the ocean mixed layer. *Journal of Marine Research* 48(3): 591 - 639

Flexas, M.M., Boyer, D.L., Espino, M., Puigdefàbregas, J., Rubio, A. and Company, J.B. (2008). Circulation ovear a submarine canyon in the NW Mediterranean. *Journal of Geophysical Research* 113(C12002), doi: 10.1029/2006JC003998.

Font, J., Salat, J. and Tintore, J. (1988). Permanent features of the circulation in the Catalan Sea. *Oceanologica Acta* 9: 51-57.

Font, J., Garcia-Ladona, E. and Gorriz, E. (1995). The seasonality of mesoscale motion in the Northern Current of the western Mediterranean: several years of evidence. *Oceanologica Acta* 18, 207-219.

Garcia-Ladona, E., Tintore, J., Pinot, J., Font, J. and Manriquez, M. (1994). Surface circulation and dynamics of the Balearic Sea. In La Violette, P., ed., The seasonal and interannual variability of the Western Mediterranean Sea, Coastal and Estuarine Studies, pp. 73-91.

Garcia-Ladona, E., Castellon, A., Font, J. and Tintore, J. (1996). The Balearic current and volume transport in the Balearic basin. *Oceanologica Acta* 19, 489-497.

Garcia-LaFuente, J., Sarhan, T., Vargas, M., Vargas, J.M. and Plaza, F. (1999). Tidal motions and tidally induced fluxes through La Línea submarine canyon, western Alboran Sea. *Journal of Geophysical Research* 104(C2):3109-3119.

Gascard, J.C. (1978). Mediterranean deep water formation baroclinic instability and oceanic eddies. *Oceanologica Acta,* 1(3): 315-330.

Gili, J.-M., Bouillon, J., Pages F., Palanques, A. and Puig, P. (1999). Submarine canyons as habitats of prolific plankton populations: three new deep-sea Hydroidomedusae in the western Mediterranean. *Zoological Journal of the Linnean Society* 125:313-329.

Gordoa, A., Illas, X., Cruzado, A., and Velásquez, Z. (2008). Spatio-temporal patterns in the north-western Mediterranean from MERIS derived chlorophyll *a* concentration. *Scientia Marina* 72: 757-767.

Hickey B.M. (1995). Coastal submarine canyons, paper presented at "Aha Huliko" A Workshop of Flow Topography Interactions, Office of Naval Research, Honolulu, 17-20 January.

Hickey, B.M. (1997). The Response of a Steep-Sided, Narrow Canyon to Time-Variable Wind Forcing. *Journal of Physical Oceanography* 27: 667-726.

Holt, J.T., Allen, J.I., Proctor, R. and Gilbert, F. (2005). Error quantification of a high-resolution coupled hydrodynamic-ecosystem coastal-ocean model: Part 1 model overview and assessment of the hydrodynamics. *Journal of Marine Systems* 57(1-2): 167 - 188. ISSN 0924-7963.

Jordi, A., Orfila, A., Basteretxea, G. and Tintoré, J. (2005). Shelf-slope exchanges by frontal variability in a steep submarine canyon. *Progress in Oceanography* 66:120-141.

Klinck, J.M., (1988). The Influence of a Narrow Transverse Canyon on Initially Geostrophic Flow. *Journal of Geophysical Research* 13(C1): 2009-515.

Klinck, J.M. (1989). Geostrophic Adjustment Over Submarine Canyons. *Journal of Geophysical Research* 94(C5): 6133-6144.

Klinck, J.M., (1996). Circulation near submarine canyons: A modeling study. *Journal of Geophysical Research* 101(C1): 1211-1223.

Kremer, J.N. and Nixon, S.W. (1978). A Coastal Marine Ecosystem: Simulation and Analysis. Ecological Studies, Vol. 24. Springer-Verlag, Heidelberg. 210 pp.

Korres, G. and Lascaratos, A. (2003). A one-way nested eddy resolving model of the Aegean and Levantine basins: implementation and climatological runs. *Annals of Geophysics* 21: 205-220.

La Violette, P. E., Tintore, J. and Font, J. (1990). The surface circulation of the Balearic Sea. *Journal of Geophysical Research* 95(C2), 1559-1568.

Lacroix, G., Ruddick, K., Park, Y., Gypens, N. and Lancelot, C. (2007). Validation of the 3D biogeochemical model MIRO&CO with eld nutrient and phytoplankton data and MERIS-derived surface chlorophyll a images. *Journal of Marine Systems* 64(1-4): 66 - 88.

Leblanc, K., Queguiner, B., Garcia, N., Rimmelin, P. and Raimbault, P. (2003). Silicon cycle in the NW Mediterranean Sea: seasonal study of a coastal oligotrophic site. *Oceanologica Acta* 26(4): 339-355. ISSN 0399-1784.

Lehmann, M.K., Fennel, K. and He, R. (2009). Statistical validation of a 3-D bio-physical model of the western North Atlantic. *Biogeosciences* 6(10): 1961-1974. ISSN 1726-4170.

López-García, M.J., Millot, C., Font, J. and García-Ladona, E. (1994). Surface circulation variability in the Balearic Basin. *Journal of Geophysical Research* 99(C2): 3285-3296.

Lotka, A. J. (1925). Elements of physical biology. Williams & Wilkins, Baltimore. [Reprinted in 1956: Elements of Mathematical Biology. Dover Publications, Inc., New York, New York].

Lucea, A., Duarte, C. M., Agustí, S. and Sondergaard, M. (2003). Nutrient (N, P and Si) and carbon partitioning in the stratified NW Mediterranean. *Journal of Sea Research* 49(3): 157 - 170. ISSN 1385-1101.

Marchesiello P., McWilliams, J. C. and Shchepetkin, A. (2001). Open boundary conditions for long-term integration of regional oceanic models. *Ocean Modelling* 3: 1-20.

Margalef, R. (1972). Interpretaciones no estrictamente estadísticas de la representación de entidades biológicas en un espacio multifactorial. *Investigación Pesquera* 36: 183-190.

Mellor, G. (2004). User's Guide for a Three-Dimensional Primitive Equation, Numerical Ocean Model. Princeton University, Princeton, N.J., 1-56

MEDOC Group (1970). Observation of Formation of Deep Water in the Mediterranean Sea, 1969. *Nature* 227:1037-1040.

Millot, C. (1990). The Gulf of Lions hydrodynamics. *Continental Shelf Research* 20: 1-9.

Millot, C. (1999). Circulation in the Western Mediterranean Sea. *Journal of Marine Systems* 20(1-4): 423 - 442. ISSN 0924-7963.

Odum, H.T. (1960). Ecological potential and analog circuits for the ecosystem. *American Scientist* 48:1-8.

Pinardi, N. and Coppini, G. (2010). Operational oceanography in the Mediterranean Sea: the second stage of development. *Ocean Science* 6: 263-267

Redfield, A. C., Ketchum, B.H., and Richards, F.A. (1963). The influence of organisms on the composition of sea water, in The Sea, edited by M. N. Hill, pp. 26– 77, Interscience, New York.

Riley, G.A., Stommel, H. M. and Bumpus, D.F. (1949). Quantitative ecology of the plankton of the western North Atlantic. *Bulletin of the Bingham Oceanographic Collection* 12(3):1–169.

Rio, M.-H., Poulain, P.-M., Pascual, A., Mauri, E., Larnicol, G. and Santoleri, R. (2007). A Mean Dynamic Topography of the Mediterranean Sea computed from altimetric data, in-situ measurements and a general circulation model. *Journal of Marine Systems* 65(1-4): 484-508. ISSN 0924-7963.

Salat, J. (1995). The interaction between the Catalan and Balearic currents in the southern Catalan Sea. *Oceanologica Acta* 18(2):227-234.

Sardà, F., Company, J. B., Bahamon, N., Rotllant, G., Flexas, M.M., Sánchez, J.D., Zúñiga, D., Coenjaerts, J., Orellana, D., Jordà, G., Puigdefábregas, J., Sánchez-Vidal, A., Calafat, A., Martín, D. and Espino, M. (2009). Relationship between environment and the occurrence of the deep-water rose shrimp *Aristeus antennatus* (Risso, 1816) in the Blanes submarine canyon (NW Mediterranean). *Progress in Oceanography* 82 (4): 227-238.

Schott, F., Visbeck, M., Send, U., Fischer, J., Stramma, L. and Desaubies, Y. (1996). Observations of Deep Convection in the Gulf of Lions, Northern Mediterranean, during the winter of 1991/92. *Journal of Physical Oceanography* 26:505-524.

Skliris, N., Goffart, A., Hecq, J.H. and Djenidei, S. (2001). Shelf-slope exchanges associated with a steep submarine canyon off Calvi (Corsica, NW Mediterranean Sea): A modeling approach. *Journal of Geophysical Research* 106(C9):19,883-19,901.

Steele, J.H. (1970). Marine food chains (Ed). University of California Press, Berkeley, 552 pp.

Sorgente, R., Drago, A.F. and Ribotti, A. (2003). Seasonal variability in the Central Mediterranean Sea circulation. *Annals of Geophysics* 21: 299-322.

Thingstad, T.F., Zweifel, U.L. and Rassoulzadegan, F. (1998). P limitation of heterotrophic bacteria and phytoplankton in the northwest Mediterranean. *Limnology and Oceanography* 43: 88–94.

Tonani, M., Pinardi, N., Dobricic, S., Pujol, I. and Fratianni, C. (2008). A high-resolution free-surface model of the Mediterranean Sea. *Ocean Science* 4, 1-14

Tonani, M., Pinardi, N., Fratianni, C., Pistoia, J., Dobricic, S. Pensieri, S., de Alfonso, M. and Nittis, K. (2009). Mediterranean Forecasting System: forecast and analysis assessment through skill scores. *Ocean Science* 5, 649-660.

Varela, R.A., Cruzado, A., Tintoré, J. and García-Ladona, E. (1992), Modelling the deep-chlorophyll maximum: A coupled physical-biological approach. *Journal of Marine Research* 50:441-463

Varela, R.A., Cruzado, A., Tintoré, J. (1994). A simulation analysis of various biological and physical factors influencing the deep chlorophyll maximum structures in oligotrophic areas. *Journal of Marine Systems* 5: 143-157.

Varela, R.A., Cruzado, A., Gabaldon, J.E. (1995). Modelling primary production in the North Sea using the European Regional Seas Ecosystem Model. *Netherlands Journal of Sea Research* 33(3): 337-361.

Vargas-Yáñez, M., Zunino, P., Benali, A., Delpy, M., Pastre, F., Moya, F., García-Martínez, M.C. and Tel, E. (2010). How much is the western Mediterranean really warming and salting? *Journal of Geophysical Research* 115: doi:10.1029/2009JC005816.

Velasquez, Z.R. (1997). Phytoplankton in the NW Mediterranean. *Ph.D. Thesis*, UPC, Barcelona, pp. 272.

Vetter E.W., Smith, C.R. and De Leo, F.C. (2010). Hawaiian hotspots: enhanced megafaunal abundance and diversity in submarine canyons on the oceanic islands of Hawaii. *Marine Ecology* 31: 183-199. ISSN 0173-9565.

Vichi, M. (2000). The influence of high-frequency surface forcing on productivity in the euphotic layer. Biological Observations in Operational Oceanography. EuroGOOS Publication N° 15. EuroGOOS Office, Southampton.

Volpe, G., Nardelli, B.B., Cipollini, P., Santoeri, R., and Robinson, I.S. (2011). Seasonal to interannual phytoplankton response to physical processes in the Mediterranean Sea from satellite observations. *Remote Sensing of Environment*, doi:10.1016/j.rse.2011.09.020.

Volterra, V. (1926). Variazioni e fluttuazioni del numero d'individui in specie animali conviventi. Mem. R. Accad. Naz. Dei lincei. Ser. VI (2). [Translated into English by M. E. Wells 1928, *Journal du Conseil International pour l'Exploration de la Mer* 3: 1–51.]

Wroblewski, J.S., Sarmiento, J.L., Flierl, G.R. (1988). An ocean basin scale model of plankton in the North Atlantic. Solutions for the climatological oceanographic conditions in May. *Global Biogeochemical* Cycles 2: 199-218.

Wroblewski, J.S. (1989). A model of the spring bloom in the North Atlantic and its impact on ocean optics. *Limnology and Oceanography* 34: 1565-1573.

Zavatarelli, M. and Pinardi, N. (2003). The Adriatic Sea modelling system: a nested approach. *Annals of Geophysics* 21: 345-364.

3

The Marine Ecosystem of the Sub-Antarctic, Prince Edward Islands

I. J. Ansorge[1], P. W. Froneman[2] and J. V. Durgadoo[3]
[1]*Oceanography Department, Marine Research Institute, University of Cape Town,*
[2]*Southern Ocean Group, Department of Zoology and Entomology, Rhodes University,*
[3]*Helmholtz Center for Ocean Research Kiel (GEOMAR), Kiel,*
[1,2]*South Africa*
[3]*Germany*

1. Introduction

Straddled between the northern and southern boundaries of the Antarctic Circumpolar Current (ACC), Sub-antarctic islands are typically oceanic; experiencing moist, cool and windy climates. They are classified as regions, in which the terrestrial and marine ecosystems are relatively simple and extremely sensitive to perturbations. One such example are the Prince Edward Islands - the most southerly part of South Africa's official territory. The islands are located in the Indian sector of the Southern Ocean at approximately 46°50'S and 37°50'E (Figure 1). The nearest landfall is the Crozet Island Group 950 km to the east, while South Africa lies over 2 000 km northwest. The islands consist of Marion and Prince Edward Island (Figure 1 - insert), two volcanic outcrops approximately 250 000 years old, but still active. Marion Island covers an area of 270 km^2; whereas Prince Edward Island – 19 km to the north-east – is only about 45 km^2 in extent. The islands rise steeply from a region of complex bottom topography with a shallow saddle, between 40 and 200 m deep, separating Prince Edward from Marion Island. Intensive investigations carried out on the oceanic frontal systems south of Africa (Lutjeharms & Valentine, 1984; Duncombe Rae, 1989 a,b; Belkin & Gordon, 1996) have shown that the Prince Edward Islands lie directly in the path of the ACC, sandwiched between the Sub-antarctic Front (SAF) and the Antarctic Polar Front (APF). As such, these islands provide an ideal ecological laboratory for studying how shifts in atmospheric and oceanic circulation patterns in the Southern Ocean will increase the ease in which these islands, their ecosystems and their ocean surrounds can be invaded by alien species (Smith, 2002).

The Prince Edward Islands, like many other oceanic islands within the Southern Ocean, are seasonally characterised by vast populations of marine organisms and a diversity and abundance of seabirds that use the islands as breeding grounds (Bergstrom & Chown, 1999; Ryan & Bester, 2008). It is estimated that the islands support over 5 million breeding pairs of top predators including flying seabirds, penguins and seals during the peak in breeding season. The energy necessary to sustain these top predators is derived from the surrounding marine environment. Changes in the marine ecosystem in response to global climate change are therefore, likely to dramatically influence the populations of top predators that seasonally occur on the islands (Ryan & Bester, 2008).

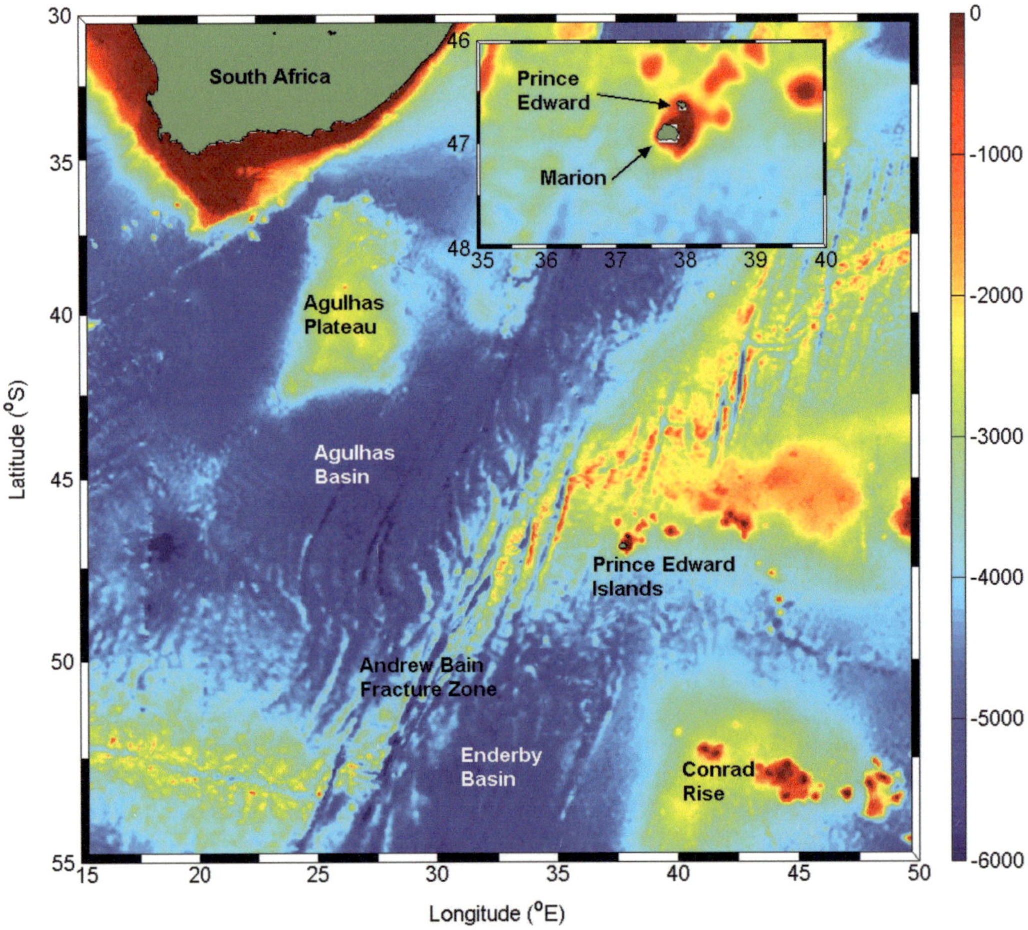

Fig. 1. Map showing the bathymetry for the south-west Indian Ocean from ETOPO2 data. Isobaths are in metres. The insert is a zoom-in of the Prince Edward and Marion Island group and the surrounding bathymetry.

1.1 Historical setting

The terrestrial nature of the islands - geological, biological and meteorological - has been studied since the South African government claimed sovereignty to them in 1947 and when Marion Island became host to a meteorological station. A detailed description of the history of the islands is presented in Cooper (2008). The oceanographic setting of the islands has received attention only since the late 1970s, when pioneering studies on the physical oceanography, primary productivity, plankton, fish and seabirds of the direct ocean environment were carried out by South African and French scientists aboard the French research vessel *M.S. Marion Dufresne* (Cooper, 2008). The presence of over 5 million birds and seals on the islands raised important questions about their relationship to the physical environment. The nesting success of many birds is critically dependent on food availability suggesting that either the islands create their own enhanced biological ecosystem in their direct vicinity, the so-called 'island mass effect', (Doty & Oguri, 1956) or that biological productivity of the ambient waters is affected by changes in the oceanic environment through frontal dynamics such as eddy generation and meanders. Other suggestions supporting this island support system, through zooplankton species, that upwelling of deep Antarctic water, as a result of predominant north-westerly winds, is the primary mechanism responsible for high productivity in the vicinity of the islands. The upwelling of nutrients in this water would favour increased phytoplankton production. Indeed, Grindley and Lane (1979) reported the presence of a predominantly Antarctic copepod fauna in this region, confirming the presence of water of Antarctic origin during the period of the 1979 cruise. El-Sayed et al., (1979) and Deacon (1983) however, argued against this hypothesis on the basis of low silica concentrations in the surface waters. Miller (1984) suggested that frontal variability may cause foreign water masses from south of the APF to intrude into the vicinity of the Prince Edward Islands. These protrusions may then explain reported appearances of Antarctic planktonic species in what is usually considered a Sub-antarctic environment. There is now growing evidence that the geographical position of the SAF in the proximity of the Prince Edward Islands plays a key role in forming local macro- and mesoscale oceanographic conditions in the region of the islands (Ansorge & Lutjeharms, 2002; Pakhomov et al., 2000).

During the past 15 years as part of the South African National Antarctic Programme (SANAP), two intensive oceanographic programmes - Marion Island Oceanographic Study (MIOS) and Dynamics of Eddy Impacts on Marion's Ecosystem (DEIMEC) have been carried out to establish the nature of the physical and biological environment south of Africa and in particular the environment in which the Prince Edward Islands are embedded. Results indicate an unusually high degree of spatial and temporal variability for this region in contrast to comparable regions of the PFZ elsewhere in the Southern Ocean (Ansorge and Lutjeharms, 2002; Durgadoo et al., 2010). The dynamics associated with this variability have only recently been investigated and described.

2. Physical oceanographic setting

Results from numerous measurements ranging from early ships data (Sultan et al., 2007), numerical model studies (Gille, 1997; Sun & Watts, 2002), remote sensing (Sandwell & Zhang, 1989; Hughes & Ash, 2001), surface drifters (Harris & Stravopoulos, 1978; Hofmann et al., 1985) as well as recent profiling ARGO data (Sokolov & Rintoul, 2009) have shown that the mean eddy kinetic energy associated with the ACC is almost non-existent over the deep ocean basins where topographic constraint is weak. Instead, levels of mesoscale variability surge around prominent topographic features and choke points such as the Drake Passage (Joyce & Patterson, 1977), the Crozet and Kerguelen Plateaux (Gille, 2003) and south of Australia (Phillips & Rintoul, 2000). Past investigations (Park et al., 1997; Pollard & Read, 2001; Kostianoy et al., 2004) have shown that the South-West Indian sector is characterised by explicit regions of extremely high mesoscale variability (Figure 2). To the north, an enhanced band of variability corresponds to the confluence of the warm Agulhas Return Current (Lutjeharms & Ansorge, 2003) and the Subtropical Convergence (Boebel et al., 2003) forming one of the strongest and fastest flowing (>1.5 ms^{-1}) frontal systems of the world ocean (Park et al., 1993). Directly south of this band, overlying the South-West Indian Ridge and immediately upstream of the Prince Edward Islands, is an isolated region of enhanced sea surface height (SSH) variability. This 'hotspot' seems to coincide with the southward deflection and intensification of the ACC at 30°E (Figure 3).

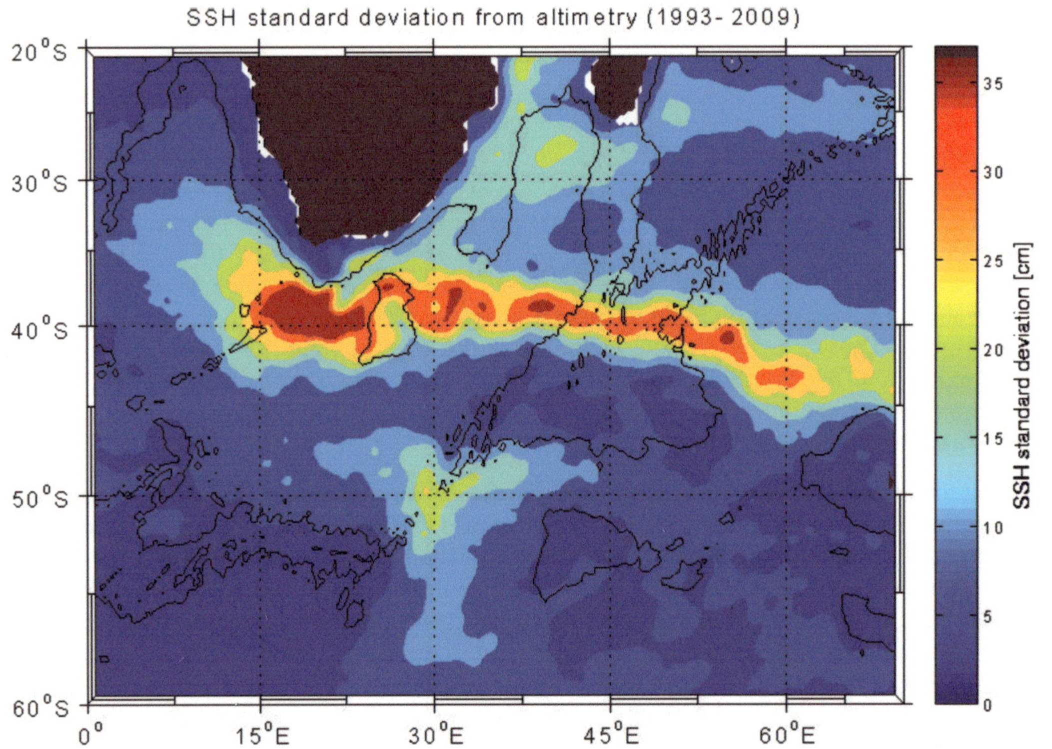

Fig. 2. Map showing the altimetry derived sea surface height variability for the south-west Indian Ocean. The isolated band of variability centred at 50°S, 30°E lies directly upstream of the Prince Edward Islands. *(courtesy Samuel Eberenz).*

Hydrographic data collected during the South-West Indian Ocean Experiment (SWINDEX) (Pollard & Read, 2001) have shown that the South-West Indian Ridge exerts a strong influence on the location and dynamics of the ACC and its associated fronts (Moore et al., 1999) resulting in substantial fragmentation of the jets downstream of the ridge. A recent examination of SST (Hughes & Ash, 2001) and SSH gradients (Sokolov & Rintoul, 2009) on either side of the South-West Indian Ridge provide an intricate examination of the ACC's multiple structure (Figure 3), confirming that the ACC narrows to a width of 5° of latitude as it is channelled through the ridge region. Downstream (i.e. east of 30°E) there is a noticeable separation in the two branches of the ACC with the SAF topographically deflected north-eastwards (Belkin & Gordon 1996; Sultan et al., 2007), thus widening the Antarctic Polar Frontal Zone (APFZ) by up to 5° of latitude.

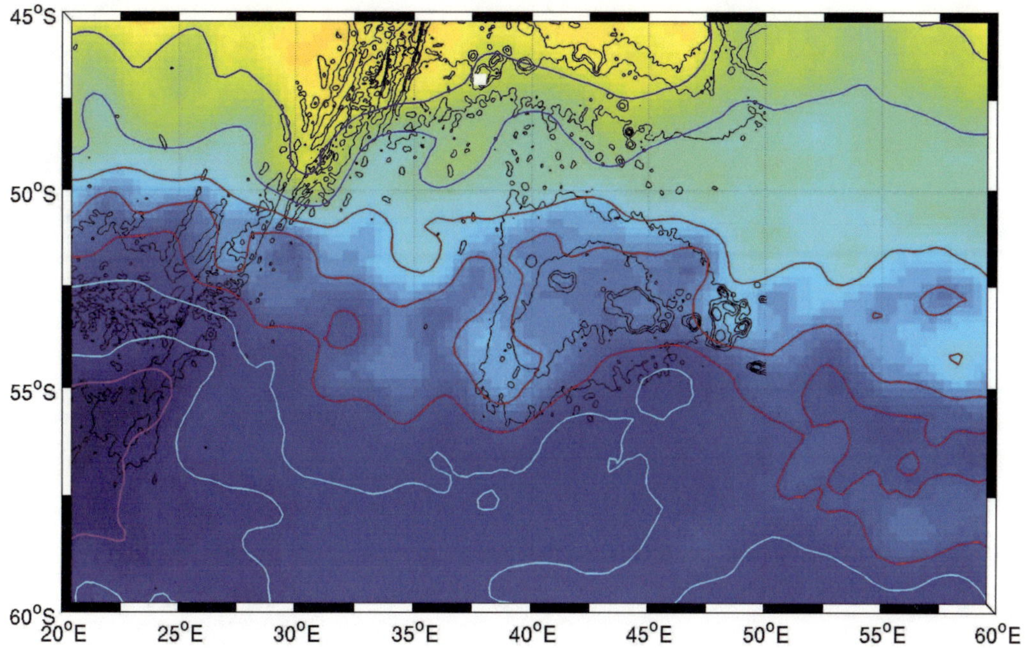

Fig. 3. Map showing the position of the frontal jets associated with the Sub-antarctic Front (blue), Antarctic Polar Front (red), Southern ACC Front (light blue) and Southern Boundary (pink). The positions have been defined from surface gradients obtained from altimetry data. The position of the Prince Edward Islands is denoted by the white square at approximately 47°S, 38°E. Isobaths are contoured at 1000 m intervals. *(courtesy Sebastiaan Swart)*

Extensive oceanographic surveys have shown that the Prince Edward Islands are sandwiched between the SAF to the north and the APF to the south (Ansorge & Lutjeharms, 2002). These fronts separate warm Sub-antarctic Surface Water (SASW) from cooler Antarctic Surface Water (AASW), with a zone of transition known as the Antarctic Polar Frontal Zone (APFZ) between the two. The SAF and APF have been shown to demonstrate a high degree of latitudinal variability in this region and it is thought that the complexity of the ACC in the vicinity of these islands (Ansorge & Lutjeharms, 2003) results in an increase in the interchange of Antarctic and Sub-antarctic surface and intermediate water masses (Deacon, 1983). Recent investigations have demonstrated conclusively that an extensive eddy train extends eastwards from the South-West Indian Ridge into the Prince Edward Island vicinity (Ansorge & Lutjeharms, 2003, 2005; Durgadoo et al., 2010, 2011). These eddies have a noticeable biological influence (Pakhomov et al., 2000; Bernard et al. 2007; Ansorge et al., 2010) by transporting physical and biological characteristics typical of the Antarctic northwards into the island vicinity (Figure 4) thus the possibility of providing an important foraging grounds for grey-headed albatrosses (Nel et al., 2001) and elephant seals (de Bruyn et al., 2009).

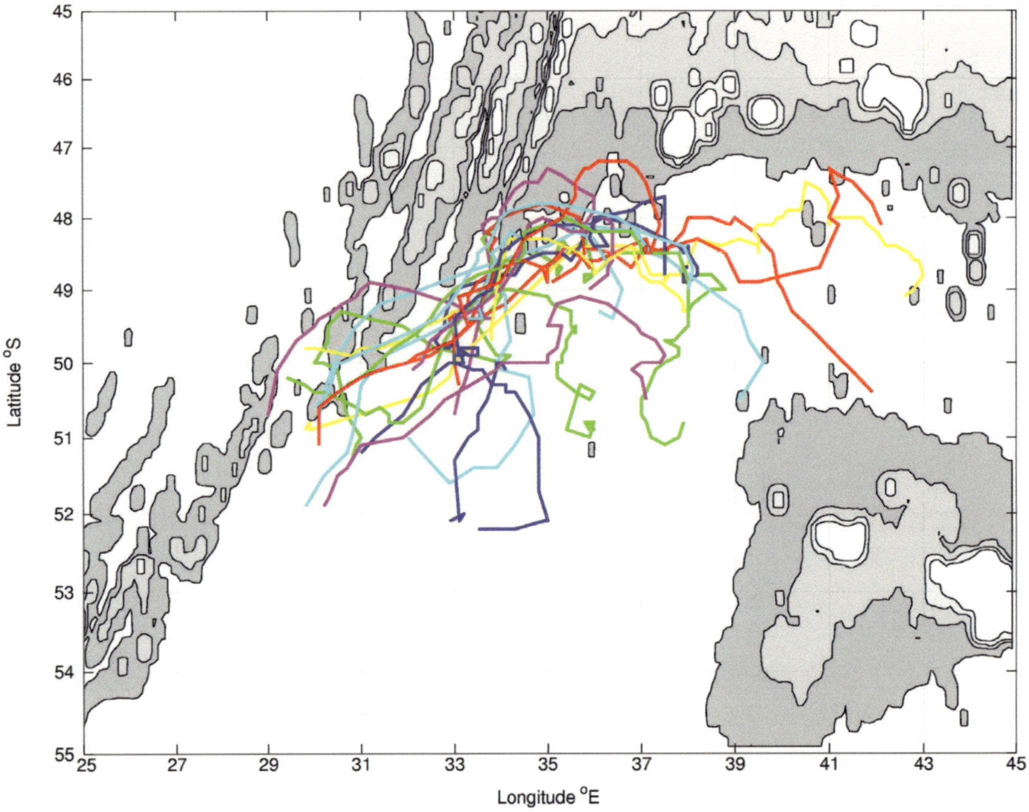

Fig. 4. Map showing the trajectories of 20 eddies tracked over 10 years from their point of generation at the South-west Indian Ridge north-eastwards towards the Prince Edward Island vicinity. Magnitude of f/h contours are shown (0.1 - 0.4 rad/s.m) at 0.1 rad/s.m intervals (Durgadoo et al., 2011).

Based on the geographic distribution of these eddies it has been surmised that their origin is as a direct result of the interaction of the ACC with the South-West Indian Ridge and in particular the series of fractures; notably the Du Toit, Andrew Bain, Marion and Prince Edward, which intersect this ridge between 25° - 35°E and 45° - 55°S and divide the South-West Indian Ridge into two almost equally extensive sections (Sclater et al., 2005). The Andrew Bain Fracture Zone is the largest of these fracture zones with a length of 750 km, and the greatest width (120 km) of any transform fault in the oceans and extends to >6000 m (Fisher & Goodwillie, 1997). It is therefore, not surprising that the highest SSH levels observed in the region (Figure 2) correlate directly to the location of this particular fracture. The clear implication of these findings is that the Prince Edward Island region has an enhanced anomaly presence not so much because of the interaction of the flow with the islands themselves, as has been inferred previously but as a consequence of the fact that they are situated at the north-eastern border of a region of unusually high mesoscale variability in the Southern Ocean (Durgadoo et al., 2010). What effect does this zone of variability - through the generation of transient eddies or latitudinal shifts in the APFZ - have on forming the macro- and mesoscale oceanographic environment of the islands themselves?

3. Biology of the island ecosystem

3.1 Phytoplankton studies

Macronutrient concentrations in the open waters of the APFZ are moderate to low with surface silicate, nitrate and phosphate concentrations ranging from 0.2 to 16.5 mmol m^{-3}, from 9.5 to 97.5 mmol m^{-3} and from 0.1 to 16.6 mmol m^{-3}, respectively (Allanson *et al.* 1985; Balarin, 2000). Shifts in the surface macronutrient concentrations within the open waters of the region generally coincide with the intrusion of warmer Sub-antarctic waters from the north and cooler Antarctic waters from the south. Additionally, eddies generated by the interaction of the ACC with the South-West Indian Ridge may locally also contribute to the spatial and temporal variations in suface concentrations of macronutrients in the open waters of the APFZ.

The total phytoplankton biomass and production within the APFZ is generally < 2.0mg chl-a m^{-3} <300 mg C m^{-2} d^{-1}, and is dominated by small the nano- (2-20μm) and picophytoplankton (< 2.0μm) size fractions (Table 1) (Balarin, 2000; Bernard & Froneman, 2005; McQuaid & Froneman, 2008). The low phytoplankton stocks recorded in the open waters can be attributed to low phytoplankton growth rates conferred by the high wind activity which contributes to deep water mixing (Balarin, 2000). The small nano- and picophytoplankton are better adapted to persist in poor light environments and low macronutrient waters. The contribution of the smaller phytoplankton size classes to the total phytoplankton biomass and production is generally > 90% (Balarin, 2000). Notable exceptions are recorded in the vicinity of the frontal systems that delimit the APFZ which demonstrate increased phytoplankton concentrations (Balarin, 2000; McQuaid & Froneman, 2008). The elevated phytoplankton stocks in the vicinity of the fronts reflect the increased contribution of the larger microphytoplankton (>20μm, mainly diatoms) to the total phytoplankton production and biomass. The shallow shelf waters of the Prince Edward Islands also periodically demonstrate the, 'island mass effect' of increased phytoplankton concentrations (Pakhomov & Froneman, 1999). Here, the phytoplankton biomass and productivity may exceed that of the open waters by 2-3 times (Pakhomov & Froneman, 1999). The elevated phytoplankton stocks periodically recorded in the vicinity of the islands can be ascribed to increased water column stability, macronutrient and trace metal (Fe) concentrations derived from the freshwater runoff from the islands, which are retained on the shallow shelf waters of the islands by anti-cyclonic eddies of the Taylor Cone type (Perrissinotto & Duncombe Rae, 1990). The combination of water column stability and increased macronutrient concentrations generate phytoplankton blooms dominated by large chain forming diatom species, particularly of the genera *Chaetoceros* and *Fragilariopsis*. The presence of the anti-cyclonic eddies in the immediate vicinity of the islands is linked to the geographic position of the SAF (Ansorge et al., 2009). When the SAF lies far to the north of the Prince Edward Islands, current speeds of the ACC are comparatively low resulting in a weak interaction between the islands and the prevailing current (Ansorge & Lutjeharms, 2002). Under these conditions, frictional forces dominate over advective forces resulting in the formation of eddies. Conversely, when the front lies in close proximity of the islands, advective forces prevail resulting in the islands acting as a flow through system. Under these conditions, phytoplankton stocks in the vicinity of the islands are in the range found in the open waters of the APFZ. There are virtually no seasonal studies on phytoplankton biomass and productivity in the region of the Prince Edward Islands (Table 1). It is worth noting that estimates of total phytoplankton concentration and productivity in APFZ in the other sectors of the Southern Ocean during summer are typically 1-2 times higher than the values recorded in the region of the islands during winter (Laubscher et al., 1993). This would suggest a strong seasonal pattern in the phytoplankton biomass and productivity in the APFZ.

Source	**Season**	**Region**	**Phytoplankton biomass (mg chl-a m^{-3})**	**Phytoplankton production (mg C m^{-2} d^{-1})**
El-Sayed et al. (1979)	Autumn	Inter-island region	0.09-1.88	211
Allanson et al. (1985)	Autumn	Inter-island region	0.06-0.87	84-2100
Allanson et al. (1985)	Spring	Inter-island region	0.06-0.87	ND
Perissinotto & Duncombe Rae (1990)	Autumn	Inter-island region	0.10-2.80	70-3000
Balarin (2000)	Autumn	Inter-island region	0.20-0.81	119-353
Bernard & Froneman (2005)	Autumn	Open waters	0.15-0.28	ND
Froneman & Balarin (1998)	Autumn	Open waters	0.29-0.52	ND
Bernard (2006)	Autumn	Inter-island region	0.24-0.71	ND
Allan (2011)	Autumn	Inter-island region	0.13-0.29	ND

Table 1. Estimates of total chlorophyll-a concentration and primary production in the open waters and inter-island region of the Prince Edward Islands. ND = no data.

3.2 Zooplankton studies

The zooplankton community structure (>200μm) within the open waters of the APFZ has been described on several occasions. Results of these studies indicate that there is no endemism among the holoplankton of the APFZ and that the region demonstrates extreme variability in the zooplankton species composition (Pakhomov & Froneman, 1999; McQuaid & Froneman, 2008). The variability in the zooplankton can be ascribed to the mesoscale variability in the oceanographic environment including cross frontal mixing, the intrusion of tongues of warm Subtropical water to the north and of cold Antarctic surface water to the south (Bernard & Froneman, 2002; 2003). Additionally, the formation of warm and cold eddies generated by the interaction of the ACC with the South-West Indian Ridge may also contribute to the transport of species from different water masses into the APFZ waters (Bernard et al., 2007). The extreme variability in the oceanographic environment within the APFZ contributes to the zooplankton comprising species with different biogeographic affinities including species which are Subtropical, Sub-antarctic and Antarctic in origin (Bernard & Froneman 2002; 2003; Hunt et al., 2001; McQuaid & Froneman, 2008).

The zooplankton community structure within the APFZ is numerically dominated by mesozooplankton (200-2000µm) comprising mainly copepods (*Oithona, Calanus* and *Metridia* spp.), pteropods (mainly *Limacina retroversa*), amphipods (*Themisto gaudichaudi*) and chaetognaths (*Eukrohnia hamata* and *Sagitta gazellae* (Pakhomov & Froneman, 1999; McQuaid & Froneman, 2008). Estimates of the contribution of the mesozooplankton to the total zooplankton abundance are highly variable and range from 52-88% of the total. The larger macrozooplankton (> 2000µm) may, however, contribute substantially to the total zooplankton biomass within the region (up to 45% of the total) although their contribution to the total zooplankton counts is generally < 15%. Among the macrozooplankton, the most important groups by numbers are the euphuasiids (*Euphausia vallentini, Nematoscelis megalopes* and *Thysanoessa* spp.), chaetognaths (*Sagitta gazellae* and *S. maxima*) and tunicates (*Salpa thompsonii*). The contribution of these groups to the total macrozooplankton counts typically demonstrates a high degree of both temporal and spatial variability reflecting the variable oceanographic environment of the APFZ.

Estimates of the total zooplankton abundance and biomass in the region of the Prince Edward Islands are highly variable, and range between 5 and 4850 ind m^{-3} and between 0.6 and 62.7 mg dwt m^{-3}, respectively (Table 2) (McQuaid & Froneman, 2008). Although there are no clear spatial patterns evident in the total zooplankton abundance and biomass within the APFZ, the frontal systems that delimit the APFZ, the SAF to the north, and the APF to the south, typically demonstrate increased zooplankton numbers which can attributed to the increased contribution of the larger macrozooplankton to the total zooplankton biomass (Pakhomov & Froneman, 1999; McQuaid & Froneman, 2008). Additionally, there is some evidence in the literature to suggest that the periodic intrusion of colder Antarctic Surface Waters in the APFZ is associated with elevated zooplankton abundances and biomass values (McQuaid & Froneman, 2008). There are currently limited seasonal data available on the zooplankton community structure available in the region of the Prince Edward Islands. It is worth noting, however, that the estimates of the total zooplankton abundance and biomass within the APFZ water during autumn are nearly an order of magnitude lower that estimates obtained in the APFZ in other sectors of the Southern Ocean during summer. This would suggest a strong seasonal pattern in the total zooplankton abundance and biomass within the region. Nonetheless, the zooplankton species composition appears, however, to be broadly similar between the different seasons.

Source	**Season**	**Abundance (ind.m^{-3})**	**Biomass (mg Dwt m^{-3})**
Grindley & Lane (1979)	Autumn	400-4850	8.7-28.4
Grindley & Lane (1979)	Spring	1575-1854	14.6-34.9
Boden & Parker (1986)	Autumn	22-594	12.9-53.0
Froneman et al. (1998)	Autumn	5-263	0.6-15.7
Ansorge et al. (1999)	Autumn	10-312	2.47-62.70
Bernard & Froneman (2002)	Autumn	49-1512	0.7-25.0
Bernard & Froneman (2003)	Autumn	78-1034	9.8-27.9
Bernard (2006)	Autumn	230-1004	ND

Table 2. Estimates of the total zooplankton abundance and biomass in the open waters of the Polar Frontal Zone and in vicinity of the Prince Edward Islands. ND = no data presented.

3.3 Nekton studies

Only a few nekton studies have been conducted in the region of the Prince Edward Islands. Results of these investigations suggest that the total nekton abundance and biomass in the region are generally low, < 2 ind $1000m^{-3}$ and < 0.1 mg dwt $1000m^{-3}$ (McQuaid & Froneman, 2008). It should be noted, however, that these studies have largely employed sampling gear that would likely underestimate the nekton abundances and biomass values. There appear to be no significant spatial patterns in the nekton abundance and biomass evident although values in the region of the fronts tend to be higher than those in the open waters.

4. Benthic community studies

The shallow waters of the Prince Edward Islands support a diverse (up to 550 species) and biomass rich benthic community which are numerically and by biomass dominated by suspension-feeders comprising mainly polychaetes, bivalves and brachiopods (Branch et al., 1993). The benthic community is thought to be sustained by the mass sedimentation of phytoplankton cells generated by the 'island mass effect' (Perissinotto & McQuaid, 1990; Pakhomov & Froneman, 1999; Allan, 2011). Locally, the kelp, *Durvillea antarctica*, also appears to contribute to the supply of food to the benthic community of the islands (Kaehler et al., 2006; Allan, 2011). A key component of the benthic community is the caridian shrimp, *Nauticaris marionis*, which represents the second most abundant component of the benthos in the vicinity of the islands (Branch et al., 1993). The sub-adults and adult consume mainly benthic and suspension feeders while their larvae feed mainly on phytoplankton (Vumazonke et al., 2003; Allan, 2011). The adult shrimp represent a key component in the diets of a number of top predators, including penguins and flying seabirds, found on the islands and thus serve as a link between the plankton, benthos and land- based predators (Perissinotto & McQuaid, 1990)

5. Terrestrial-marine interactions

The energy necessary to sustain the large numbers of top predators found seasonally on the islands is obtained from both allochthonous and autochthonous sources. The allocthonous source is derived from the advection of zooplankton and nekton towards the islands via the easterly flowing ACC (Pakhomov & Froneman, 1999). The zooplankton and nekton trapped in the shallow island shelf waters are vulnerable to predation by the top predators during the daytime. The depleted stocks are subsequently replenished during the night-time. This mechanism has been termed 'The replenishing hypothesis' by McQuaid and Froneman (2008). The periodic development of dense phytoplankton bloom associated with the so called, 'island mass effect' which sustains the benthic rich community within the shallow shelf waters of the Prince Edward Islands represents the main autochthonous source of energy necessary to sustain the top predators on the islands. Collectively, these two food delivery mechanisms are termed, "*the life support system of the Prince Edward Islands*" (Pakhomov & Froneman, 1999; McQuaid & Froneman, 2008). It is now well understood that the geographical position of the SAF in the proximity of the Prince Edward Islands plays a crucial role in forming local macro- and mesoscale oceanographic conditions (Pakhomov et al., 2000, Ansorge & Lutjeharms, 2002) and that any changes in its position may have dire consequences to the functioning of the island's "*life support system*".

6. Global climate change and the Prince Edward Island ecosystem

Recent studies have shown that since the 1950's, the ACC has strengthened and migrated southwards by 50–70 km (Gille, 2002). Changes in the intensity and geographic position within these frontal systems are likely to coincide with dramatic changes in the distribution of species and total productivity within the Southern Ocean and in particular at the Prince Edward Islands. The impact a southward migration of the ACC will have on the Prince Edward Islands ecosystem over the next century is indeed complex. It has been suggested (Ansorge et al., 2009) that shifts in the ACC may alter the intensity and frequency of eddies spawned at the South-West Indian Ridge while closer to the islands a more southern position of the SAF may result in an increase in the through-flow regime as can be seen from recent investigations (Pakhomov & Chown, 2003; Ansorge et al., 2009). Physical data further confirm that the mean sea surface temperatures at the Prince Edward Islands have increased by >1°C over the past 60 years (Melice et al., 2003). Mirroring this is a decrease of nearly 500 mm in precipitation, an increase of over 200 hours in sunshine and an increase in winds from the warmer sector in the north-west (Melice et al., 2003). The warming of the surface waters in the region of the islands has been coupled with an elevated contribution of warmer Subtropical Zone zooplankton species to the total zooplankton counts over the last three decades (Figure 5). A recent review of their composition around the Prince Edward Islands indicates that over the past two decades the contribution of Antarctic species decreased by ~20%, whereas the number of subtropical species found in the areas had increased from 6% to 26% (Ansorge et al., 2009). This is also supported by the incidental catches of subtropical fish species during the long-line fishery in the proximity of these

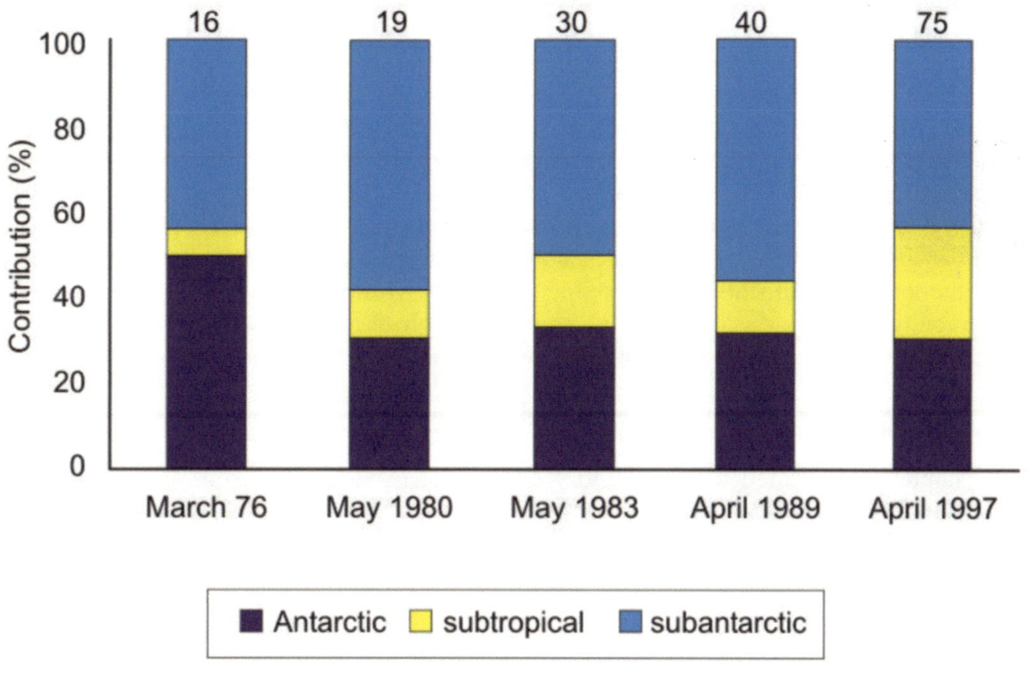

Fig. 5. Long-term changes in the composition of zooplankton species in the vicinity of the Prince Edward Islands since March 1976 (modified from Pakhomov et al., 2000).

islands. Although short-term variability and eddy transport cannot be completely discounted, it may be postulated that warmer water species have intruded into the APFZ more frequently during the past decades (Pakhomov et al., 2004). The most direct effect of a meridional shift in the SAF can be seen from changes within the species composition of the zooplankton.

Most recently, Allan (2011) demonstrated that the isotope ratios (carbon and nitrogen) of the numerically dominant suspension feeders of benthos and the caridian shrimp, *N. marionis,* have become significantly depleted since the 1980's. The observed depletion in isotope signatures was linked to the increased contribution of allochthonous food sources in the diets of these organisms due to the decreased frequency of occurrence of the so called, "island mass effect" (Allan, 2011). Indeed, a decline in stable isotope carbon values of a bottom dwelling shrimp *Nauticaris marionis* tissue indirectly postulates a decrease in the occurrence of bloom conditions in the inter-island region between 1980s and 2000s (Pakhomov et al., 2004). Lastly, a decrease in chlorophyll concentrations near the islands since 1976 provide further support that a variation in the position of the SAF has occurred during the past 30 years (Pakhomov & Chown, 2003).

It is unclear whether this change is expected to continue and at what rate. However, studies using coupled ocean-atmosphere climate models suggest that the westerly wind belt, which drives the ACC, is intensifying (Oke & England, 2004) and shifting polewards (Large & Yeager, 2004) in response to global warming. This shift has been associated with a southward migration of the SAF towards the islands. The decreased contribution of autochthonous production in the diets of the benthos in the region of the islands is therefore, the result of large scale changes in the prevailing oceanographic conditions in the region of the islands in response to global warming. It is likely that global climate change may in the future become associated with the disruption of the "*Life support system of the Prince Edward Islands*"and further investigations are required to understand better what impact these changes will have on the system. Furthermore, the impact of this shift on the top predators found seasonally on the islands remains largely unknown. However, it is worth noting that the populations of top predators that feed predominantly in the vicinity of the islands have decreased over the past two decades, possibly as a result of decreased food availability (Ryan & Bester, 2008). This represents a fundamental shift in the balance between allochthonous and autochthonous trophic pathways within the system and confirms the vulnerability of marine ecosystems to changes in physical conditions. Importantly, this indicates that the more dramatic consequences of climate change may be indirect ones.

7. Acknowledgments

This chapter is dedicated to the memory of the late Professor Emeritus JRE Lutjeharms. Funds for this study were obtained from the University of Cape Town, Rhodes University and the South African National Antarctic Programme (SANAP).

8. References

Allan, EL. (2011). Trophodynamics of the benthic and hyperbenthic communities inhabiting the sub-Antarctic Prince Edward islands: stable isotope and fatty acid signatures. *PhD thesis*, Rhodes University, South Africa. Pp 134

Ansorge, IJ., Froneman, PW., Pakhomov, EA., Lutjeharms, JRE., Van Ballegooyen, R., Perissinotto, R. (1999). Physical-biological coupling in the water surrounding the Prince Edward Islands (Southern Ocean). *Polar Biology,* 21, 135-145.

Ansorge, IJ., Lutjeharms, JRE. (2000). Twenty-five years of physical oceanographic research at the Prince Edward Islands. *South African Journal of Science*, 2000, 96, 557–565.

Ansorge, IJ., Lutjeharms, JRE. (2002). The hydrography and dynamics of the ocean environment of the Prince Edward islands (Southern Ocean). *Journal of Marine Systems,* 37, 107-127.

Ansorge, IJ., Lutjeharms, JRE. (2003). Eddies originating from the South-West Indian Ridge. *Journal of Marine Systems*, 39, 1-18.

Ansorge, IJ., Lutjeharms, JRE. (2005). Direct observations of eddy turbulence at a ridge in the Southern Ocean. *Geophysical Research Letters,* 32, L14603, doi: 10.1029/ 2005GL022588.

Ansorge, IJ., Durgadoo, JV., Pakhomov, EA. (2009). Dynamics of physical and biological systems of the Prince Edward Islands in a changing climate. *Papers and Proceedings of the Royal Society of Tasmania,* 143, (1), 15-18.

Ansorge, IJ., Pakhomov, EA., Kaehler, S., Lutjeharms JRE., Durgadoo, JV. (2010). Physical and biological coupling in eddies in the lee of the South-West Indian Ridge. *Polar Biology,* DOI 10.1007/s00300-009-0752-9

Balarin, MG. (2000). Size fractionated phytoplankton biomass and production in the Southern Ocean. *MSc thesis.* Rhodes University, Pp 132.

Belkin, IM., Gordon, AL. (1996). Southern Ocean fronts from the Greenwich meridian to Tasmania. *Journal of Geophysical Research,* 101, 3675–3696.

Bergstrom, D., Chown, SL. (1999). Life at the front: history, ecology and change on southern ocean islands. *Trends in Ecology Evolution,* 14, 472-477.

Bernard, KS. (2006). The role of the euthecosome pteropd, Limacina retroversa, in the Polar Frontal Zone, Southern Ocean. *PhD thesis*, Rhodes University, Pp 154.

Bernard, KS, Froneman PW (2002). Mesozooplankton community structure in the Southern ocean upstream of the Prince Edward Islands. *Polar Biology* 25, 597-604.

Bernard, KS., Froneman, PW. (2003). Mesozooplankton community structure and grazing impact in the Polar Frontal Zone during austral autumn 2002. *Polar Biology,* 26, 268-275.

Bernard, KS., Froneman, PW. (2005). Trophodynamics of selected mesozooplankton in the west-Indian sector of the Polar Frontal Zone, Southern Ocean. *Polar Biology,* 28, 8, 594-606.

Bernard, ATF., Ansorge, IJ., Froneman, PW., Bernard, KS., Lutjeharms, JRE. (2007). Entrainment of Antarctic euphausiids into the Sub-antarctic by a cold eddy. *Deep Sea Research Part I,* Oceanographic Research Papers, 54, 10, 1841 - 1851.

Boebel, O., Rossby, T., Lutjeharms, JRE., Zenk, W., Barron, C. (2003). Path and variability of the Agulhas Return Current. *Deep Sea Research Part II*: Topical Studies in Oceanography, 50, 1, 35-56.

Boden, BP., Parker, LD. (1986). The plankton of the Prince Edward Islands. *Polar Biology,* 5, 81-93.

Branch, GM., Attwood, CG., Gianakouras, D., Branch, ML., (1993). Patterns in the benthic communities on the shelf of the sub-Antarctic Prince Edward Islands. *Polar Biology,* 13, 23-34.

Cooper, J. (2008). Human history. In: *The Prince Edward Islands: and-sea interactions in a changing ecosystem*, eds

Deacon, GER. (1982). Physical and biological zonation in the Southern Ocean. *Deep Sea Research*, 29, 1, 1-15.

de Bruyn, PJN., Tosh, CA., Oosthuizen, WC., Bester, MN., Arnould, JPY. (2009). Bathymetry and frontal system interactions influence seasonal foraging movements of lactating Subantarctic fur seals from Marion Island. *Marine Ecology Progress Series*, 394, 263–276, doi: 10.3354/meps08292.

Doty, MS., Oguri, M. (1956). The Island Mass Effect. *Journal du Conseil*, 22, 1, 33-37, doi:10.1093/icesjms/22.1.33.

Duncombe Rae, CM. (1989a). Frontal systems encountered between southern Africa and the Prince Edward Islands during April/May 1987. *South African Journal of Antarctic Research*. 19, 21–25.

Duncombe Rae, CM.(1989b). Physical and chemical marine environment of the Prince Edward Islands (Southern Ocean) during April/May 1987. *South African Journal of Marine Science*, 8, 301–311.

Durgadoo, JV., Ansorge, IJ., Lutjeharms, JRE. (2010). Oceanographic observations of eddies impacting the Prince Edward Islands, South Africa. *Antarctic Science*, doi: 10.1017/S0954102010000088.

Durgadoo, JV., Ansorge, IJ., de Cuevas BA, Lutjeharms, JRE., Coward, AC. (2011) Decay of eddies at the South-West Indian Ridge. *South African Journal of Science*. 107(11/12), Art. #673, 10 pages. http://dx.doi.org/10.4102/sajs.v107i11/12.673

El-Sayed, SZ., Bennon, DP., Grindley, JR., Murail, JF. (1979). Some aspects of the biology of water column studies during the "Marion- Durfresne" cruise 08. *CNRFA 44*, 127-134.

Fisher, RL., Goodwillie, AM. (1997). The physiography of the Southwest Indian Ridge. *Marine Geophysical Research*, 19, 6, 451–455.

Froneman, PW., Pakhomov, EA., Meaton, V. (1998). Surface distribution of microphytoplankton of the southwest Indian Ocean along a repeat transect between Cape Town and the Prince Edward Islands. *South African Journal of Science*, 94, 124-128.

Gille, ST. (1997). The Southern Ocean Momentum Balance: Evidence for Topographic Effects from Numerical Model Output and Altimeter Data. *Journal of Physical Oceanography*, 27, 2219–2232.

Gille, ST. (2002). Warming of the Southern Ocean since the 1950's. *Science*, 295, 1275-1277.

Gille, ST. (2003). Float observations of the Southern Ocean. Part II: Eddy Fluxes, *Journal of Physical Oceanography*, 33, 1182-1196.

Gordon, AL. (1986). Interocean exchange of thermocline water. *Journal of Geophysical Research*, 91, 5037-5046.

Grindley, JR., Lane, SB. (1979). Zooplankton around Marion and Prince Edward Islands. *CNFRA 4*, 111-125

Harris, TFW., Stavropoulos, CC. (1978). Satellite-tracked drifters between Africa and Antarctica. *Bulletin of American Meteorological Society*, 59, 51-59.

Hofmann, EE. (1985). The large-scale horizontal structure of the Antarctic Circumpolar Current from FGGE drifters. *Journal of Geophysical Research*, 90, 7087-7097.

Hughes, CW., Ash, E. (2001). Eddy forcing of the mean flow in the Southern Ocean. *Journal of Geophysical Research*, 106, 2713–2722.

Hunt, BVP., Pakhomov, EA., McQuaid, CD. (2001). Short-term variation and long terms changes in the oceanographic environment and zooplankton community in the vicinity of a sub-Antarctic archipelago. *Marine Biology*, 138, 369-381.

Joyce, TM., Patterson, SL., Millard, RC. (1981). Anatomy of a cyclonic ring in the Drake Passage. *Deep Sea Research*, 28, 1265–1287.

Kaehler, S., Pakhomov, EA., Kalin, RM., Davis, S. (2006). Trophic importance of kelp-derived suspended particulate matter in through- a flow sub-Antarctic system. *Marine Ecology Progress Series*, 316, 17-22.

Kostianoy, AG., Ginzburg, AI., Frankignoulle, M., Delille, B. (2004). Fronts in the Southern Indian Ocean as inferred from satellite sea surface temperature data. *Journal of Marine Systems*, 45, 1-2, 55-73.

Large, WG., Yeager, SG. (2004). Diurnal to Decadal Global Forcing for Ocean and Sea-Ice Models: the Data Sets and Flux Climatologies. *NCAR Technical Note NCAR/TN-460+STR*, National Center for Atmospheric Research.

Laubscher, RK., Perissinotto, R., McQuaid, CD. (1993). Phytoplankton production and biomass at frontal zones in the Atlantic sector of the Southern Ocean. *Polar Biology*, 13, 471-481.

Lutjeharms, JRE., Ansorge, IJ. (2001). The Agulhas Return Current. *Journal of Marine Systems*, 30, 1-2, 115-138.

Lutjeharms, JRE., Valentine, HR. (1984). Southern Ocean thermal fronts south of Africa. *Deep Sea Research*, 31, 1461-1476.

McQuaid, CD., Froneman, PW. (2008). Biology in the oceanographic environment. In: *The Prince Edward Islands: and-sea interactions in a changing ecosystem*, eds Chown, SL., Froneman, PW., Sun Press, Stellenbosch, South Africa, 97-121.

Mélice, JL., Lutjeharms, JRE., Rouault, M., Ansorge, IJ. (2003). Sea-surface temperatures at the sub-Antarctic islands Marion and Gough during the past 50 years. *South African Journal of Science*, 99, 363-366.

Meredith, MP., Hogg, AM. (2006). Circumpolar response of Southern Ocean eddy activity to a change in the Southern Annular Mode. *Geophysical Research Letters*, 33, L16608, doi:10.1029/2006GL026499.

Miller, DGM., Boden BP., Parker, L. (1984). Hydrology and bio-oceanography of the Prince Edward islands. *South African Journal of Science*, 100, 29-32.

Moore, JK., Abbott, M., Richman, J. (1999). Location and dynamics of the Antarctic Polar Front from satellite sea surface temperature data. *Journal of Geophysical Research*, 104, 3059–3073.

Nel, DC., Pakhomov, EA., Lutjeharms, JRE., Ansorge, IJ., Ryan, PG., Klages, NTW. (2001). Exploitation of mesoscale oceanographic features by grey-headed albatrosses Thalassarche chrysostoma in the southern Indian Ocean. *Marine Ecology Progress Series*, 217, 15-26.

Oke, PR. England, MH. (2004). Oceanic response to changes in the latitude of the Southern Hemisphere subpolar westerly winds. *Journal of Climate*, 17, 1040-1054.

Pakhomov, EA., Froneman, PW. (1999). The Prince Edward Islands pelagic ecosystem, south Indian Ocean: a review of achievements. *Journal of Marine Systems* 18, 355-367.

Pakhomov, EA., Froneman, PW., Ansorge, IJ., Lutjeharms, JRE. (2000). Temporal variability in the physico-biological environment of the Prince Edward Islands (Southern Ocean). *Journal of Marine Systems,* 26, 75-95.

Pakhomov, EA., Chown, SL. (2003). The Prince Edward Islands: Southern Ocean oasis. *Ocean Yearbook,* University of Chicago Press, Chicago, USA, Vol. 17, 348-379.

Pakhomov, EA., McClelland, JW., Bernard, KS., Kaehler, S., Montoya, JP (2004). Spatial and temporal shifts in stable isotope values of the bottom-dwelling shrimp *Nauticaris marionis* at the sub-Antarctic archipelago. *Marine Biology,* 144, 317-325.

Perissinotto, R., Duncombe Rae, CM. (1990). Occurrence of anti-cyclonic eddies on the Prince Edward plateau (Southern Ocean): effects on phytoplankton productivity and biomass. *Deep Sea Research* 37, 777-793.

Park, YH., Gamberoni, L., Charriaud, E. (1993). Frontal structure, water masses and circulation in the Crozet Basin. *Journal of Geophysical Research,* 98, 12361–12385.

Park, YH., Charriaud, E., Poisson, A. (1997). Hydrography and Baroclinic transport between Africa and Antarctica on WHP Section I6. *International WOCE newsletter,* 29, 13-16.

Phillips, HE., Rintoul, SR. (2000). Eddy Variability and Energetics from Direct Current Measurements in the Antarctic Circumpolar Current South of Australia. *Journal of Physical Oceanography,* 30, 3050–3076.

Pollard, RT., Read, JF. (2001). Circulation pathways and transports of the Southern Ocean in the vicinity of the Southwest Indian Ridge. *Journal of Geophysical Research,* 106, 2881-2898.

Ryan, PG., Bester, MN. (2008) Pelagic predators. In: *The Prince Edward Islands: and-sea interactions in a changing ecosystem,* eds Chown, SL., Froneman, PW., Sun Press, Stellenbosch, South Africa, 121-164.

Sandwell, DT., Zhang, B. (1989). Global Mesoscale Variability From the Geosat Exact Repeat Mission: Correlation With Ocean Depth, *Journal of Geophysical Research,* 94(C12), 17,971–17,984, doi:10.1029/JC094iC12p17971.

Sclater, JG., Grindlay, NR., Madsen, JA., Rommevaux-Jestin, C. (2005). Tectonic interpretation of the Andrew Bain transform fault: Southwest Indian Ocean, *Geochemistry, Geophysics, Geosystems,* 6, Q09K10, doi:10.1029/2005GC000951.

Smith, VR. (2002). Climate change in the sub-Antarctic: an illustration from Marion Island. *Climatic Change,* 52, 345-357.

Sokolov, S., Rintoul, SR. (2009). Circumpolar structure and distribution of the Antarctic Circumpolar Current fronts: 1. Mean circumpolar paths. *Journal of Geophysical Research,* 114, C11018 (2009).

Sultan, E., Mercier, H., Pollard, RT. (2007). An inverse model of the large scale circulation in the South Indian Ocean. *Progress In Oceanography,* 74, 1, 71-94.

Sun, C., Watts, DR. (2002). Heat flux carried by the Antarctic Circumpolar Current mean flow, *Journal of Geophysical Research,* 107(C9), 3119, doi:10.1029/2001JC001187.

Swart, NC., Ansorge, IJ., Lutjeharms, JRE. (2008) Detailed characterization of a cold Antarctic eddy. *Journal of Geophysical Research,* 113, C01009, doi: 10.1029/2007JC004190.

Vumazonke, L., Pakhomov, EA., Froneman, PW., McQuaid, CD. (2003). Diet and daily ration of male and female caridian shrimp, Nauticaris marionis, at the Prince Edward Archipelago. *Polar Biology* 26, 420-432.

Meiofauna as a Tool for Marine Ecosystem Biomonitoring

Maria Balsamo, Federica Semprucci,
Fabrizio Frontalini and Rodolfo Coccioni
*Department of Earth, Life and Environmental Sciences (DiSTeVA),
University of Urbino,
Italy*

1. Introduction

Meiofauna are the metazoan component of the benthos, and also include large protozoans (e.g. foraminifera). They are defined by their body size (44-1000 µm) and are the most diversified element of the marine biota: as many as 24 of the 35 animal phyla have meiobenthic representatives which live in meiofauna, whether for all their life or just temporarily. It is the most abundant benthic group in the marine realm, and is thought to be closely connected to other faunal compartments of the benthic system. The function of meiofauna in marine benthic systems seems to be much more complex than previously supposed, and requires investigation to clarify their ecological importance in the benthic domain (see Balsamo et al., 2010 for review).

The aims of this paper are: to review advances in the use of meiofauna as a bio-indicator for the monitoring of marine ecosystems; and to highlight future perspectives of this approach. In particular, the use of the two most abundant and diverse meiofaunal groups (Foraminifera and Nematoda) will be considered.

2. Meiofauna as an environmental bio-indicator in marine ecosystems

The use of meiofauna as a biological indicator is a more recent development than the utilization of macrofauna in the assessment and monitoring of aquatic ecosystems (Coull & Chandler, 1992). The advantages of the former are numerous and strongly emphasized by Kennedy & Jacoby (1999), while some of the arguments traditionally advanced against their use underline difficulties in identification, the high rate of sampling frequency and the microscopic size of these organisms. However, new technologies and tools, such as standardized methodologies, electronic identification keys, molecular approaches and the creation of new indices, currently allow for and promote the use of meiofauna in ecological studies (see Giere, 2009, for review).

3. Two representative meiofaunal groups: Foraminifera and Nematoda

Among the meiobenthic protozoans, Foraminifera (class Foraminifera, *phylum* Granuloreticulata) are the most abundant and diverse of the shelled microorganisms in the oceans (Sen Gupta, 1999). The *phylum* Nematoda, meanwhile, is the most plentiful (often >50% of the total meiofauna, up to >90% in deep-sea sediments) and diverse metazoan meiofauna taxon (Boucher & Lambshead, 1995; Giere, 2009).

Foraminifera play a significant role in global biogeochemical cycles of inorganic and organic compounds, making them one of the most important groups on Earth (Yanko et al., 1999). Furthermore, many foraminiferal taxa secrete a carbonate shell that is readily preserved, and so record evidence of environmental stresses and changes over time. They are commonly small and abundant compared to other hard-shelled taxa and easy to collect, providing a highly reliable database for statistical analysis, even when only a limited volume of samples is available. Because of their widespread distribution, short life and reproductive cycles, high biodiversity, and specific ecological requirements, foraminifera may respond to environmental changes (e.g. Alve, 1995; Murray & Alve, 2002; Yanko et al., 1994). Moreover, with their high number of species and genera - around three to four thousand of the former (Murray, 2007) - benthic foraminifera are more likely to contain a variety of specialists that are sensitive to environmental change. For all of these reasons, they are particularly sensitive and can thus be successfully used for their value as bio-indicators of environmental change in a wide range of marine environments (Armynot du Châtelet & Debenay, 2010; Frontalini & Coccioni, 2011). The use of benthic foraminifera as bio-indicators of environmental quality can be investigated in terms of population density and diversity, assemblage structure, reproduction capability, test morphology - including size (dwarfism), prolocular morphology, ultrastructure, pyritization, abnormality, and the chemistry of the test. The study of pollution effects on benthic foraminifera and their use as proxies began in the 1960s (Boltovskoy, 1965; Resig, 1960; Watkins, 1961), and has been increasingly developed in recent decades as a result of environmental research (for reviews, see Alve, 1995; Boltovskoy et al., 1991; Frontalini & Coccioni, 2011; Murray & Alve, 2002; Nigam et al., 2006; Yanko et al., 1994).

The ecological value of nematodes is related not only to their notable quantitative importance in the benthic domain, but also to their pivotal role within the trophic chains of aquatic ecosystems and the stabilizing effects of shores (Platt & Warwick, 1980). The advantages cited above of using foraminifera as bio-indicators could be extended to nematodes (see Heip et al., 1985; Vanaverbeke et al., 2011), and it is for this reason that this *phylum* was recently proposed as an indicator with which to assess the ecological quality of marine ecosystems according to the Water Framework Directive (WFD, Directive 2000/60/EC) (Moreno et al., 2011). The nematode assemblage is generally studied in terms of density, diversity, assemblage structure, trophic guilds, life history strategies, body size and biological trait analysis. The Index of Trophic Diversity (ITD; Heip et al. 1984) and the Maturity index (MI), which is based on the ecological characteristics and reproductive strategies of nematodes (Bongers, 1990; Bongers et al., 1991), are the two indices that are more commonly applied in ecological studies of nematode assemblages.

Moreno et al. (2011), in analyzing the most frequently used indices in the ecological assessment of nematodes, have suggested that the taxonomic approach (presence/absence of specific indicator genera) reveals the best correspondence between environmental status and biological response, whereas, among the synthetic descriptors, c-p % composition and diversity (Shannon index, H') can be used to evaluate ecological quality status efficiently.

4. Foraminiferal and nematode responses to different pollution sources and disturbances

Human activities, including industry, agriculture, mining, dredging, and dumping introduce large amounts of pollutants into marine areas, causing permanent and significant disturbance to and a major impact on ecosystems. Pollution may also occur in offshore environments, such as drilling rigs and oil platforms. When present in sufficient quantities, and under certain conditions, pollutants influence the biota living within and at the sediment interface. The benthic community generally responds to adverse ecological conditions, primarily by undergoing: i) local extinctions; ii) compositional biocenosis and trophic group changes; iii) assemblage modifications, which include changes in abundance and diversity; iv) dwarfism (Lilliput effect); v) changes in reproduction capability; and vi) cytological, biological and morphological variations (Fig. 1). Indeed, many studies of the effects on foraminiferal and nematode communities of a variety of disturbances have been carried out in different parts of the world (see, Fig. 2).

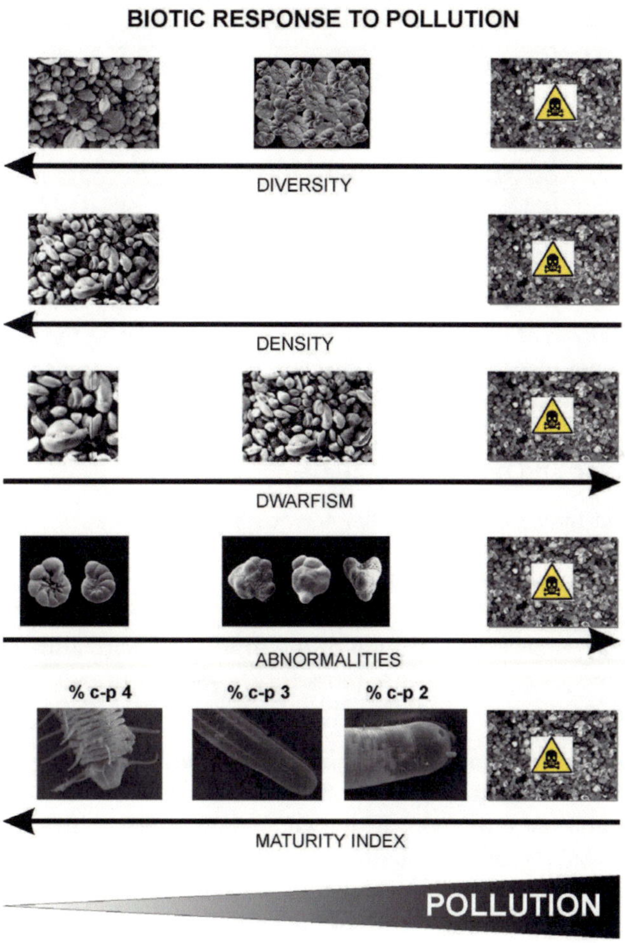

Fig. 1. Schematic response of Foraminifera and Nematoda to pollution.

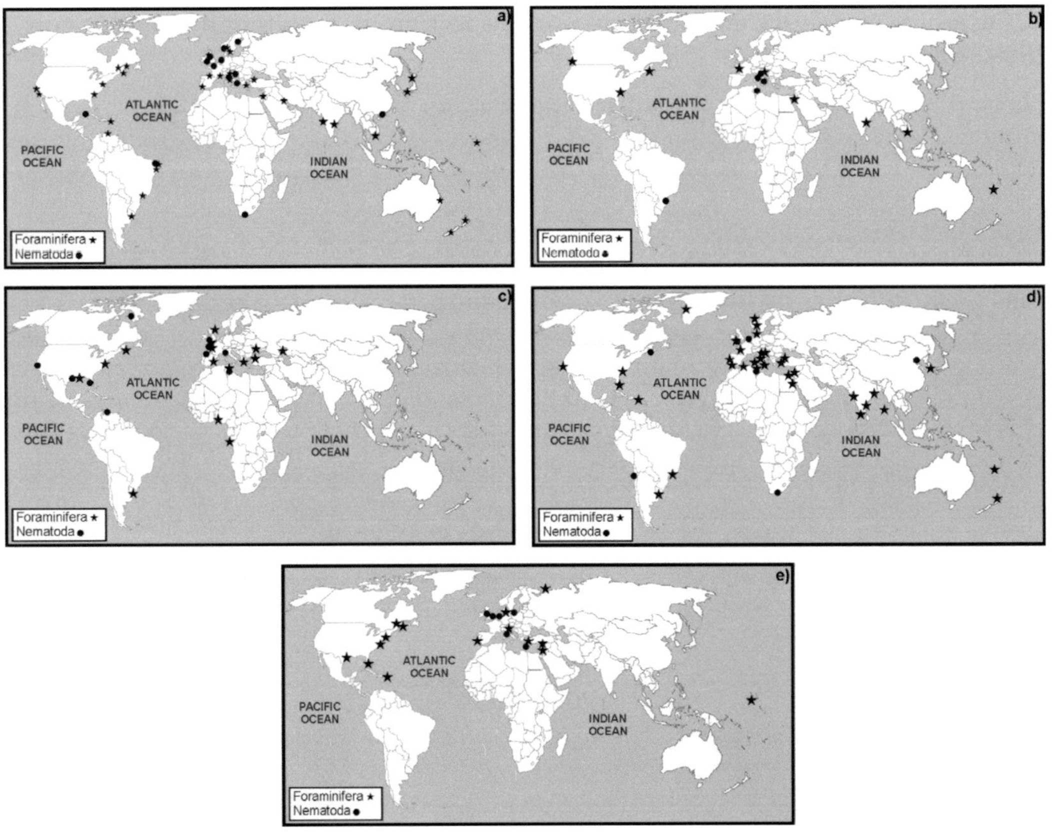

Fig. 2. Global distribution of the most significant studies involving the responses of benthic foraminifera and nematodes to: a) sewage discharge; b) aquaculture; c) hydrocarbons; d) trace elements; and e) other types of stress (e.g. thermal pollution and physical disturbance)

4.1 Sewage discharge and organic enrichment

The input of organic matter may have deleterious effects on marine ecosystems and their benthic community by inducing high nutrient levels. This can ultimately lead to oxygen deficiency, also known as eutrophication. Eutrophication in shelf environments is mainly due to the nutrient load of rivers, which is in turn related to river flow, the catchment area, and the industrialization and urbanization of the drainage basin. Both the quantity and the quality (labile - more degradable; and refractory - more difficult to metabolize) of the organic material have an impact on the benthic communities.

The main food sources (not taking into account the endosymbiotic species) for benthic foraminifera are organic matter (mainly the labile fraction) and the entire bacterial community that proliferates thereon. Moreover, when present in low quantities, like in deep-sea environments, the organic matter acts as a limiting factor. In contrast, it exerts a strong influence in marine coastal areas, where periodically enhanced production and the riverine input of nutrients may lead to an increase in the standing crop of the benthic foraminifera (e.g. Watkins, 1961). According to Alve (1995), the increase in organic matter in the so-called defined 'hyper-trophic' zone may stimulate the growth of large populations of benthic foraminifera. Alternatively, an excess of organic material can negatively influence the benthic community. In particular, dysoxic-anoxic conditions can be established where the oxygen demand to metabolize the organic matter exceeds its supply. In fact, most studies have revealed that around the areas of organic matter supply (e.g. outfall and pipe), a barren-abiotic zone might occur in response to the development of anoxic conditions (Alve, 1995). An additional benthic foraminiferal response to organic matter pollution includes the modification of the original assemblages. In particular, an increased dominance of agglutinated foraminiferal taxa at pollution sites has been found. This effect is probably the result of the taphonomic process, which is related to the dissolution induced by the input of large quantities of organic matter that have altered the properties of the sediments. It has been inferred that both increasing the concentration of the organic matter and the lower availability of oxygen may progressively exclude or favour some taxa. As a consequence, a species-specific response and a change in assemblage composition are to be expected according to the degree of eutrophication (Van der Zwaan, 2000). In fact, some taxa, including *Buliminella elegantissima* and *Bulimina marginata denudata,* were overwhelmingly dominant in the living assemblages close to the outfall area for the Los Angeles County sewage system (Bandy et al., 1964). It has also been documented that some taxa migrate upwards in sediment when oxygen concentrations diminish, apparently mirroring the oxygen gradient (Alve & Bernhard, 1995). More recently, on the basis of bibliographic analyses, Frontalini & Coccioni (2011) synthetized the degree of tolerance of some benthic foraminiferal taxa, particularly *Hopkinsina pacifica, Nonionella turgida, Valvulineria bradyana* and *Uvigerina peregrina,* along with several species belonging to the genera *Bulimina, Buliminella, Fursenkoina, Bolivina* and *Epistominella,* which were regarded as being the most tolerant to organic matter enrichment. Changes in density and assemblage composition are not the only response by benthic foraminifera to organic matter, with a reduction of species diversification also being reported in the vicinity of many outfalls (e.g. Resig, 1960). The development of an azoic area, high concentrations of sulphides, and reducing conditions are documented around the point sources (mainly pulp and paper mills) of refractory organic matter. In more restricted environments (e.g. fjords), where the connection with the open sea is limited, anoxia may occur very rapidly. The effect of increased human-induced organic flux was documented in short cores in Drammensfjord and Frierfjord (Norway), where marked changes in assemblage composition, diversity and abundance were reported (Alve, 1991a, 1999). An additional, important contribution, which exemplifies the effect of eutrophication over time, was provided by Barmawidjaja et al. (1995), who accurately described the distribution of benthic foraminifera in a 57-long core drilled in front of the Po Delta. Significant benthic foraminiferal changes and steps were observed. These changes were ascribed to: human-induced alterations of the main outflow canals of the Po River (1840 and 1880); a steady increase of the nutrient load (from 1900 onwards); the intensification of eutrophication (1930); and the first signs of anoxic events (1960). Since 1880, the original assemblages have been gradually replaced by more stress tolerant versions (e.g. dominated by *N. turgida*). Another substantial change occurred in 1930, when opportunistic stress tolerant species (*H. pacifica, Bolivina seminuda* and *Quinqueloculina stalkeri*) became dominant. The authors regarded *H. pacifica* (peaking in 1960) as the most stress tolerant taxon. Foraminiferal assemblages changes over both time and space have been reported in the Osaka Bay by Tsujimoto et al. (2006a,b), who reported for the eutrophicated inner part of the bay an assemblages dominated by agglutinated forms (*Trochammina hadai* and *Eggerella advena*). The same authors documented marked foraminiferal assemblages over the past 50 years, where the dominance of agglutinants, which was linked to the increase of eutrophication from the 1960s to the 1970s, was followed by a decrease of these forms in response to the imposition of improved environmental regulations. On the basis of a comparison of foraminiferal assemblages from samples collected in 1983 and 2001 in the northern sector of the Lagoon of Venice, Albani et al. (2010) reported unchanged conditions for 50% of the lagoon. They also documented the effectiveness of the purification plant that had been operating since the 1980s, as well as improvements in the water quality in the area near Porto Marghera. The enrichment of organic matter has also been regarded as responsible for increasing the number of abnormal tests and the emergence of large and protruding proloculi (e.g. Seiglie, 1971). It has been also reported that adult specimens are smaller in nutrient-rich, but oxygenated, environments, probably in response to rapid growth and high reproductive rates (Yanko et al., 1999).

In nematode assemblages, a general increase in abundance (in particular of deposit feeders) and decreases in diversity can be observed in correspondence with sewage outfalls or organic enrichment, although responses are not always unequivocal (Armenteros et al., 2010; Austen & Widdicombe, 2006; review by Coull & Chandler, 1992; Fraschetti et al., 2006; Sandulli & De Nicola-Giudici, 1990, 1991; Schratzberger & Warwick, 1998; Schratzberger et al., 2008; Somerfield et al., 2003) (Fig. 2a). In a laboratory experiment, Schratzberger & Warwick (1998) tested the response of two estuarine nematode assemblages (from organic-poor sandy and organic-rich muddy estuaries, respectively) to the intensity and frequency of organic enrichment. In the sand microcosms, the nematode assemblage changed more drastically than in the muddy versions: diversity and species' richness decreased significantly, including in response to low doses, whereas in the muddy estuaries this only occurred at medium and high levels of organic loading. The authors concluded that sand nematodes inhabiting a less organically enriched environment have a lower ecological tolerance to this type of stress than their mud counterparts. Accordingly, communities of different sediment types may be affected differently by organic disturbance.

A multivariate approach has revealed significant differences between the controls and both of the organically enriched treatments. Some indicator species of the different disturbance levels have also been identified. In sand microcosms with low dose levels, there was an initial increase in abundance of *Chromadora nudicapitata, Daptonema normandicum* and *D. hirsutum*, followed by an increase of *D. hirsutum* and a contextual decrease of *D. setosum* and *Odontophora longisetosa.* High levels of organic enrichment were characterized by a reduction in abundance of *O. longisetosa, D. setosum* and *Chromaspirina inglisi*. In contrast, in the mud microcosms, the discriminating low-abundance species were *Ptycholaimellus ponticus* and *D. procercum,* whereas *Terschellingia communis, T. longicaudata, Metachromadora vivipara* and *Sabatieria breviseta* decreased only in high dose levels. Some of these species are well-known for their physiological and behavioural adaptations to poorly oxygenated environments. These changes include a low respiratory rate, slow movements and the presence of intracellular inclusions of insoluble metal sulphide depositions (Nicholas et al., 1987; Warwick & Gee, 1984; Warwick & Price, 1979). Schratzberger & Warwick (1998) suggested that many small doses of a same amount of organic matter have a milder effect on assemblage structure than when administered in fewer but larger doses. The presence of some of the species' indicators of organic enrichment referred to by Schratzberger & Warwick (1998) has also been reported by other authors in relation to organically disturbed sediment. A recent experiment in Cienfuegos Bay (Armenteros et al., 2010) highlighted nematode abundance and diversity decreases as well as alterations in taxonomic and trophic structure. The authors suggested that the accumulation of by-products of bacterial metabolism (i.e. ammonia and hydrogen sulphide) is more deleterious for nematodes than oxygen *per se*. This is because nematodes have developed various behavioural (e.g. migration to 'oxygen islands') and/or physiological mechanisms (e.g. symbiosis with bacteria, low metabolic rate) to cope with hypoxic and even temporal anoxic events. Species-specific responses of nematode assemblages to organic enrichment, already documented by Schratzberger & Warwick (1998) and Schratzberger et al. (2008), have also been recorded by Armenteros et al. (2010), who report *S. pulchra, S. parasitifera, T. communis, T. gourbaultae* and *T. longicaudata* to be discriminant species of different levels of organic enrichment. The genera *Sabateria* and *Terschellingia* are among the dominant taxa in both the Pina Basin, an urbanized embayment on the coast of Pernambuco, Brazil (Somerfield et al., 2003), and some Ligurian harbours that are characterized by organic enrichment (Moreno et al., 2008, 2009). Fraschetti et al. (2006) documented a decrease in nematode abundance in a rocky subtidal area influenced by sewage outfall, but there was no reduction in the number of genera or changes to their taxonomic composition. Interestingly, the MI did not change significantly between the impacted and the control sites: its values were high overall, as was the detected percentage of c-p 4 and the epistrate feeders. This finding suggested that sewage discharge did not affect nematode assemblages in terms of favouring colonizers.

4.2 Aquaculture

Aquaculture has undergone a dramatic expansion worldwide, inducing a growing interest in and concern for its potential impact on coastal marine environments. The first effect is the build-up of faeces or pseudofaeces (biodeposition) on the benthic habitat just below the culture area. This can alter particle-size, organic content, and nitrogen-cycling, which can easily induce eutrophication and a decrease in oxygen penetration in the sediment (see, for review, Mirto et al., 2002; Netto & Valgas, 2010).

Only a few studies have been carried out on the foraminiferal assemblages that are specifically affected by aquaculture. These were mainly conducted in the Atlantic Ocean (Bouchet et al., 2007; Clark, 1971; Schafer et al., 1995; Scott et al., 1995), the Pacific Ocean (Debenay et al., 2009), the Red Sea (Angel et al., 2000), the China Sea (Debenay & Luan, 2006), the Japan Sea (Tarasova & Preobrazhenskaya, 2007), the Adriatic Sea (Vidović et al., 2009) and the Bay of Bengal (Jayaraju et al., 2008) (Fig. 2b). The high input of easily degradable organic matter and the wide variety of chemical and biological products used in aquaculture may introduce into the water body persistent and potentially toxic residues. Accordingly, benthic foraminifera may be negatively affected by aquacultural activities. In particular, foraminiferal population dynamics were investigated at an outfall site of land-based salmonid aquaculture (Clam Bay, Nova Scotia) by Clark (1971), who reported a strongly inverse relationship between local foraminiferal density and the discharge of fish meal. Moreover, a reduction of foraminiferal abundance, but no marked compositional changes in the assemblages, has been identified under salmon cages (finfish aquaculture) as a response to increased organic matter sedimentation (Scott et al., 1995). Aquaculture facilities have also produced localized anoxic areas which occur under fish cages, as seen in eastern Canada. This leads to less diversity and an increase of non-calcareous foraminifera (Schafer et al., 1995). The mechanisms responsible for these changes are probably not linked directly to an increased food supply, but can be found in the changing of sediment properties, the sedimentation rate and the low values of oxygen concentrations. In particular, a strongly positive correlation between the Foraminiferal Abnormality Index (FAI) (*sensu* Coccioni et al. (2005)) and: (1) the quantity of easily oxidized material deposited at the bottom of shrimp ponds; and (2) the sediment oxygen demand has been documented in New Caledonia (Debenay et al., 2009). These findings were corroborated by the poorly diversified assemblages, which were mainly dominated by *Ammonia tepida* and *Quinqueloculina seminula*, indicating very restricted conditions and major environmental stress. It has also been suggested that a very high FAI could be a potential indicator of great accumulations of native organic matter, leading to a high sediment oxygen demand. Meanwhile, changes in assemblage composition and decreased density have been observed under scallop cages (Minonosok Bay, Sea of Japan) by Tarasova & Preobrazhenskaya (2007). Higher proportions of agglutinated species as a response to shrimp farming and rice culture were also documented in the Mekong Delta (Vietnam) by Debenay & Luan (2006). Moreover, two sediment transects were investigated below a commercial fish farm in the Gulf of Eilat (Red Sea) by Angel et al. (2000), who found higher abundances of agglutinated species in the ''hypertrophic'' zone adjacent to the fish cages. The same author documented abnormal specimens, mainly *Peneroplis planatus*, in the zone, but it was unclear whether they had formed as a result of adverse conditions related to the fish farm. The impact of fish farming on benthic foraminiferal assemblages was evaluated in Drvenik Veliki Island (Croatia) by Vidović et al. (2009), who reported alterations in their composition but no changes in diversity. The same authors suggested that the presence, absence or relative abundance of species could be a possible indicator of organic enrichment due to fish farm activities.

Although free-living nematodes were assumed to tolerate sediment organic enrichment, field observations pointed to the general impact of fish farms on nematode abundance, namely a decrease in the levels thereof (La Rosa et al., 2001; Mazzola et al., 2002; Mirto et al., 2002). When it comes to the biomass, however, Duplisea & Hargrave (1996) did not find any variations below the cages. In contrast, Mirto et al., (2002) documented an increase in body weight as an effect of fish farm biodeposition, which was due to the dominance of some large, but tolerant, Comesomatidae genera (mainly *Pierrickia, Dorylaimopsis, Sabatieria*). The MI from fish farm sediment provided more congruent results, revealing an index decrease which was parallel to the biodeposition increase. This result also means that there is a good tool with which to detect the resilience of nematode assemblages. In contrast, the ITD did not highlight any clear impact of this type of disturbance (Mirto et al., 2002), with the same authors also reporting unambiguous responses to fish-farm disturbance, including in respect of the diversity parameters of the nematode community (*k*-dominance curves, richness, H′ and J). Different results were documented in a bluefin tuna fish farm in Vibo Marina, Italy (Western Mediterranean Sea) by Vezzulli et al. (2008), who recorded unclear variations in H′, J and the MI between cages and control stations, probably due to the limited impact of this type of farm. However, the analysis of the community at the genus level gave more consistent results in both of the investigations. Mirto et al. (2002) found that some nematode genera were highly sensitive to biodeposition (*Setosabatieria, Latronema* and *Elzalia*), and disappeared almost completely in farm sediments, whereas other tolerant genera mainly increased their dominance (*Sabatieria, Dorylaimopsis* and *Oxystomina*). Vezzulli et al. (2008) documented a dominance of *Tricoma, Desmoscolex, Quadricoma* and *Halalaimus* at their control station, and *Daptonema, Marylynnia, Sabatieria* and *Terschellingia* at the fish farm stations. Consequently, given that the indices cited above are not always sensitive enough to detect fish farm disturbance, the authors suggested that the identification of sensitive/tolerant nematode genera is the best indicator when it comes to noting even early benthic community changes below fish cages. The impact of intensive fish farming on the benthic environment seems to be greater than that of mussel farming, since mussels feed on natural resources (suspended particles) and are not sustained by any additional intensive feeding. In this respect, Danovaro et al. (2004) have documented the limited effect of a mussel farm on meiofaunal structure. This was due to the farm's minimum impact on sediment oxygen penetration and the biochemical compositions of the sediment organic matter. Yet several other investigations have reported that mussel biodeposition can have a serious effect on farm sediments, leading to severe organic matter accumulation and consequential reducing conditions, possibly also inducing significant changes in meiofaunal structure and, in particular, nematode assemblages (Mahmoudi et al., 2008; Mirto et al., 2000; Netto & Valgas, 2010). Many factors can influence these seemingly conflicting results, such as the culturing method used, the density of the cultivated mussels, the water depth and the hydrographical conditions in the system under investigation (Danovaro et al., 2004). In particular, H', richness and J significantly decreased in the mussel farm, which was probably related to changes in the sediment beneath the cages, which is characterized by a higher silt-clay percentage (Mahmoudi et al., 2008; Netto & Valgas, 2010). Indeed, nematode diversity is significantly affected by even slight granulometric variations (see Semprucci et al., 2010a and references therein). Multivariate analyses have highlighted that *Mesacanthion diplechma* was highly sensitive to the impact of a mussel farm, whereas *Paracomesoma dubium*, *T. longicaudata* and *T. communis* were very resistant thereto (Mahmoudi et al., 2008). A comparable opportunistic assemblage was reported by Netto & Valgas (2010), with an increase being mainly found in the tolerant genera of *Terschellingia*, *Sabatieria* and *Daptonema*. Among the functional parameters of the assemblage analyzed, the trophic diversity resulted lower below the cages, and selective deposit feeders or microvorous species were dominant (Netto & Valgas, 2010), probably because of the high microbial densities (Mirto et al. 2000). Meanwhile, the MI values were low due to a significant increase in opportunistic species, in particular the c-p 2 class (Netto & Valgas, 2010).

4.3 Hydrocarbons

In marine systems, the major sources of hydrocarbon contamination are oil exploration and production, natural seeps, atmospheric input, tanker accidents, industrial discharge, and urban run-off (Beyrem et al., 2010).

The effect of hydrocarbons on benthic foraminiferal assemblages in field and experimental studies has been evaluated in only a few papers (e.g. Alve, 1995; Denoyelle et al., 2010, in press; Ernst et al., 2006; Lockin & Maddocks, 1982; Mayer, 1980; Mojtahid et al., 2006; Morvan et al., 2004; Murray, 1985; Sabean et al., 2009; Vénec-Peyre, 1981; Whitcomb, 1978; Yanko & Flexer, 1991; Yanko et al., 1994) (Fig. 2c). A culture experiment on *Ammonia beccarii* and *Allogromia latilocollaris* exposed to crude oil revealed an inhibition in both reproduction and growth, while exposure to oil distillates induced narcosis, resulting in the death of all specimens (Whitcomb, 1978). The effect of a crude oil spill on the benthic foraminiferal assemblages on the coast of Brittany (France), which was caused by the Amoco Cadiz accident, was evaluated by Vénec-Peyre (1981), who reported morphological abnormalities but no changes in diversity or density. Meanwhile, minor or no negative effects of petroleum operations were in evidence in several platforms in Louisiana and the North Sea (Lockin & Maddocks, 1982; Murray, 1985). In contrast, a marked negative effect on benthic foraminiferal assemblages and a reduction of density and diversity were recorded in the Caspian Sea (Mayer, 1980) and the Odessa Bay (Yanko & Flexer, 1991). In order to address the response of benthic foraminifera to the 'Erika' oil spill, Morvan et al. (2004) evaluated at monthly/bimonthly intervals a site situated on the tidal mudflat in the Bay of Bourgneuf (France). Although a clear link between the occurrence of abnormalities and oil pollution was not found, an impoverished fauna in terms of density and species' richness was documented. The exposure of *A. tepida* at different concentrations of oil mixed in seawater revealed that the number of juveniles per reproduction event was lower in contaminated than in control cultures (Morvan et al., 2004). This experiment clearly documented the potential impact of oil on foraminiferal test shape, cytology and reproduction processes. Accordingly, in a laboratory experiment on the impact of oil on intertidal faunas, it was shown that the toxicity of oil components may lead to an increased mortality of benthic foraminiferal faunas Ernst et al. (2006). The effect of the discharge of oily drill cuttings on benthic foraminiferal assemblages was evaluated in the outer continental shelf off Congo (Africa) by Mojtahid et al. (2006). Different foraminiferal zones were determined according to pollution intensity. Low foraminiferal densities were recognized at the immediate vicinity of the discharge point, whereas the zone slightly further away from the disposal sites was characterized by very high foraminiferal densities and dominated by opportunistic taxa like *Bulimina aculeata*, *Bulimina marginata*, *Spiroplectinella sagittula* (reported as *T. sagittula*), *Trifarina bradyi* and *Bolivina* spp. Furthermore, foraminiferal densities seem to decrease along with the dominance of the opportunistic taxa. The environmental impact of weathered crude oil on benthic foraminifera in an Atlantic coastal salt marsh was evaluated by Sabean et al. (2009), who demonstrated the negative impact of oil on foraminiferal assemblages, as testified by a dramatic increase in abnormalities in *Miliammina fusca* when compared with the non-oiled control. Meanwhile, benthic foraminifera and macrofauna were evaluated as bio-indicators of an oil-based drill mud disposal site off Congo (Denoyelle et al., 2010). Poor faunas, dominated by some very tolerant taxa at the most polluted sites, were found, whereas greatly increased densities thereof, dominated by opportunistic taxa, were identified slightly further away from the disposal site. A chronic bioassay method has been developed by Denoyelle et al. (in press), who incubated the foraminifera *A. tepida* for 30 days in natural seawater with different concentrations of Fuel Oil no. 2. It was found that increased concentrations of this pollutant induced a significant decrease in the percentage of individuals displaying both pseudopodal activity and newly built chambers. This response clearly varies in terms of the function of the concentrations of the added pollutants. Although the impact of hydrocarbon pollution is far from being fully understood, increased mortality and abnormality and decreased density and diversity are among the effects on benthic foraminiferal assemblages.

Variable responses in terms of nematode abundance after oil contamination have been reported in the relevant literature: abundance may be relatively unaffected (Boucher, 1980; Elmgren et al., 1983; Fricke et al., 1981; Gee et al., 1992), or, in contrast, can significantly fall (Beyrem et al., 2010; Carman et al., 1995; Danovaro et al., 1995, 1999; Elmgren et al., 1980; Giere, 1979; Mahmoudi et al., 2002, 2005). The phenomenon of rapid recovery is present only in some cases (Fig. 2c). These controversial results may be due to: different dosages and the toxicity of hydrocarbons; variable responses to the disturbance of the species; the different bioavailability of hydrocarbons with sediment type; and the hydrodynamic conditions. Furthermore, different responses to fuel oil contamination may also be a consequence of community chronic exposure, which alters biota sensitivity to these contaminants (Mahmoudi et al., 2005). Accordingly, a benthic community from a contaminated area is more tolerant to hydrocarbon contamination than a comparable community from a less contaminated environment (Carman et al., 2000). Beyrem et al. (2007) documented that if the presence of diesel alone does not seem to affect total nematode abundance, the diesel-metal combination may cause significant abundance decreases as a result of the higher production of mucus-exopolymers by diesel. Exopolymers are in fact known to have a strong affinity for a variety of metals, and are readily consumed by benthic organisms, thus increasing the exposure of animals to metals (see Beyrem et al., 2007 and references therein). A reduction of species' diversity has been also reported by several other authors (Boucher, 1980; Boucher et al., 1981; Danovaro et al., 1995; Mahmoudi et al., 2002, 2005), probably due to a general increase in the mortality of the most sensitive species (Beyrem & Aïssa, 2000; Carman et al., 2000; Mahmoudi et al., 2002, 2005). For instance, a few days after the Amoco Cadiz oil spill, no significant impact of hydrocarbons on nematode assemblages was detected (Renaud-Mornant et al. 1981). However, after several months, a significant decline in nematode diversity was both revealed and related to a change in the structure of the community: pristine species associated with sandy sediments were clearly replaced by species which were typical of muddy sediments. Little data on the impact of hydrocarbon contamination on nematode trophic structure or trophic diversity are available, although significant alterations have rarely been found, probably due to the limited influence of oil pollution on nematode trophic structure (Danovaro et al., 1995, 2009; Schratzberger et al., 2003). Nevertheless, an alteration of single trophic guilds may be recorded. For instance, Danovaro et al. (1995) reported a decrease of group 1B after oil contamination. This group is likely to be directly affected by oil toxicity as its members ingest tar particles and oil emulsion during feeding (Danovaro, 2000). Surprisingly, group 2B was found to be dominant, or even increasing in abundance, after the oil spill (Danovaro, 2000; Schratzberger et al., 2003). Moreover, the importance of group 2A may rise as a result of the increased microphytobenthic biomass (Carman et al., 1995; Danovaro, 2000). Species - or genus - specific responses were reported by several authors. Danovaro et al. (1995) found that genera such as *Chromaspirina*, *Hypodontolaimus*, *Oncholaimellus*, *Paracanthonchus*, *Setosabatieria* and *Xyala* disappeared immediately after the Agip Abruzzo oil spill, although they recovered rapidly, thus appearing to be opportunists. In contrast, genera such as *Daptonema* and *Viscosia* appeared to be more tolerant to hydrocarbons. In their microcosm experiment, Mahmoudi et al. (2005) found that *Chaetonema* sp. was highly sensitive to diesel contamination; *Pomponema* sp. and *Oncholaimus campylocercoides*, meanwhile, were diesel-sensitive, while *Hypodontolaimus colesi*, *D. trabeculosum* and *D. fallax* were opportunistic and *Marylynnia stekhoveni* diesel-resistant. However, two of these species, *M. stekhoveni* and *O. campylocercoides*, may present with variable levels of sensitivity (see Beyrem et al., 2010), highlighting the fact that the impact of hydrocarbon on marine assemblages requires further investigation. As for *O. campylocercoides*, it is worth noting that Boufahja et al. (2011a) have recently demonstrated that the biometry, life cycle and fecundity of this species may be useful indices for the biomonitoring of hydrocarbon pollution in marine ecosystems.

4.4 Trace elements

The term 'heavy metal' is widely utilized but inadequately described in the scientific literature (see Duffus, 2002). In this paper, the term 'trace elements' will be used instead of 'heavy metals'. 'Trace elements' is a collective term, which refers to any metallic element that has a relatively high density. While most trace elements are biologically essential at very low concentrations, they become toxic to marine organisms above a specific threshold (Kennish, 1992). Although trace elements can be introduced into the environment by natural causes, the major input thereof are anthropogenic in origin, e.g. mining sites, foundries and smelters, the purification of metals, combustion by-products, traffic and coal, natural gas, paper, and chloro-alkali activities. One of the chief problems associated with trace elements is their persistence; unlike organic pollutants they do not decay, thus proving that they have a high potential for bioaccumulation and biomagnification.

Over the last four decades, many studies, conducted in different environmental settings, have focused on the response of benthic foraminifera to trace element pollution (e.g. Alve, 1991b, 1995; Armynot du Châtelet et al., 2004; Coccioni, 2000; Coccioni et al., 2005, 2009; Frontalini & Coccioni, 2008, 2011; Frontalini et al., 2009, 2010; Yanko et al., 1998, 1999) (Fig. 2d). These studies have revealed that this kind of pollution, which may cause pathological processes in the foraminiferal cell, plays an important role in the development of abnormal (teratological) tests. It may also lead to: changes in foraminiferal density and diversity; alterations in assemblage composition; size variation; and structural modification, including in megalospheric and dwarf specimens (for a review, Alve, 1995; Frontalini & Coccioni, 2011; Yanko et al. 1994). It has also been suggested that the presence of morphological abnormalities in benthic foraminiferal tests could be a powerful *in situ* bio-indicator of trace element pollution. Accordingly, Coccioni et al. (2005) developed the FAI to index and compare the levels of morphological abnormality occurring at different sites. More recently, and on the basis of bibliographic analysis, Frontalini & Coccioni (2011) were able to synthetize the inferred sensitive or tolerant response of many foraminiferal species and genera to pollution. Although most of these inferences are not yet supported by the results of laboratory experimentation, a high degree of tolerance to trace elements has been inferred for several taxa, including *Ammonia, Cribroelphidium, Haynesina, Brizalina* and *Bolivina,* as also reported by other researchers (Armynot du Châtelet & Debenay, 2010; Armynot du Châtelet et al., 2004; Carnahan et al., 2008, 2009; Yanko et al., 1999). In recent years, several studies have focused on the response of particular benthic foraminiferal species to selected trace elements in controlled laboratory conditions (e.g. Gustafson et al., 2000; Le Cadre & Debenay, 2006; Nigam et al., 2009; Saraswat et al., 2004). Laboratory culture experiments, through which the benthic foraminiferal response to various elements and concentrations of pollutants can be observed over time, represent the most effective and direct method with which to assess the effect of a single parameter on benthic foraminiferal assemblages. In fact, these experiments provide continuous and accurate observations on the benthic foraminiferal response under controlled conditions, whereby a single parameter can be altered, keeping the rest constant. In this way, the benthic foraminiferal response to specific parameters can be directly characterized. This leads to benthic foraminiferal culture studies wherein foraminifers are subjected to specific pollutants and their particular response thereto is thus documented. In particular, Gustafson et al. (2000) reported a decrease in benthic foraminiferal density when exposed to Tri-n Butyltin. Meanwhile, in a monospecific experiment involving the near-shore benthic foraminiferal species *Rosalina leei,* which was subjected to different concentrations of Hg, Saraswat et al. (2004) documented that the growth rate, as well as the maximum size, decreased considerably in the specimens subjected to gradually higher concentrations of mercury. Test abnormalities also developed at the same time. Moreover, the specimens were subjected to progressively increasing concentrations of Hg to see the further effects thereof: although their growth ceased, the specimens were still living at concentrations as high as 260 ng/l. The effect of graded concentrations of Cu on *A. beccarii* and *A. tepida* has been analyzed by Le Cadre & Debenay (2006) at the morphological level. The two species were sensitive to low, but survived high, concentrations of this trace element. Increasing such concentrations leads to greater delay before the production of new chambers, thus lengthening the period of time before reproduction and reducing the number of juveniles. The proportion of abnormal tests also rose. Moreover, cellular ultrastructure modifications of abnormal specimens exposed sub-lethal contaminations, while thickening of the organic lining, the proliferation of fibrillar vesicles, increases in the number and volume of lipidic vesicles, the disruption of the plasma membrane, increased numbers of residual bodies, and the detection of sulfur within the cells were also found. It has been suggested that the latter is the result of a detoxification mechanism involving the production of a metallothionein-like protein. More recently, Nigam et al. (2009) documented the response of the benthic foraminifera *R. leei* to the sudden or gradual addition of Hg into the media. They reported that when Hg was added suddenly, the specimens did not show any change in morphology during the initial 40 days. However, later on, test deformities developed at higher concentrations, or complete mortality occurred within 20 days at concentrations as high as 300 ng/l. Additional changes were documented in the rate of reproduction, the number of juveniles produced and the survival rate of the juveniles, with growth found to be inversely proportional to mercury concentration. When Hg concentration was increased gradually, irregularities in the newly added chambers were noticed only in the specimens subjected to very high levels of mercury, while growth was found to be inversely proportional to the concentration of this trace element.

Nematodes appear to be good biological indicators of trace element contamination, with it being documented that they may be even more sensitive than other meiofaunal taxa, such as copepods (Somerfield et al., 1994). Several field investigations and laboratory experiments on the effects of trace elements within nematode assemblages have been carried out and reported on the literature (e.g. Austen & McEvoy, 1997; Austen & Somerfield, 1997; Beyrem et al., in press; Boufahja et al., 2011b; Derycke et al., 2007; Guo et al., 2001; Gyedu-Ababio et al., 1999; Hedfi et al., 2007, 2008; Hermi et al., 2009; Mahmoudi et al., 2002, 2007; Millward & Grant, 1995; Somerfield et al., 1994) (Fig. 2d). Nematode abundance may be altered in different ways by different trace elements (Austen & McEvoy, 1997; Hedfi et al., 2007, 2008; Heip et al., 1984; Hermi et al., 2009; Boufahja et al., 2011b; Mahmoudi et al., 2002). In contrast, nematode diversity is very sensitive to these metals and significantly decreases after exposure (Austen & Somerfield, 1997; Boufahja et al., 2011b; Hedfi et al., 2007, 2008; Mahmoudi et al., 2002; Millward & Grant, 1995; Somerfield et al., 1994; Tietjen, 1980). The chemical form, as well as the type of trace element, is important in determining the toxicity effect on nematode assemblages (see Coull & Chandler, 1992 for review). The uptake of metals (e.g. Cu and Zn) by nematodes primarily occurs through cuticular mucous secretions, and may be very different in different species, even of the same genus (Howell, 1982, 1983). However, laboratory studies have shown that the effect of trace elements depends not only on the nature of the element, but also on some environmental factors e.g. temperature, salinity and trophic availability (Coull & Chandler, 1992). Nematode assemblage responses to different types of contaminants have been reported by several authors. Somerfield et al. (1994) and Austen & Somerfield (1997) documented that some species, such as *Ptycholaimellus ponticus*, *S. pulchra*, *Molgolaimus demani* and *Axonolaimus paraspinosus*, can tolerate a wide range of trace elements and may thus have evolved some tolerance adaptations. *Terschellingia* species survived well in all the microcosm treatments including those containing the highest metal doses (Austen & Somerfield, 1997). *Tripyloides gracilis* increased its presence in the most affected sediments (Austen & Somerfield, 1997; Tietjen, 1980). In accordance with the work of Somerfield et al. (1994), toxicity tests on the entire nematode assemblage from the severely contaminated estuary of Restronguent Creek (UK) (Millward & Grant, 1995) revealed that it was more resistant to Cu due to an enhanced Cu tolerance in some dominant species (e.g. *T. marinus*, *A. spinosus*, *D. setosum*, *Eleutherolaimus* sp., *Theristus acer*). Millward & Grant (1995) demonstrated for the first time that pollution-induced community tolerance (PICT) may be used as a tool to evaluate the biological impact of a chronic pollutant on the marine benthic system. Interesting results are also

Trace Elements	**Intolerant/sensitive**	**Opportunistic**	**Resistant**
Ni	*Leptonemella aphanothecae*	*Daptonema normandicum*, *Neochromadora trichophora*, *Odontophora armata*	*Oncholaimus campylocercoides*, *Bathylaimus capacosus*
Cu	*Microlaimus affinis*, *Monoposthia mirabilis*	-	-
Hg	*Araeolaimus bioculatus*	*Marylynnia stekhoveni*	*Prochromadorella neapolitana*
Cr	*Leptonemella aphanothecae*	*Daptonema normandicum*, *Sabatieria longisetosa*	*Bathylaimus capacosus*, *Bathylaimus tenuicaudatus*
Pb	*Calomicrolaimus honestus*	*Oncholaimus campylocercoides*	-
Zn	*Hypodontolaimus colesi*	*Xyala* sp., *Viscosia franzii*	*Oncholaimus campylocercoides*
Co	*Oncholaimellus mediterraneus*, *Oncholaimus campylocercoides*, *Neochromadora trichophora*	*Spirinia gerlachi*, *Viscosia franzii*, *Promonhystera* sp.	*Marylynnia stekhoveni*
Co/Zn	-	*Viscosia franzii*, *Sabatieria pulchra*	*Marylynnia stekhoveni*

Table 1. Some nematode sensitive/tolerant species proposed as possible preventive indicators of a contaminated sea (Beyrem et al., in press; Boufahja et al., 2011b; Hedfi et al., 2007, 2008; Hermi et al., 2009; Mahmoudi et al., 2007)

reported in some field surveys in the Mediterranean Sea: Mahmoudi et al. (2002) found significant, negative correlations between Cu, Pb and Zn and the biomass and diversity of nematodes, even though the trace element concentrations in the sediments of the Bou Ghrara Lagoon (Tunisia) were lower than envisaged in international norms. Similarly, Semprucci et al. (2010b) found an MI decrease in relation to a peak of Pb in concentration values between the effect-range low (ERL) and the median (ERM) criteria proposed by the NOAA. Nematodes from sediments from the Swartkops estuary in Port Elizabeth, South Africa were investigated by Gyedu-Ababio et al. (1999); the relationship between density, genera, community structure and environmental parameters, including the sediment concentrations of seven heavy metals (Mn, Ti, Cr, Pb, Fe, Sn and Zn), were analyzed. A combination of H' and the MI proved to be very useful for assessing polluted or stressed sites. The nematode community structure at polluted and low/no pollution sites, respectively, differed significantly. *Monhystera* and *Theristus* were found to be colonizers, and these genera were thus regarded as indicators of polluted sediments. Extensive microcosm experiments carried out along the Tunisian coastal zone have tested the nematode assemblage in respect of different trace elements, thus highlighting a set of nematode sensitive/tolerant species (Table 1). The application of the microcosm bioassay approach in these studies has enabled a clear range of sensitivity of nematode assemblages to several trace element doses to be depicted. In a laboratory study on the individual and combined effects of Co and Zn, Beyrem et al. (in press) inferred that there was an antagonistic interaction between these metals: the simultaneous presence of Co and Zn seems to have a lesser impact on nematode species' composition. The clear species-specific responses to the different types of contaminants found in these studies support the possible use of these taxa in the effective biomonitoring of trace elements in coastal marine ecosystems.

4.5 Other types of anthropogenic disturbance

Although the impact induced by organic matter and trace element pollution is a major concern, marine and transitional environments can actually also be influenced by other kinds of anthropogenic disturbance. This could be the discharge of radioactive waste, thermal pollution or physical disturbance (e.g. dredging disposal), which may lead to partial or complete defaunation. Physical disturbance, i.e. processes that lead to the disruption of sediments (see Boyd et al., 2000 and references therein), is a key factor in controlling the spatial and temporal composition of benthic communities. Anthropogenic activities, including coastal development, dredged material disposal and bottom trawling, may cause widespread physical disturbance of the seabed and changes in sedimentation patterns in shelf seas.

At the present time, only a very few studies have addressed the effect of these disturbances on benthic foraminiferal assemblages (e.g. Arieli et al., 2011; Bartlett, 1972; Hechtel et al., 1970; Reish, 1983); (Fig. 2e). A decrease in foraminiferal diversity and a rapid increase of density were documented in response to heated effluent from a power plant in Long Island Sound (northeastern US) (Hechtel et al., 1970). Although it is difficult to separate the effect of thermal pollution from other kinds of stresses, Bartlett (1972) suggested that there was impoverished diversity and lower density in areas along the Atlantic Provinces (eastern Canada). Morphological abnormalities and low density were among the effects on benthic foraminifera in the eurythermal environment of Guayanilla Bay (Puerto Rico).

The most comprehensive investigation of the impact of thermal pollution from a power plant on benthic foraminifera has been carried out in Hadera (Israel) (Arieli et al., 2011). This *in situ* monthly monitoring has revealed that thermal pollution has a detrimental effect on benthic foraminiferal assemblages. In particular, decreased density and diversity values were documented as the sea surface temperature (SST) increased. It was also suggested that 30°C is the critical threshold above which foraminifera' growth and reproduction are severely retarded, and some species are better adapted to tolerating a high SST than others.

The effects of dredged material on the structure of nematode assemblages have been investigated by several authors (e.g. Boyd et al., 2000; Schratzberger et al., 2000, 2006; Somerfield et al., 1995) (Fig. 2e). Boyd et al. (2000) reported a decrease in all of the diversity indices, but not in the abundance values. In accordance with the experiments carried out by Schratzberger et al. (2000, 2009), the most remarkable effect of dredging disposal on the structure of nematode assemblages was the proliferation within the disposal sites of the non-selective deposit feeders *S. pulchra* group (both *breviseta* and *punctata*) and *D. tenuispiculum*. Somerfield et al. (1995) also found *D. tenuispiculum* and *S. punctata* (part of the *pulchra* group) to be numerically abundant in the same dredged material disposal site of Liverpool Bay. As a consequence, the authors suggested that these species were indicators of dredging disturbance. Schratzberger & Jennings (2002) documented the impact of trawling on nematodes. Their data revealed that trawling had a significant impact on the composition of nematode assemblages: diversity and species' richness were significantly lower at high levels of trawling disturbance than at low or medium levels thereof. Recently, Schratzberger et al. (2009) have reviewed the effects of several types of physical disturbance on nematodes. They noted that epigrowth feeding genera such as *Spirinia* and *Desmodora* were highly susceptible to seabed disturbance, thus leading to genus diversity reduction at the most affected stations. A decreased trophic diversity at these stations was primarily due to the increased dominance of non-selective deposit feeders. In contrast, a decrease in genus diversity as a result of bottom trawling did not lead to any changes in trophic diversity. *Sabatieria*, meanwhile, proliferated at high levels of anthropogenic disturbance as a result of coastal development, dredged material disposal and bottom trawling.

5. Conclusions

According to the European Marine Strategy Framework Directive (2008/56/EC), seafloor integrity should be at a level ensuring the safeguarding of the structure and function of ecosystems. Consequently, monitoring the quality of the environment appears to be essential for devising effective protection strategies and appropriate forms of management of marine systems. In particular, the Water Framework Directive (WFD, Directive 2000/60/EC) highlights the importance of biological descriptors when it comes to evaluating and monitoring environmental conditions. Among these, benthic foraminifera and nematodes are highly suitable and sensitive biological organisms through which our comprehension of marine and transitional marine environments can be further explored. In particular, benthic foraminifera have been demonstrated to be particularly sensitive microorganisms and they have been successfully utilized for their value as bioindicators of environmental change in a wide range of marine environments. Since the complex interplay of different biological, chemical, ecological and physical element, and the synergic and antagonistic relationships among them operating at different times and scale the identification of foraminiferal response to specific environmental factor(s) is sometimes hampered. The complex behavior is reflected in shifting patterns of parameters like density, diversity, assemblages composition and percentages of abnormality that are far to be completely understood. The covariance among the nature of the substrate, and the oxygen and organic matter contents within it, which affect the benthic foraminiferal density, clearly, represents a case. The lower foraminiferal diversification of transitional marine environments (i.e., lagoon, coastal lake, salt marsh), when compared to more open water, cannot be readily used as pollution indicator if not associate with similar environments. In the same environmental settings, the percentages of foraminiferal abnormalities would be used with care when additional disturbance factors (i.e., rapid salinity changes) are involved. Moreover, the comparison of different studies is further complicated by the adoption of different techniques and methods. In order to compare these studies, the same set of techniques must be used from the initial sampling to the final treatment of data at least for the same environmental setting. This can only be guaranteed if there is an agreement among scientists and a flexible protocol(s) is developed. Although a definitive scientific agreement has not yet been reached, the development of a protocol represents a milestone for foraminiferal application within governmental and international programs which regulate environmental surveys on marine and transitional marine environments. Notwithstanding nematodes are particularly suitable as bio-indicators in all the types of environments (freshwater, marine and terrestrial ones), a general limitation of ecological investigations of nematodes is related to their difficult taxonomic identification. In this respect, a new and important challenge for the future is the implementation of molecular technique applications, which could be an important tool for making nematode identification easier and faster. The taxonomic approach is certainly the best and more sensitive way for the evaluation of the nematode species response to the pollution effects. However, even if the functional roles of nematodes may be highly species-specific, the application of the taxonomic sufficiency (at genus level) may overall give excellent results. In particular, a multivariate analysis of the assemblage structure and the detection of specific bio-indicators give the best results in all the types of pollution discussed. Also taxonomic diversity may be a good tool for detecting the responses of nematodes, especially to sewage, aquaculture and trace elements pollutions. However, the results from the diversity index should be treated with caution because of the influence of some natural environmental parameters (e.g. sediment grain size). At the present time, the functional trait approach (e.g. trophic groups, life history strategies) seems to be no more powerful than the traditional taxonomic methods, but it can provide additional ecological information on the ecosystem functioning. In particular, Maturity Index and colonizer-persister class percentages are successfully applied in literature on the effects of aquaculture and trace elements. Furthermore, the c-p percentage has been recently suggested as the best descriptor of the quality ecological status in marine ecosystems, together with the taxonomic approach. Although significant variations of single trophic groups are usually observed in several types of pollutions, the Index of Trophic Diversity shows controversial responses, especially in aquaculture and hydrocarbon pollutions, and in physical disturbance. This might suggest a non-selective impact of these stressors on the global trophic structure of the nematode assemblage or it may be related to the limits of the Wieser classification, which does not perfectly reflect the trophic position of all the nematode species. A good alternative to feeding types appears to be based on the life history of nematodes or on the integration of both of these functional traits. However, the ecological and practical advantages of using Foraminifera and Nematoda in benthic ecological studies are basic elements for choosing these taxa as descriptor groups in the assessment of the quality status of marine sediments.

6. References

Albani, A.D., Donnici, S., Serandrei Barbero, R., & Rickwood, P.C. (2010). Seabed sediments and foraminifera over the Lido Inlet: comparison between 1983 and 2006 distribution patters. *Continental Shelf Research,* Vol. 30, pp. 847-858.

Alve, E. (1991a). Foraminifera, climatic change, and pollution: a study of Late Holocene sediments in Drammensfjord, Southeast Norway. *The Holocene,* Vol. 1, pp. 243–61.

Alve, E. (1991b). Benthic foraminifera reflecting heavy metal pollution in Sørljord, Western Norway. *Journal of Foraminiferal Research,* Vol. 34, pp. 1641-1652.

Alve, E. (1995). Benthic foraminifera response to estuarine pollution: a review. *Journal of Foraminiferal Research,* Vol. 25, pp. 190-203.

Alve, E. (1999). Colonisation of new habitats by benthic foraminifera: a review. *Earth-Science Reviews,* Vol. 46, pp. 167-185.

Alve, E., & Bernhard, J.M. (1995). Vertical migratory response of benthic foraminifera to controlled oxygen concentrations in an experimental mesocosm. *Marine Ecology Progress Series,* Vol. 116, pp. 137–51.

Angel, D. L., Verghese, S., Lee, J. J., Saleh, A. M., Zuber, D., Lindell, D., & Symons, A. (2000). Impact of a net cage fish farm on the distribution of benthic foraminifera in the northern Gulf of Eilat (Aqaba, Red Sea). *Journal of Foraminiferal Research,* Vol. 30, pp. 54-65.

Arieli, R.N., Almogi-Labin, A., Abramovich, S., & Herut, B. (2011). The effect of thermal pollution on benthic foraminiferal assemblages in the Mediterranean shoreface adjacent to Hadera power plant (Israel). *Marine Pollution Bulletin,* Vol. 62, pp. 1002-1012.

Armenteros, M., Pérez-García, J. A., Ruiz-Abierno, A., Díaz-Asencio, L., Helguera, Y., Vincx, M., & Decraemer W. (2010). Effects of organic enrichment on nematode assemblages in a microcosm experiment. *Marine Environmental Research,* Vol. 70, pp. 374-82.

Armynot du Châtelet, E., Debenay, J.P., & Soulard, R. (2004). Foraminiferal proxies for pollution monitoring in moderately polluted harbors. *Environmental Pollution,* Vol. 127, pp. 27-40.

Armynot du Châtelet, E., & Debenay, J.P. (2010). Anthropogenic impact on the western French coast as revealed by foraminifera: a review. *Revue de Micropaléontologie,* Vol. 53, pp. 129-137.

Austen, M. C., & McEvoy, A.J. (1997). The use of offshore meiobenthic communities in laboratory microcosm experiments: response to heavy metal contamination. *Journal of Experimental Marine Biology and Ecology,* Vol. 211, pp. 247-261.

Austen, M.C., & Somerfield, P.J. (1997). A community level sediment bioassay applied to an estuarine heavy metal gradient. *Marine Environmental Research,* Vol. 43, pp. 315-328.

Austen, M.C., & Widdicombe, S. (2006). Comparison of the response of meio- and macrobenthos to disturbance and organic enrichment. *Journal of Experimental Marine Biology and Ecology,* Vol. 330, pp. 96-104.

Balsamo, M., Albertelli, G., Ceccherelli, V.U., Coccioni, R., Colangelo, M.A., Curini-Galletti, M., Danovaro, R., D'Addabbo, R., Leonardis, C., Fabiano, M., Frontalini, F., Gallo, M., Gambi, C., Guidi, L., Moreno, M., Pusceddu, A., Sandulli, R., Semprucci, F., Todaro, M.A., & Tongiorgi, P. (2010). Meiofauna of the Adriatic Sea: current state of knowledge and future perspective. *Chemistry and Ecology*, Vol. 26, pp. 45 - 63.

Bandy, O.L., Ingle, J.C., & Resig, J.M. (1964). Foraminiferal trends, Laguna Beach outfall area, California. *Limnology and Oceanography*, Vol. 9, pp. 112–23.

Barmawidjaja, D. M., Van der Zwaan, G. J., Jorissen, F. J., & Puskaric, S. (1995). 150 years of eutrophication in the northern Adriatic Sea: evidence from a benthic foraminiferal record. *Marine Geology*, Vol. 122, pp. 367-384.

Bartlett, G.A. (1972). Ecology and the concentration and effect of pollutants in nearshore marine environments. *International Symposium on Identification and Measurement of Environmental Pollutants*, National Research Council, Canada, Abstracts, p. 277.

Beyrem, H., & Aïssa, P., 2000. Les nématodes libres, organismes sentinelles de l' évolution des concentrations d'hydrocarbures dansla baie de Bizerte (Tunisie). *Cahiers de Biologie Marine*, Vol. 41, pp. 329–342.

Beyrem, H., Mahmoudi, E., Essid, N., Hedfi, A., Boufahja, F., & Aïssa, P. (2007). Individual and combined effects of cadmium and diesel on a nematode community' in a laboratory microcosm experiment. *Ecotoxicology and Environmental Safety*, Vol. 68, pp. 412–418.

Beyrem, H., Louati, H., Essid, N., Aïssa, P., & Mahmoudi E. (2010). Effects of two lubricant oils on marine nematode assemblages in a laboratory microcosm experiment. *Marine Environmental Research*, Vol. 69, pp. 248-253.

Beyrem, H., Boufahja, F., Hedfi, A., Essid, N., Aïssa, P., & Mahmoudi, E. (in press). Laboratory study on individual and combined effects of cobalt- and zinc-spiked sediment on meiobenthic nematodes. *Biological Trace Elements Research*, DOI 10.1007/s12011-011-9032-y

Boltovskoy, E. (1965). *Los Foraminıferos Recientes*. Editorial Universitaria de Buenos Aires (EUDEBA), Buenos Aires.

Boltovskoy, E., Scott, D.B., & Medioli, F.S. (1991). Morphological variations of benthic foraminiferal test in response to changes in ecological parameters: a review. *Journal of Paleontology*, Vol. 65, pp. 175-185.

Bongers, T. (1990). The maturity index: an ecological measure of environmental disturbance based on nematode species composition. *Oecologia*, Vol. 83, pp. 14-19.

Bongers, T., Alkemade, R., & Yeates, G.W. (1991). Interpretation of disturbance-induced maturity decrease in marine nematode assemblages by means of the Maturity Index. *Marine Ecology Progress Series*, Vol. 76, pp. 135-142.

Boucher, G. (1980). Impact of Amoco Cadiz oil spill on intertidal and sublittoral meiofauna. *Marine Pollution Bulletin*, Vol. 11, pp. 95–101.

Boucher, G,. & Lambshead, P.J.D. (1995). Ecological biodiversity of marine nematodes in samples from temperate, tropical, and deep-sea regions. *Conservation Biology*, Vol. 9, pp. 1594-1604.

Boucher, G., Chamroux S., LeBorgne L., & Mevel G. (1981). Ėtude expérimentale d'une pollution par hydrocarbures dans un microecosystème sédimentaire. I: Effect de la contamination du sediment sur la méiofaune. In, Amoco Cadiz, Consequences d'un

Pollution Accidentelle par les Hydrocarbures. Actes Coll. Intern. C.O.B. Brest (France) 19-22 Nov. 1919, Ed. CNEXO, Paris, pp. 229-243.

Bouchet, V. M. P., Debenay, J. P., Sauriau, P.-G., Radford-Knoery, J., & Soletchnike, P. (2007). Effects of short-term environmental disturbances on living benthic foraminifera during the Pacific oyster summer mortality in the Marennes-Oleron Bay (France). *Marine Environmental Research,* Vol. 64, pp. 358–383.

Boufahja, F., Hedfi, A., Amorri, J., Aïssa, P., Beyrem, H., & Mahmoudi E. (2011b). An assessment of the impact of chromium-amended sediment on a marine nematode assemblage using microcosm bioassays. *Biological Trace Elements Research,* Vol. 142, pp. 242-255.

Boufahja, F., Hedfi, A., Amorri, J., Aïssa, P., Beyrem, H., & Mahmoudi E. (2011a). Examination of the bioindicator potential of *Oncholaimus campylocercoides* (Oncholaimidae, Nematoda) from Bizerte bay (Tunisia). *Ecological Indicators,* Vol. 11, pp. 1139–1148.

Boyd, S.E., Rees, H.L. & Richardson C.A. (2000). Nematodes as sensitive indicators of change at dredged material disposal sites. *Estuarine, Coastal and Shelf Science,* Vol. 51, pp. 805–819

Carman, K.R., Fleeger, J.W., Means, J.C., Pomarico, S.M., & McMillin, D.J. (1995). Experimental investigation of polynuclear aromatic hydrocarbons on an estuarine sediment food web. *Marine Environmental Research,* Vol. 40, pp. 298–318.

Carman, K.R., Fleeger, J.W. & Pomarico, S.M. (2000). Does historical exposure to hydrocarbon contamination alter the response of benthic communities to diesel contamination? *Marine Environmental Research,* Vol. 49, pp. 255–278.

Carnahan, E.A., Hoare, A.M., Hallock, P., Lidz, B.H., & Reich, C.D. (2008). Distribution of heavy metals and foraminiferal assemblages in sediments of Biscayne Bay, Florida, USA. *Journal of Coastal Research,* Vol. 24, pp. 159-169.

Carnahan, E.A., Hoare, A.M., Hallock, P., Lidz, B.H., & Reich, C.D. (2009). Foraminiferal assemblages in Biscayne Bay, Florida, USA: Response to urban and agriculture influence in a subtropical estuary. In: Romano E. & Bergamin L. (Eds.), Foraminifera and marine pollution. *Marine Pollution Bulletin,* Vol. 59, pp. 221-233.

Clark, D.F. (1971). Effects of aquaculture outfall on benthonic foraminifera in Clam Bay, Nova Scotia. *Maritime Sediments,* Vol. 7, pp. 76–84.

Coccioni, R., Frontalini F., Marsili A., & Troiani F. (2005). Foraminiferi bentonici e metalli in traccia: implicazioni ambientali, In: *La dinamica evolutiva della fascia costiera tra le foci dei fiumi Foglia e Metauro: verso la gestione integrata di una costa di elevato pregio ambientale,* Coccioni, R. (Ed.), pp. 57-92, Quaderni del Centro di Geobiologia, Urbino University, Italy.

Coccioni, R., Frontalini, F., Marsili, A., & Mana, D. (2009). Benthic foraminifera and trace element distribution: A case-study from the heavily polluted lagoon of Venice (Italy). In: Romano E., Bergamin L. (Eds.), Foraminifera and marine pollution, *Marine Pollution Bulletin,* Vol. 56, pp. 257-267.

Coccioni, R. (2000). Benthic foraminifera as bioindicators of heavy metal pollution - a case study from the Goro Lagoon (Italy). In: *Environmental Micropaleontology: The Application of Microfossils to Environmental Geology,* Martin, R.E. (Ed.), pp. 71-103, Kluwer Academic/Plenum Publishers, New York.

Coull, B.C., & Chandler, G.T. (1992). Pollution and meiofauna: field, laboratory, and mesocosms studies. *Oceanography and Marine Biology An Annual Review*, Vol. 30, pp. 191–271.

Danovaro, R., 2000. Benthic microbial loop and meiofaunal response to oil-induced disturbance in coastal sediments: a review. *International Journal of Environment and Pollution*, Vol. 13, pp. 380– 391.

Danovaro, R., Fabiano, M., & Vincx, M. (1995). Meiofauna response to the Agip Abruzzo oil spill in subtidal sediments of the Ligurian Sea. *Marine Pollution Bulletin*, Vol. 30, pp. 133–145.

Danovaro, R., Gambi, C., Luna, G. M., & Mirto, S. (2004). Sustainable impact of mussel farming in the Adriatic Sea (Mediterranean Sea): evidence from biochemical, microbial and meiofaunal indicators. *Marine Pollution Bulletin*, Vol. 49, pp. 325–333.

Danovaro, R., Della Croce, N., Marrale, D., Martorano, D., Parodi, P., Pusceddu, A., & Fabiano, M. (1999). Biological indicators of oil induced disturbance in coastal sediments of the Ligurian Sea. In: *Assessment & Monitoring of Marine Science*, Lokman, S., Shazili, N.A.M., Nasir, M.S., & Borowtizka, M.A. (Eds.), pp. 75–85, University Putra Malaysia Terengganu, Kuala Terengganu, Malaysia.

Danovaro, R., Gambi, C., Höss, S., Mirto, S., Traunspurger, W., & Zullini, A. (2009). Case studies using nematode assemblage analysis in aquatic habitats. In: *Nematodes as Environmental Indicators*, M. J. Wilson, & Kakouli-Duarte, T. K. (Eds.), pp. 146-171, CAB Internationals. Wallingford, UK.

Debenay, J.P., & Luan B.T. (2006). Foraminiferal assemblages and the confinement index as tools for assessment of saline intrusion and human impact in the Mekong delta. *Revue de Micropaléontologie*, Vol. 49, pp. 74-85

Debenay, J.P., Della Patrona, L., Herbland, A., & Goguenheim, H. (2009a). The impact of Easily Oxidized Material (EOM) on the meiobenthos: Foraminifera abnormalities in Shrimp ponds of New Caledonia, implications for environment and paleoenvironment survey. *Marine Pollution Bulletin*, Vol. 59, pp. 323-335.

Denoyelle, M., Jorissen, F.J., Martin, D., Galgani, F., & Miné, J. (2010). Comparison of benthic foraminifera and macrofaunal indicators of the impact of oil-based drill mud disposal. *Marine Pollution Bulletin*, Vol. 60, pp. 2007–2021.

Denoyelle, M., Geslin, E., Jorissen, F.J., Cazesc, L., & Galgani, F., & Miné, J. (in press). Innovative use of foraminifera in ecotoxicology: A marine chronic bioassay for testing potential toxicity of drilling muds. *Ecological Indicators*, doi:10.1016/j.ecolind.2011.05.011.

Derycke, S., Hendrickx, F., Backeljau, T., D'Hondt, S., Camphijn, L., Vincx, M., & Moens, T. (2007). Effects of sublethal abiotic stressors on population growth and genetic diversity of *Pellioditis marina* (Nematoda) from the Westerschelde estuary. *Aquatic Toxicology*, Vol. 82, pp. 110–119.

Duffus, J.H. (2002). 'Heavy Metals' A meaningless term? IUPAC, *Pure and Applied Chemistry*, Vol. 74, pp. 793-807.

Duplisea, D.E., & Hargrave, B.T. (1996). Response of meiobenthic size-structure, biomass and respiration to sediment organic enrichment. *Hydrobiologia*, Vol. 339, pp. 161–170.

Elmgren, R., Vargo. G.A., Grassle, J.F., Grassle, J.P., Heinle, D.R., Langlois, G., & Vargo, S.L. (1980). Trophic interactions in experimental marine ecosystems perturbed by oil. In: *Microcosmos in Ecological Research*, Giesy J.P. (Ed.), pp. 779-800, U.S. Tech. Info. Cen., Dept. Energy, Symposium Series 52 (CONF-781101).

Elmgren, R., Hasson, S., Larsson, U., Sundelin, B., & Boehm, P.D. (1983) The Tsesis oil spill: Acute and long-term impact on the benthos. *Marine Biology*, Vol. 73, pp. 51–65.

Ernst, S.R., Morvan, J., Geslin, E., Le Bihan, A., & Jorissen, F.J. (2006). Benthic foraminiferal response to experimentally induced Erika oil pollution. *Marine Micropaleontology*, Vol. 61, pp. 76–93.

Fraschetti, S., Gambi, C., Giangrande, A., Musco, L., Terlizzi, A., & Danovaro, R. (2006). Structural and functional response of meiofauna rocky assemblages to sewage pollution. *Marine Pollution Bulletin*, Vol. 52, pp. 540–548.

Fricke, A.H., Henning, H.F.H.O., & Orren, M.J. (1981). Relationship between pollution and psammolittoral meiofauna density of two South African beaches. *Marine Environmental Research*, Vol. 5, pp. 57–77.

Frontalini, F., & Coccioni, R. (2008). Benthic foraminifera for heavy metal pollution monitoring: a case study from the central Adriatic Sea coast of Italy. *Estuarine, Coastal and Shelf Science*, Vol. 76, pp. 404–417.

Frontalini, F., & Coccioni, R. (2011). Benthic foraminifera as bioindicators of pollution: A review of Italian research over the last three decades. *Revue de Micropaléontologie*, Vol. 54, pp. 115-127.

Frontalini, F., Coccioni, R., & Bucci, C. (2010). Benthic foraminiferal assemblages and trace element contents from the lagoons of Orbetello and Lesina. *Environmental Monitoring and Assessment*, Vol. 170, 245-260.

Frontalini, F., Buosi, C., Da Pelo, S., Coccioni, R., Cherchi, A., & Bucci, C. (2009). Benthic foraminifera as bio-indicators of trace element pollution in the heavily contaminated Santa Gilla lagoon (Cagliari, Italy). *Marine Pollution Bulletin*, Vol. 58, pp. 858-877.

Gee, J.M., Austen, M., De Smet, G., Ferraro. T., McEvoy, A., Moore, S., Van Gausbeki, D., Vincx, M., & Warwick, R.M. (1992). Soft sediment meiofauna community responses to environmental pollution gradients in the German Bight and at a drilling site off the Dutch coast. *Marine Ecology Progress Series*, Vol. 91, pp. 289–302.

Giere, O. (1979). The impact of oil pollution on intertidal meiofauna. Field studies after the La Coruna spill, May, 1976. *Cahiers de Biologie Marine*, Vol. 20, pp. 231–251.

Giere, O. (2009). *Meiobenthology: the microscopic motile fauna of aquatic sediments*. 2nd edition. Springer, Berlin Heidelberg.

Guo, Y., Somerfield, P.J., Warwick, R.M., & Zhang, Z. (2001). Large-scale patterns in the community structure and biodiversity of freeliving nematodes in the Bohai Sea, China. *Journal of Marine Biology Association of United Kingdom*, Vol. 81, pp. 755-763.

Gustafson, M., Ingela, D.F., Blanck, H., Hall, P., Molander, S., & Nordberg, K. (2000). Benthic foraminiferal tolerance to Tri-n-Butyltin (TBT) pollution in an experimental mesocosm. *Marine Pollution Bulletin*, Vol. 40, pp. 1072–1075.

Gyedu-Ababio, T.K., Furstenberg, J.P., Baird, D., & Vanreusel, A. (1999) Nematodes as indicators of pollution: a case study from the Swartkops river system, South Africa. *Hydrobiologia*, Vol. 397, pp. 155–169.

Hechtel, I.G., Ernst, E.J., & Kalin, R. (1970). Biological effects of thermal pollution, Northport, New York. State University of New York, Marine Science Research Contribution, Report Series, Vol. 3, pp. 83–85.

Hedfi, A., Mahmoudi, E., Boufahja, F., Beyrem, H., & Aïssa, P. (2007). Effects of increasing levels of nickel contamination on structure of offshore nematode communities in experimental microcosms. *Bulletin of Environmental Contaminant and Toxicology,* Vol. 79, pp. 345–349.

Hedfi, A., Mahmoudi, E., Beyrem, H., Boufahja, F., Essid, N. & Aïssa, P. (2008). Réponse d'une communauté de nematodes libres marins à une contamination par le cuivre: étude microcosmique. *Bulletin de la Société zoologique de France,* Vol. 133, pp. 97-106.

Hermi, M., Mahmoudi, E., Beyrem, H., Aïssa, P., & Essid N. (2009) Responses of a free-living marine nematode community to mercury contamination: results from microcosm experiments. *Archives of Environmental Contamination and Toxicology,* Vol. 56, pp. 426–433.

Heip, C., Herman, R., & Vincx, M. (1984). Variability and productivity of meiobenthos in the Southern Bight of the North Sea. *Rapport et process-verbaux des reunions. Conseil international pour l'Exploration de la Mer,* Vol. 183, pp. 507-521.

Heip, C., Vincx, M. & Vranken, G. (1985). The ecology of marine nematodes. *Oceanography and Marine Biology An Annual Review,* Vol. 23, pp. 399-489.

Howell, R. (1982). Levels of heavy metal pollutants in two species of marine nematodes. *Marine Pollution Bulletin,* Vol. 13, pp. 396–398.

Howell, R. (1983). Heavy metals in marine nematodes: uptake, tissue distribution and loss of copper and zinc. *Marine Pollution Bulletin,* Vol. 14, pp. 263–268.

Jayaraju, N., Raja Reddy, B.C.S., & Reddy, K.R. (2008). The response of benthic foraminifera to various pollution sources: a study from Nellore coast, East Coast of India. *Environmental Monitoring and Assessment,* Vol. 142, pp. 319–323.

Kennedy, A. D., & Jacoby, C. A. (1999). Biological indicators of marine environmental health: meiofauna - a neglected benthic component? *Environmental Monitoring and Assessment*, Vol. 54, pp. 47-68.

Kennish, M.J. (1992). *Ecology of Estuaries: Anthropogenic Effects.* CRC Press Inc., Boca Raton, FL.

La Rosa, T., Mirto, S., Mazzola, A., & Danovaro, R. (2001). Differential responses of benthic microbes and meiofauna to fish-farm disturbance in coastal sediments. *Environmental Pollution,* Vol. 112, pp. 427–434.

Le Cadre, V., & Debenay, J.P. (2006). Morphological and cytological responses of *Ammonia* (foraminifera) to copper contamination: implication for the use of foraminifera as bioindicators of pollution. *Environmental Pollution,* Vol. 143, pp. 304–317.

Lockin, J.A., & Maddocks. R.F. (1982). Recent foraminifera around petroleum production platforms on the southwest Louisiana Shelf. *Gulf Coast Association of Geological Societies Transactions,* Vol. 32, pp. 377–97.

Mahmoudi, E., Beyrem, H., Baccar L., & Aïssa, P. (2002). Response of free-living Nematodes to the quality of water and sediment at Bou Chrara Lagoon (Tunisia) during winter 2000. *Mediterranean Marine Science,* Vol. 3/2, pp. 133-146.

Mahmoudi, E., Essid, N., Beyrem, H., Hedfi, A., Boufahja, F., Vitiello, P. & Aïssa, P. (2005). Effects of hydrocarbon contamination on a free living marine nematode community: results from microcosm experiments. *Marine Pollution Bulletin*, Vol. 50, pp. 1197–1204.

Mahmoudi, E., Essid, E., Beyrem, H., Hedfi, A., Boufahja, F., Vitiello, P., & Aïssa, P. (2007). Individual and combined effects of lead and zinc of a free living marine nematode community: results from microcosm experiments. *Journal of Experimental Marine Biology and Ecology*, vol. 343, pp. 217–226.

Mahmoudi, E., Essid, N., Beyrem, H., Hedfi, A., Boufahja, F., Aissa, P., & Vitiello, P. (2008). Mussel-farming effects on Mediterranean benthic nematode communities. *Nematology*, Vol. 10, pp. 323-333.

Mayer, E.M. (1980). *Foraminifera of the Caspian and Aral Seas*, Unpubl. Ph.D. thesis, Moscow University.

Mazzola, A., Mirto, S., La Rosa, T., Fabiano, M., & Danovaro, R. (2000). Fish-farming effects on benthic community structure in coastal sediments: analysis of meiofaunal recovery. *ICES Journal of Marine Science*, Vol. 57, pp. 1454–1461.

Millward, R.N., & Grant, A. (1995). Assessing the impact of copper on nematode communities from a chronically metal enriched estuary using pollution-induced community tolerance. *Marine Pollution Bulletin*, Vol. 30, 701–706.

Mirto, S., La Rosa, T., Danovaro, R., & Mazzola, A. (2000). Microbial and meiofaunal response to intensive mussel-farm biodeposition in coastal sediments of the western Mediterranean. *Marine Pollution Bulletin*, Vol. 40, pp. 244–252.

Mirto, S., La Rosa, T., Gambi, C., Danovaro, R., & Mazzola, A. (2002). Nematode community response to fish-farm impact in the western Mediterranean. *Environmental Pollution*, Vol. 116, pp. 203-214.

Mojtahid, M., Jorissen, F., Durrieu, J., Galgani, F., Howa, H., Redois, F., & Camps, R. (2006). Benthic foraminifera as bio-indicators of drill cutting disposal in tropical east Atlantic outer shelf environments. *Marine Micropaleontology*, Vol. 61, pp. 58-75.

Moreno, M., Ferrero, T.J., Gallizia, I., Vezzulli, L., Albertelli, G., & Fabiano, M. (2008). An assessment of the spatial heterogeneity of environmental disturbance within an enclosed harbour through the analysis of meiofauna and nematode assemblages. *Estuarine Coastal Shelf Science*, Vol. 77, pp. 565–576.

Moreno, M., Albertelli, G., & Fabiano, M. (2009). Nematode response to metal, PAHs and organic enrichment in tourist marinas of the mediterranean sea. *Marine Pollution Bulletin*, Vol. 58, pp. 1192-1201.

Moreno, M., Semprucci, F., Vezzulli, L., Balsamo, M., Fabiano, M. & Albertelli, G. (2011). The use of nematodes in assessing ecological quality status in the Mediterranean coastal ecosystems. *Ecological Indicators*, Vol. 11, pp. 328-336.

Morvan, J., Le Cadre, V., Jorissen, F., & Debenay, J.P. (2004). Foraminifera as potentiql bio-indicators of the 'Erika' oil spill in the Bay of Bourgneuf: fuel and experiments studies. *Aquatic Living Resources*, Vol. 17, pp. 217-322.

Murray, J.W. (1985). Recent foraminifera from the North Sea (Forties and Ekofisk areas) and the continental shelf west of Scotland. *Journal of Micropaleontology*, Vol. 4, pp. 117–25.

Murray, J.W. (2007). Biodiversity of living benthic foraminifera: how many species are there? *Marine Micropaleontology*, Vol. 64, pp. 163-176.

Murray, J.W., & Alve, E. (2002). Benthic foraminifera as indicators of environmental change: marginal-marine, shelf and upper-slope environments. In: *Quaternary Environmental Micropalaeontology*, Haslett, S.K. (Ed.), pp. 59–90, Edward Arnold (Publishers) Limited, London.

Netto, S.A., & Valgas, I., (2010). The response of nematode assemblages to intensive mussel farming in coastal sediments (Southern Brazil). *Environmental Monitoring and Assessment*, Vol. 162, pp. 81–93.

Nicholas, W.L., Goodchild, D.J., & Steward, A. (1987). The mineral composition of intracellular inclusions in nematodes from thiobiotic mangrove mud-flats. *Nematologica*, Vol. 33, pp. 167-179.

Nigam, R., Saraswat, R., & Panchang, R. (2006). Application of foraminifers in ecotoxicology: retrospect, perspect and prospect. *Environmental International*, Vol. 32, pp. 273–283.

Nigam, R., Linshy, V.N., Kurtarkar, S.R., & Saraswat, R. (2009). Effects of sudden stress due to heavy metal mercury on benthic foraminifer *Rosalina leei*: laboratory culture experiment. *Marine Pollution Bulletin*, Vol. 59, pp. 362–368

Platt, H. M. & Warwick, R. M. (1980). The significance of free-living nematodes to the littoral ecosystem. In: *The Shore Environment: Ecosystems*, J.H. Price, D.E.G. Irvine & Farnham W.F. (Eds.), pp. 729-759, Academic Press, New York.

Reish, D.J. (1983). Survey of the marine benthic infauna collected from the United States radioactive waste disposal sites off the Farallon Islands, California, Report EPA-520/1-83-006 (EPA520183006), 65 pp.

Renaud-Mornant, J., Gourbault, N., de Panafiew, J.B., & Heleouet, M.N. (1981). Effets de la pollution par hydrocarbures sur la meiofauna de la baie de Morlaix. In: *Amoco Cadiz, Consequences d'une Pollution Accidentale par les Hydrocarbures. Actes coli. Intern. C.O.B.*, Brest (France), 19–22 November 1979, pp. 551–561.

Resig, J.M. (1960). Foraminiferal ecology around ocean outfalls off southern California. In: *Disposal in the Marine Environment*. Person, E. (Ed.), pp. 104–121, Pergamon Press, London.

Sabean, J.A.R., Scott, D.B., Lee, K., & Venosa, A.D. (2009). Monitoring oil spill bioremediation using marsh foraminifera as indicators. *Marine Pollution Bulletin*, Vol. 59, pp. 352–361.

Sandulli, R., & De Nicola-Giudici, M. (1990). Pollution effects on the structure of meiofaunal communities in the Bay of Naples. *Marine Pollution Bulletin*, Vol. 21, pp. 144–153.

Sandulli, R., & De Nicola-Giudici, M. (1991) Responses of meiobenthic communities along a gradient of sewage pollution. *Marine Pollution Bulletin*,Vol. 22, pp. 463–467.

Saraswat, R., Kurtarkar, S.R., Mazumder, A., & Nigam, R. (2004). Foraminifers as indicators of marine pollution: a culture experiment with *Rosalina leei*. *Marine Pollution Bulletin*, Vol. 48, pp. 91–96.

Schafer, C.T., Winters, G.M., Scott, D.B., Pocklington, P., & Honig, C. (1995). Survey of living Foraminifera and polychaete populations at some Canadian aquaculture sites: Potential for impact mapping and monitoring. *Journal of Foraminiferal Research*, Vol. 25, pp. 236–59.

Schratzberger, M., Rees, H. L. & Boyd, S. E. (2000). Effects of the simulated deposition of dredged material on the structure of nematode assemblages—the role of burial. *Marine Biology*, Vol. 136, pp. 519–530.

Schratzberger, M., Bolam S., Whomersley P., & Warr K. (2006). Differential response of nematode colonist communities to the intertidal placement of dredged material. *Journal of Experimental Marine Biology and Ecology*, Vol. 334, pp. 244–255

Schratzberger, M., Daniel F., Wall C.M., Kilbride R., Macnaughton S. J., Boyd S. E., Rees H. L., Lee K., & Swannell R. P.J. (2003). Response of estuarine meio- and macrofauna to in situ bioremediation of oil-contaminated sediment. *Marine Pollution Bulletin*, Vol. 46, pp. 430–443.

Schratzberger, M., Forster, R.M., Goodsir, F., & Jennings, S. (2008). Nematode community dynamics over an annual production cycle in the central North Sea. *Marine Environmental Research*, Vol. 66, pp. 508-519.

Schratzberger M., & Jennings S. (2002). Impacts of chronic trawling disturbance on meiofaunal communities. *Marine Biology* Vol. 141, pp. 991–1000.

Schratzberger, M., & Warwick, R.M. (1998). Effects of the intensity and frequency of organic enrichment on two estuarine nematode communities. *Marine Ecology Progress Series*, Vol. 164, pp. 83-94.

Schratzberger, M., Lampadariou, N., Somerfield, P. J., Vandepitte, L. & Vanden Berghe, E. (2009). The impact of seabed disturbance on nematode communities: linking field and laboratory observations. *Marine Biology*, Vol. 156, pp. 709-724.

Scott, D.B., Schafer, C.T., & Honig, C. (1995). Temporal variations of benthonic foraminiferal assemblages under and near aquaculture operations: historical documentation of possible environmental impacts. *Journal of Foraminiferal Research*, Vol. 25, pp. 224-235.

Seiglie, G.A. (1971). A preliminary note on the relationship between foraminifers and pollution in two Puerto Rican bays. *Caribbean Journal of Science*, Vol. 11, pp. 93–8.

Semprucci, F., Boi, P., Manti, A., Covazzi-Harriague, A., Rocchi, M., Colantoni, P., Papa, S. & Balsamo, M. (2010b). Benthic communities along a littoral of the Central Adriatic Sea (Italy). *Helgoland Marine Research*, Vol. 64, pp. 101–115.

Semprucci, F., Colantoni, P., Baldelli, G., Rocchi, M., Balsamo, M. (2010a). The distribution of meiofauna on back-reef sandy platforms in the Maldives (Indian Ocean). *Marine Ecology: An Evolutionary Perspective*, Vol. 31, pp. 592-607.

Sen Gupta, B.K. (1999). *Modern Foraminifera*. Kluwer Academic Publisher, Dordrecht.

Somerfield, P.J., Fonseca-Genevois V.G., Rodrigues A.C.L., Castro F.J.V., & Santos G.A.P. (2003). Factors affecting meiofaunal community structure in the Pina Basin, an urbanized embayment on the coast of Pernambuco, Brazil. *Journal of Marine Biology Association of United Kingdom*, Vol. 83, pp. 1209-1213.

Somerfield, P.J., Gee, J.M., & Warwick, R.M. (1994). Soft sediment meiofaunal community structure in relation to a long-term heavy metal gradient in the Fal estuary system. *Marine Ecology Progress Series*, Vol. 105, pp. 79–88.

Somerfield, P. J., Rees, H. L., & Warwick, R. M. (1995). Interrelationships in community structure between shallow-water marine meiofauna and macrofauna in relation to dredgings disposal. *Marine Ecology Progress Series*, Vol. 127, pp. 103–112.

Tarasova, T.S., &. Preobrazhenskaya T.V. (2007). Benthic foraminifera at a scallop mariculture site in Minonosok Bay, the Sea of Japan. *Biologiya Morya*, Vol. 33, pp. 25-36.

Tietjen, J.H. (1980). Population structure and species distribution of the free-living nematodes inhabiting sands of the New York Bight apex. *Estuarine Coastal Shelf Science*, Vol. 10, pp. 61-73.

Tsujimoto, A., Nomura, R., Yasuhara, M., Yamazaki, H., & Yoshikawa, S. (2006a). Impact of eutrophication on shallow marine benthic foraminifers over the last 150 years in Osaka Bay, Japan. *Marine Micropaleontology*, Vol. 60, pp. 258–268.

Tsujimoto, A., Nomura, R., Yasuhara, M., & Yoshikawa, S. (2006b). Benthic foraminiferal assemblages in Osaka Bay, southwestern Japan: faunal changes over the last 50 years. *Paleontological Research*, Vol. 10, pp. 141–161.

van der Zwaan, G., J., (2000). Variation in natural vs. anthropogenic eutrophication of shelf areas in front of major rivers. In: *Environmental Micropaleontology: the Application of Microfossils to Environmental Geology*, Martin, R.E. (Ed.), pp. 385-404, Kluwer Academic/Plenum Publishers, New York.

Vanaverbeke, J., Merckx, B., Degraer, S., & Vincx, M. (2011). Sediment-related distribution patterns of nematodes and macrofauna: two sides of the benthic coin? *Marine Environmental Research*, Vol. 71, pp. 31-40.

Vénec-Peyré, M.T. (1981). Les Foraminifères et la pollution: etude de la microfaune de la Cale du Dourduff (Embochure de la Riviere de Morlaix). *Cahiers de Biologie Marine*, Vol. 22, pp. 25–33.

Vezzulli, L., Moreno, M., Marin, V., Pezzati, E., Bartoli, M., & Fabiano, M., 2008. Organic waste impact of capture- based Atlantic bluefin tuna aquaculture at an exposed site in the Mediterranean Sea. *Estuarine Coastal Shelf Science*, Vol. 78, pp. 369-384.

Vidović, J. Ćosović, V., Juračić, M., & Petricioli, D. (2009). Impact of fish farming on foraminiferal community, Drvenik Veliki Island, Adriatic Sea, Croatia. *Marine Pollution Bulletin*, Vol. 58, pp. 1297–1309.

Warwick, R.M., & Gee, J.M. (1984). Community structure of estuarine meiobenthos. *Marine Ecology Progress Series*, Vol. 18, pp. 97-111.

Warwick, R.M., & Price, R. (1979). Ecological and metabolic studies on free-living marine nematodes from an estuarine mud-flat. *Estuarine, Coastal and Marine Science*, Vol. 9, pp. 257-271.

Watkins, J.G. (1961). Foraminiferal ecology around the Orange County, California, ocean sewer outfall. *Micropaleontology*, Vol. 7, pp. 199–206.

Whitcomb, N.J. (1978). Effects of oil pollution upon selected species of benthic foraminiferids from the lower York River, Virginia. *Geological Society of America Annual Meeting*, Abstracts with Programs, p. A 515.

Yanko, V., & Flexer, A. (1991). Foraminiferal benthonic assemblages as indicators of pollution (an example of Northwestern Shelf of the Black Sea). *Proceedings of the Third Annual Symposium on the Mediterranean Margin of Israel*, Haifa, Israel. Abstracts Volume, pp. 5.

Yanko, V., Kronfeld, J., & Flexer, A. (1994). Response of benthic foraminifera to various pollution sources: implications for pollution monitoring. *Journal of Foraminiferal Research*, Vol. 24, pp. 1–17.

Yanko, V., Ahmad, M., & Kaminski, M. (1998). Morphological deformities of benthic foraminiferal tests in response to pollution by heavy metals: implications for pollution monitoring. *Journal of Foraminiferal Research*, Vol. 28, pp. 177–200.

Yanko, V., Arnold, A.J., & Parker, W.C. (1999). Effects of marine pollution on benthic Foraminifera. In: *Modern Foraminifera*, Sen Gupta, B.K. (Ed.), pp. 217-235, Kluwer Academic Publisher, Dordrecht.

5

Chemical Interactions in Antarctic Marine Benthic Ecosystems

Blanca Figuerola, Laura Núñez-Pons, Jennifer Vázquez, Sergi Taboada, Javier Cristobo, Manuel Ballesteros and Conxita Avila
University of Barcelona, Spanish Institute of Oceanography, Spain

1. Introduction

Antarctic marine ecosystems are immersed in an isolated, relatively constant environment where the organisms inhabiting their benthos are mainly sessile suspension feeders. For these reasons, physical and chemical biotic interactions play an essential role in structuring these marine benthic communities (Dayton et al., 1974; Orejas et al., 2000). These interactions may include diverse strategies to avoid predation (e.g. Iken et al., 2002), competition for space or food (e.g. Bowden et al., 2006) and avoiding fouling (e.g. Rittschof, 2001; Peters et al., 2010). For instance, in the marine benthos, one of the most extended effective strategies among sessile soft-bodied organisms is chemical defense, mediated by several bioactive natural products mostly considered secondary metabolites (e.g. Paul et al., 2011). The study of the "chemical network" (chemical ecology interactions) structuring the communities provides information about the ecology and biology of the involved species, the function and the structure of the community and, simultaneously, it may lead to the discovery of new compounds useful to humans for their pharmacological potential (e.g. Avila, 1995; Bhakuni, 1998; Munro et al., 1999; Faulkner, 2000; Lebar et al., 2007; Avila et al., 2008). In the last three decades, the study of marine chemical ecology has experienced great progress, thanks to the new technological advances for collecting and studying marine samples, and the possibility of identification of molecules with smaller amounts of compounds (e.g. Paul et al., 2006, 2011; Blunt et al., 2011).

Polar organisms have been less studied, compared with their temperate and tropical counterparts (Paul, 1992; Blunt et al., 2009). However, recent studies report that Antarctic benthic invertebrates are a rich and diverse source of natural products, with great interest from both the ecological and the pharmacological point of view (e.g. Avila et al., 2000, 2008; Amsler et al., 2001; Iken et al., 2002; Lebar et al., 2007; Reyes et al., 2008; Taboada et al., 2010; Paul et al., 2011). Moreover, several researches have demonstrated that some Antarctic species inhabiting shallow areas from McMurdo Sound and the Antarctic Peninsula possess chemical defenses (for review see Avila et al., 2008; McClintock et al., 2010), even if only in few cases the chemistry of the metabolites involved has been fully described and/or their ecological role has been established (e.g. Núñez-Pons et al., 2010; Núñez-Pons et al., in prep).

In the last years our research group has been studying the ecological activity of marine natural products obtained from Antarctic benthic organisms by using *in situ* experiments. Furthermore, as part of our investigations, previously unknown species for science have been described (Ballesteros & Avila, 2006; Ríos & Cristobo, 2006; Figuerola et al., in press), and new compounds have been isolated and described too (e.g. Antonov et al., 2008, 2009, 2011; Reyes et al., 2008; Carbone et al., 2009; Carbone et al., in prep). Also, we have extended the range of species from our previous analysis by studying Antarctic macroalgae, which are known to be prolific producers of secondary metabolites with pharmaceutical applications (e.g. Hoyer et al., 2002; Ankisetty et al., 2004). As a general objective our aim here is to integrate all the experimental data obtained from the assays conducted with different taxonomical groups in order to establish a preliminary ecological model of the chemically-mediated interactions in the Antarctic benthos. This model will, for the first time, consider the mechanisms that regulate the chemical interactions among the different Antarctic benthic organisms studied. Our specific objectives are trying to determine the a) feeding-deterrence activities towards sympatric predators, including a macropredator (*Odontaster validus* Koehler) and a mesograzer, *Cheirimedon femoratus* Pfeffer, b) toxicity potential against a copepod, *Metridia gerlachei* Giesbrecht, c) cytotoxicity against embryos and sperm of the Antarctic sea urchin *Sterechinus neumayeri* Meissner and d) antifouling activity against microbial biofilms.

2. Material and methods

2.1 Samples collection and identification

Marine benthic invertebrates and algal samples were collected in the Southern Ocean in four Antarctic campaigns: two in the Eastern Weddell Sea (Antarctica) and vicinities of Bouvet island (Sub-Antarctica) on board the R/V Polarstern, from the Alfred Wegener Institute for Polar and Marine Research (AWI Bremenhaven, Germany) during the ANT XV/3 (January-March 1998) and ANT XXI/2 cruises (November 2003-January 2004); a third one on board the BIO Hespérides during the ECOQUIM-2 cruise (January 2006) around the South Shetland Islands; and finally, the ACTIQUIM-1 cruise at Deception Island mainly by scuba-diving, although other sampling methods were used as well (December 2008-January 2009). Sample collection took place between 0 m and 1524 m depth by using various trawling devices: bottom trawl, Agassiz trawl, Rauschert dredge and epibenthic sledge, and also, as said, by scuba diving (0-15m). Samples were sorted and photographed on deck, frozen at -20 °C, and a voucher portion of each sample or, in some cases, whole individuals, were fixed in 10% formalin or 70% ethanol and stored at the Dept. of Animal Biology (Invertebrates), University of Barcelona (Spain), for taxonomical identification.

Individuals of the sea star *Odontaster validus*, the sea urchin *Sterechinus neumayeri*, the amphipod *Cheirimedon femoratus* and the copepod *Metridia gerlachei* were collected for *in situ* ecological experiments in Deception Island by scuba diving at Port Foster Bay (Deception Island: 62° 59,369' S, 60° 33,424' W) from 0-15 m depth (December 2008 - January 2009 and January 2010). After experimentation, these invertebrates were brought back alive to the sea.

2.2 Chemical extractions

Chemical extractions were done in the laboratories from the Faculty of Biology (University of Barcelona). Frozen animals were carefully dissected into different sections when possible, in order to locate the compounds within the body of the organisms (although this is not discussed here). The different sections were made according to the taxonomic group (e.g. internal/external, apical/basal parts in sponges, echinoderms and tunicates; poliparium/axis in cnidarian octocorals; mantle/foot in opistobranch molluscs; gill slits in ascidians; tentacles in holoturian echinoderms…). These body sections were extracted separately, and thus the total number of extracts is larger than the total number of species tested. Samples were extracted with acetone, and sequentially partitioned into diethyl ether and butanol fractions. All steps were repeated three times, except for the butanol which was done once. Organic solvents were then evaporated under reduced pressure, resulting in dry diethyl ether and butanolic extracts, and an aqueous residue. An aliquot of all the diethyl ether extracts (lipophilic fraction) was used for the bioassays at different concentrations for the different experiments. The detailed description of the extraction procedure has been reported elsewhere (Avila et al., 2000; Iken et al., 2002). Butanolic extracts and water residues were kept aside for future investigations.

2.3 Experiments of chemical ecology and statistical treatment

All experiments of chemical ecology took place in the Spanish Antarctic Base "Gabriel de Castilla" in Deception Island (South Shetland Islands, Antarctica) during the Austral Summers of 2008-2009 and 2009-2010.

2.3.1 Feeding experiments with a macropredator, the seastar *Odontaster validus*, and a mesograzer, the amphipod *Cheirimedon femoratus*

The omnivorous sea star *O. validus* occupies the top predator position that fish occupy in temperate and tropical areas (McClintock, 1994). For this reason, this ubiquitous sea star is used as putative macropredator in feeding-deterrence experiments to test the presence of chemical defenses in selected marine invertebrates and algae (e.g. Avila et al., 2000, Iken et al., 2002). The amphipod *Cheirimedon femoratus* was chosen as mesograzer consumer in feeding-preference assays because this voracious, omnivorous-scavanger crustacean is found in notably high densities in Antarctica exerting remarkable, localized ecological pressures, often underestimated (Huang et al., 2007).

The sea star experiments were carried out over 24 h. Extracts, fractions and/or isolated compounds were dissolved in the solvent carrier (diethyl ether) and slowly pipetted at their natural dry weight concentration (mg extract g^{-1} dry wt tissue) onto shrimp pieces, and the solvent was left to totally evaporate under the hood, resulting in a uniform coating of extract. Normalization of natural concentrations based on biomass using wet or dry weight are appropriate when ingredients are homogeneously distributed, and also when using biting and not-biting predators. Moreover, dry weight has been proven to be the most constant parameter for avoiding the variability caused by the water content. Control shrimp pieces were treated with solvent only. Feeding-deterrence experiments are described in detail in precedent investigations (e.g. Avila et al. 2000). The bioassays consisted on 10 replicates in which the sea stars were individually transferred into 2.5 l-buckets filled with fresh seawater (1±0.5°C), and they were offered a treatment or a control diet, respectively, by putting a shrimp piece in the centre of the bucket and the asteroid on top. A food item was considered rejected when *Odontaster validus* lost physical contact with it, and it was considered eaten when the food was ingested completely after the testing period (Fig. 1). Afterwards, eaten and uneaten shrimp pieces were counted for statistical analysis. Feeding repellence was evaluated as a contingency table 2x2, and since the number of replicates was small (n=10) by using Fisher's Exact tests for each experiment using extract-treated shrimp pieces referred to the control run simultaneously (Sokal & Rohlf, 1995).

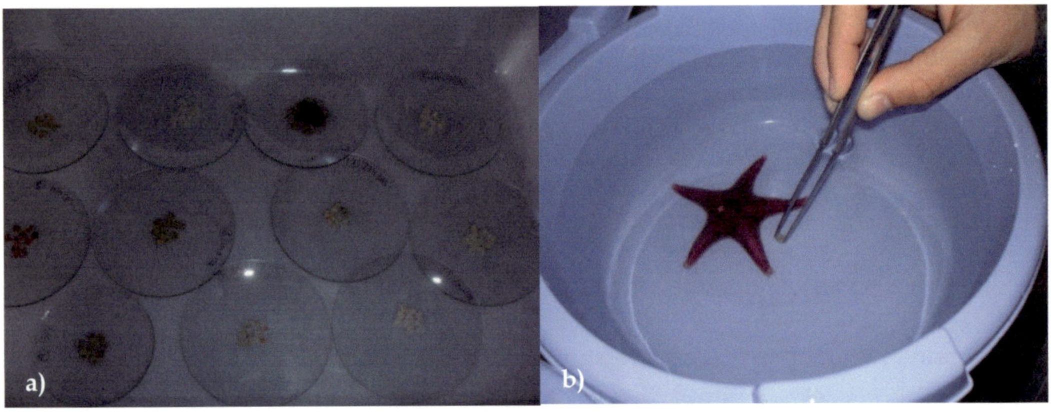

Fig. 1. *Odontaster validus* feeding-repellence experiments. a) Shrimp pieces being prepared with extract coatings for the tests; b) A sea star being offered a shrimp piece.

The generalist amphipod *Cheirimedon femoratus* was used as a potential mesograzer predator. It was presented to a simultaneous choice of two different food types, consisting of a control (extract-free) diet (which the predator readily consumed) and a treatment diet, where the extracts were included at natural concentration according to a dry weight basis (see above). Both diets consisted on alginate-based artificial foods containing a powdered commercial aquarium diet as a food attractant. Control food was prepared with only solvent, which was left to evaporate onto the food powder prior to being gelified into food pellets. For treatment diets, extracts were added into the food mixture dissolved in the carrier solvent (which was similarly evaporated). Groups of 15-20 amphipods were transferred into 1L-bottles filled with sea water, and were presented to a choice of extract-free control and extract-treated diets. The assays ran until either food type had been consumed up to one-half or more. At the end of the experiment, the consumed food was calculated for statistical analysis and determination of feeding preferences of extract-treated foods from the paired simultaneous controls to consequently establish repellent activities. The two food types were presented together, and therefore we measured separately for each replicate container and each food type the quantity of ingested food, and calculated the differences for each experimental unit (replicate). The changes in the two food types held in the same container are not independent and possess correlated errors, making it impossible to analyze them separately. Each replicate is represented by a paired result yielding two sets of data (treatments and controls), which can be compared, since assumption of normality and homogeneity of variances are not met, by non-parametric procedures, that is by applying the Exact Wilcoxon test, which was calculated using R-command software.

2.3.2 Toxicity activity against the copepod *Metridia gerlachei*

Metridia gerlachei is a common omnivorous copepod frequently found in the waters of Port Foster (King & LaCasella, 2003). For this experiments, we used plates with 2 ml seawater where 10-15 copepods were placed. Each experiment consisted of 5 replicates with the ethereal extract to be tested at natural concentration, 5 negative control assays (only filtered sea water), and 5 solvent assays (filtered seawater with solvent). During experimentation, copepods were observed over time for survival. Extracts were considered toxic when, considering the 5 replicates for each test, >50% of the copepods died.

2.3.3 Cytotoxicity activity against embryo and sperm in the Antarctic sea urchin *Sterechinus neumayeri*

Sessile organisms may prevent the settlement of sympatric organisms by displaying cytotoxic activities that may act against embryos and larvae of other invertebrates, in their attempt to colonize the surface of sessile invertebrates, such as sponges, ascidians, bryozoans and polychaetes (e.g. Heine et al., 1991; McClintock et al., 1990). In the Antarctic marine benthic environment, *Sterechinus neumayeri* is one of the most abundant and common species of sea urchin, and its biology is well known (*e.g.* Bosch et al., 1987; Brey et al., 1995). For these reasons, this species was chosen for our bioassays.

After acclimatization, sea urchins were induced to spawn by injecting 1ml of 0.5 M KCl solution into the coelomic cavity through the peristome. The cytotoxicity test was developed according to the protocol proposed by Volpi Ghirardini and collaborators (2005) for the Mediterranean sea urchin *Paracentrotus lividus*. Some modifications were introduced in the original procedure, mainly focused on the volume of sea water used and the time that embryos were exposure to extracts, in order to adapt it to the characteristics of *S. neumayeri*. Details of this modified method are described in a paper that is being prepared (Figuerola et al., in prep) (Fig. 2). The percentage of blastula stage in each treatment was determined for statistical analysis. A S regression model ($Y = \exp(0.702 + 124{,}928/X)$, $R^2 = 0.6125$) was calculated between % of the number of the blastula (Y) and the initial concentration of eggs

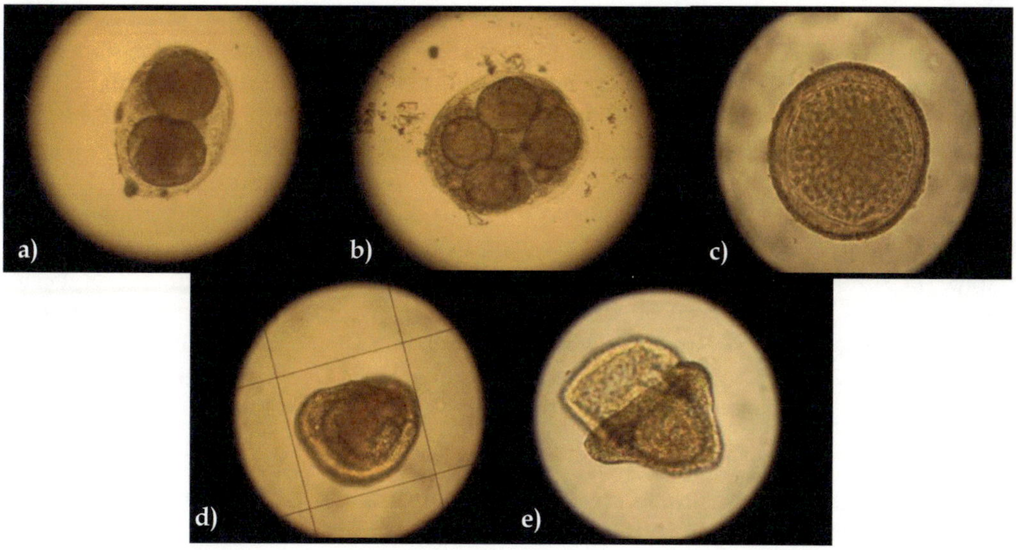

Fig. 2. Different stages of embryonic development of the sea urchin *Sterechinus neumayeri*. a) Stage of 2 cells; b) Stage of 4 cells; c) Morula stage; d) Dipleurula larva; e) Pluteus larva.

(X) only using the data from the control group and solvent to obtain a representation of the normal behavior (without the influence of the extracts) of the experimental conditions. Also, confidence intervals (CI) of prediction (upper and lower: UCL and LCL) of 95% coverage were calculated to detect extract samples outside CI.

The sperm test was developed following the procedure of similar experiments conducted in the past using *S. neumayeri* (Heine et al., 1991; McClintock et al., 1992). Previous sperm assays utilized 25-ml volumes of test solutions in 25 x 150 mm glass test tubes. The refined method used here was based on 0.25 ml test volumes. The use of smaller wells also allowed an increase in the number of samples to be tested at one time. *S. neumayeri* sperm was obtained as described above for the cytotoxicity tests. Every day that a sperm toxicity test was conducted, a blank control (sperm with filtered sea water) and a positive control (sperm in filtered sea water with ethereal extract) were run simultaneously. Ten replicates using extracts at different concentrations were tested for each of the samples. Sperm mobility was checked under a light microscope (40x) 20 min after the sperm solution was added to each well. Extracts were considered toxic when <25% of the sperm was active. Moreover, a binomial test of proportions ($p<0{,}001$) was calculated in order to evaluate the effects of different concentrations of extracts.

2.3.4 Antifouling activity

Different marine organisms such as bacteria, algae and invertebrates colonize surfaces underwater. However, many sessile marine invertebrates possess chemical or physical defenses to prevent the settlement of epibionts (e.g. Kelly et al., 2003; Sivaperumal et al., 2010). The presence of different kinds of compounds may influence the growth of other species which could settle near or over marine invertebrates. We evaluated the antibacterial activity of different extracts using the methods described in the literature with Antarctic bacteria (e.g. Jayatilake et al., 1996; De Marino et al., 1997, Mahon et al., 2003). Selected bacteria from the sea water were collected during the campaign, cultured on marine agar Difco brand (DMA 2216), and later sent to specialists for further identification. Filter paper discs impregnated with 20 µL of solution were placed on the surface of inoculated plates. Each test consisted in one disc without any additive (negative control), one disc with chloramphenicol (positive control), one disc impregnated with the solvent (diethyl ether, negative control) and one disc impregnated with the extract at natural concentration. Each culture of microorganisms was inoculated for triplicate on the surface of marine agar with the paper discs. Diffusion methods were based on the homogeneous distribution of the extract on solid culture media. The amount of the extract, as the number of bacteria (inoculum), was carefully controlled. After incubation, we measured the diameters of the inhibition halos and the results were interpreted using cut points as established internationally. Zones of growth inhibition larger than 2 mm were considered active.

3. Results

3.1 Feeding experiments with a macropredator, the seastar *Odontaster validus*, and a mesograzer, the amphipod *Cheirimedon femoratus*

In feeding-deterrence experiments using the seastar *O. validus*, 160 extracts (139 species) were tested belonging to different Phyla: Porifera (43 species), Cnidaria (17), Tunicata (15),

Bryozoa (17), Echinodermata (5), Annelida (7), Algae (8), and other groups (11). A total of 76 deterrent extracts (66 species) were found, revealing significant differences in food consumption between simultaneous control and treatment tests ($p<0.05$), with control as the preferred food (Fisher's exact test). The deterrent extracts were thus 48,2% of the tested species, from organisms belonging to the taxa Porifera (22), Cnidaria (10), Tunicata (11), Bryozoa (10), Echinodermata (3), Annelida (4) and Algae (1), and others (5) (Avila et al. in prep) (Fig. 3).

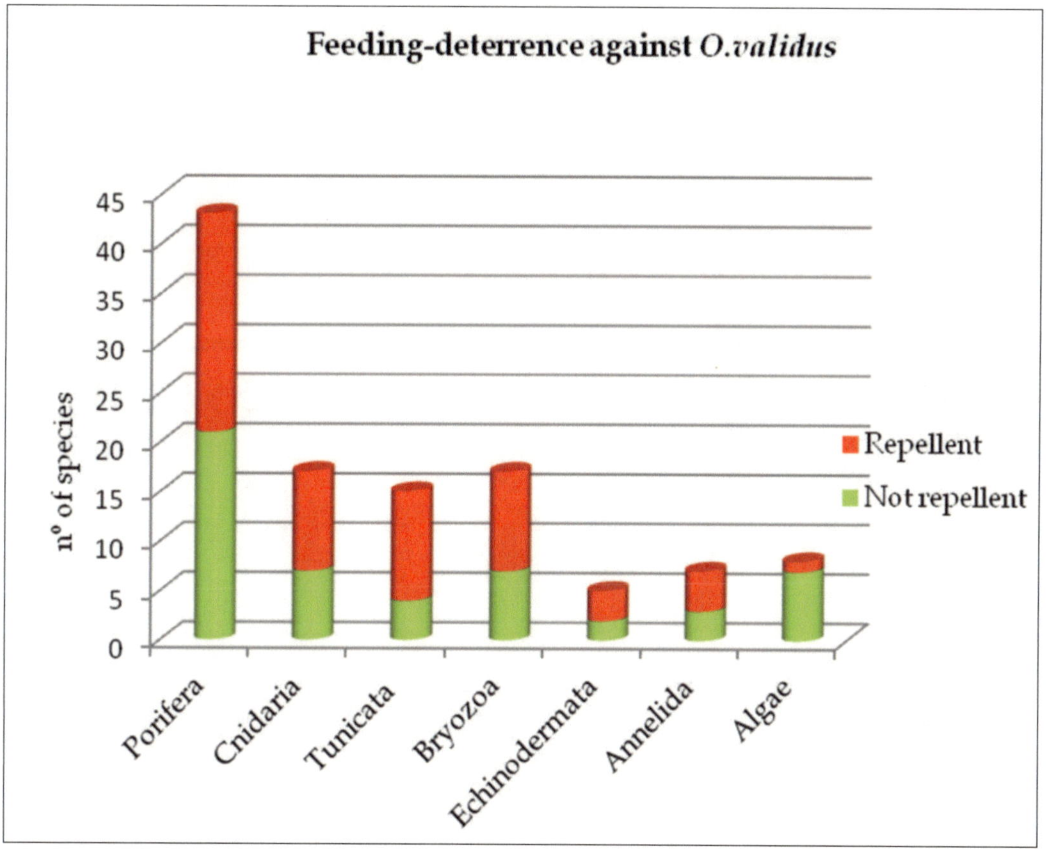

Fig. 3. Feeding-deterrence activity results against the seastar *Odontaster validus* in the different phyla tested.

In the experiment of feeding-preference using *C. femoratus*, 52 extracts were tested from Porifera (15), Cnidaria (14), Tunicata (12), Bryozoa (1), Echinodermata (1) and several extracts from macroalgae (8) (Núñez-Pons et al., in prep) A total of 36 extracts (33 species) out the 52 tested (40 species) were active (88,8% of the tested species) against the amphipod, revealing significant differences in food ingestion ($p<0.05$), being the control food preferred respect to the paired, unpreferred extract-treated, diet (Exact Wilcoxon test). These extracts corresponded to the taxa Porifera (7), Cnidaria (12), Tunicata (8) and Algae (6) (Fig.4).

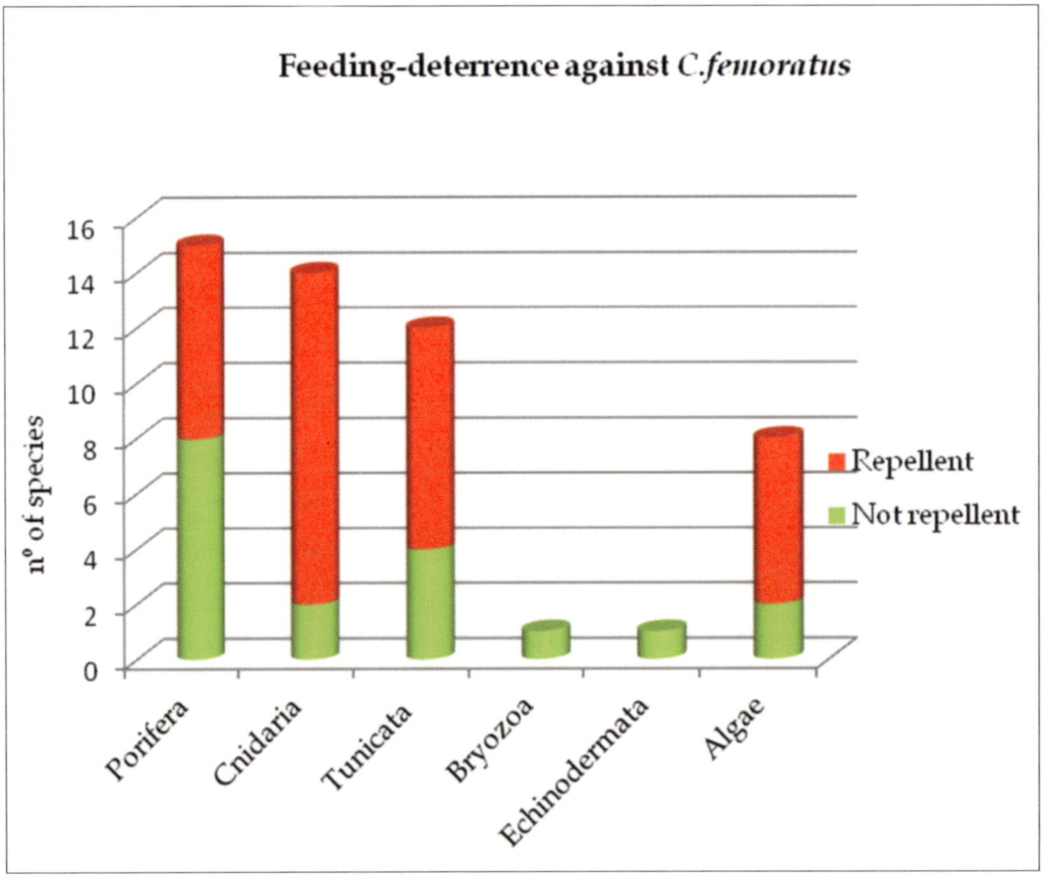

Fig. 4. Feeding-preference towards the amphipod *Cheirimedon femoratus* in the different Phyla tested.

3.2 Toxicity activity against the copepod *Metridia gerlachei*

We tested 24 species (32 extracts) belonging to the taxa Porifera (9), Cnidaria (3), Tunicata (1), Bryozoa (6), Echinodermata (4), and Hemichordata (1) and 14 of them (58, 3%) were toxic against copepods. A total of 14 active extracts (12) were detected (50% of the tested species) from organisms belonging to the Phyla Porifera (4), Cnidaria (1), Tunicata (1), Bryozoa (2), Echinodermata (4) (Fig. 5).

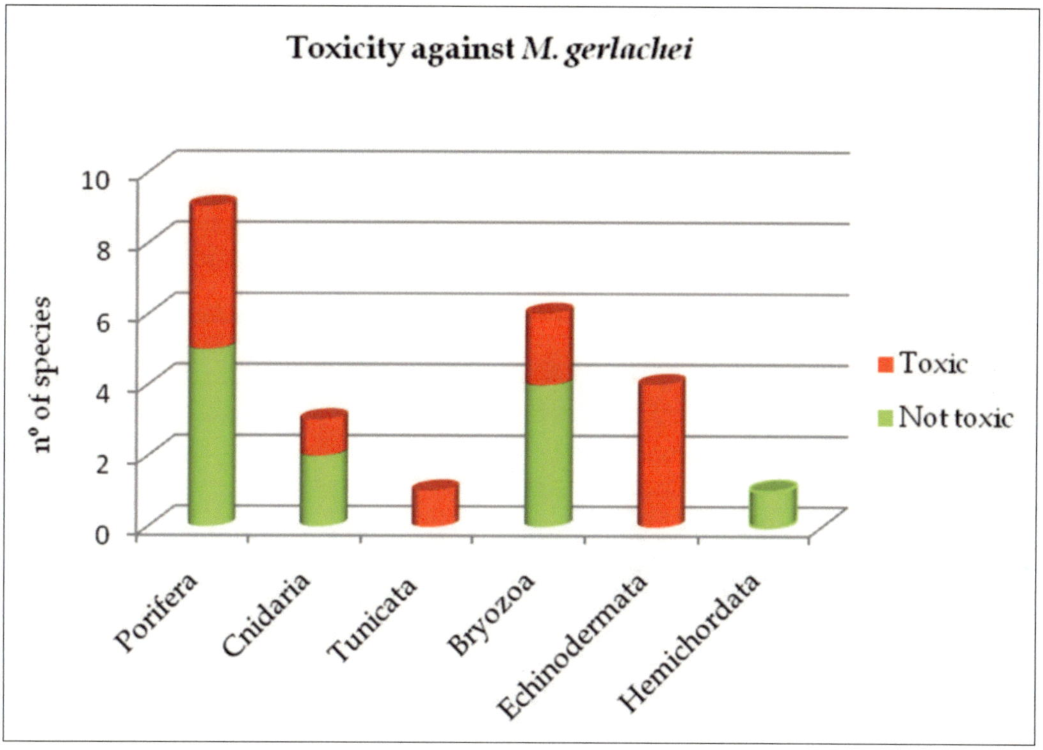

Fig. 5. Toxicity activity against *M. gerlachei* in different taxa.

3.3 Cytotoxicity against sea urchin embryos

A total of 17 species were tested, belonging to the Phyla Porifera (9), Cnidaria (1), Tunicata (2), Bryozoa (2), Annelida (1), Nemertea (1) and Algae (1). The toxic extracts (extracts outside confidence intervals described above) belong to Porifera (4), Cnidaria (1), Tunicata (2), Annelida (1), Nemertea (1) and Algae (1) (Fig. 6).

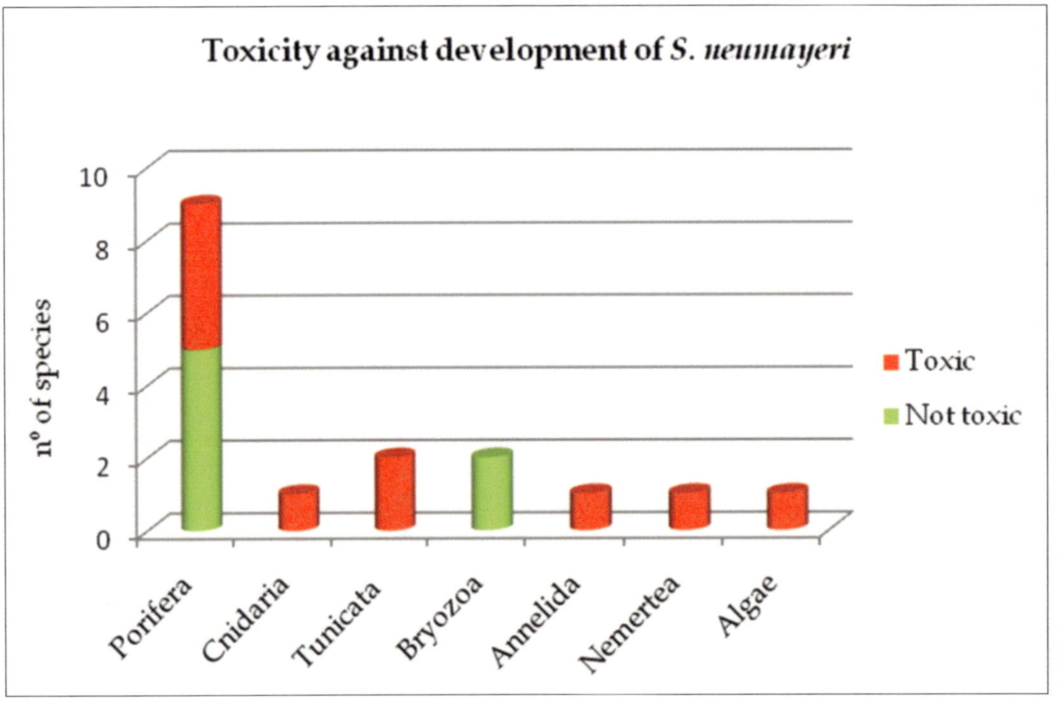

Fig. 6. Toxicity against development of *Sterechinus neumayeri* in different Phyla.

3.4 Cytotoxicity against sea urchin sperm

A total of 20 species (24 extracts) were tested. All the extracts except one were toxic to sperm (Fig.7) (<25% of the sperm was active) at the maximum concentration (1 mg ml^{-1}) and about 90% of the samples tested were active at the intermediate concentration. These extracts belonged to the taxa Porifera (4), Cnidaria (3), Bryozoa (2), Echinodermata (2), Annelida (6), Nemertea (1) and Hemichordata (3). Finally, 13 (15 extracts) out the 20 tested species (65%) were toxic to sperm at the lowest concentration.

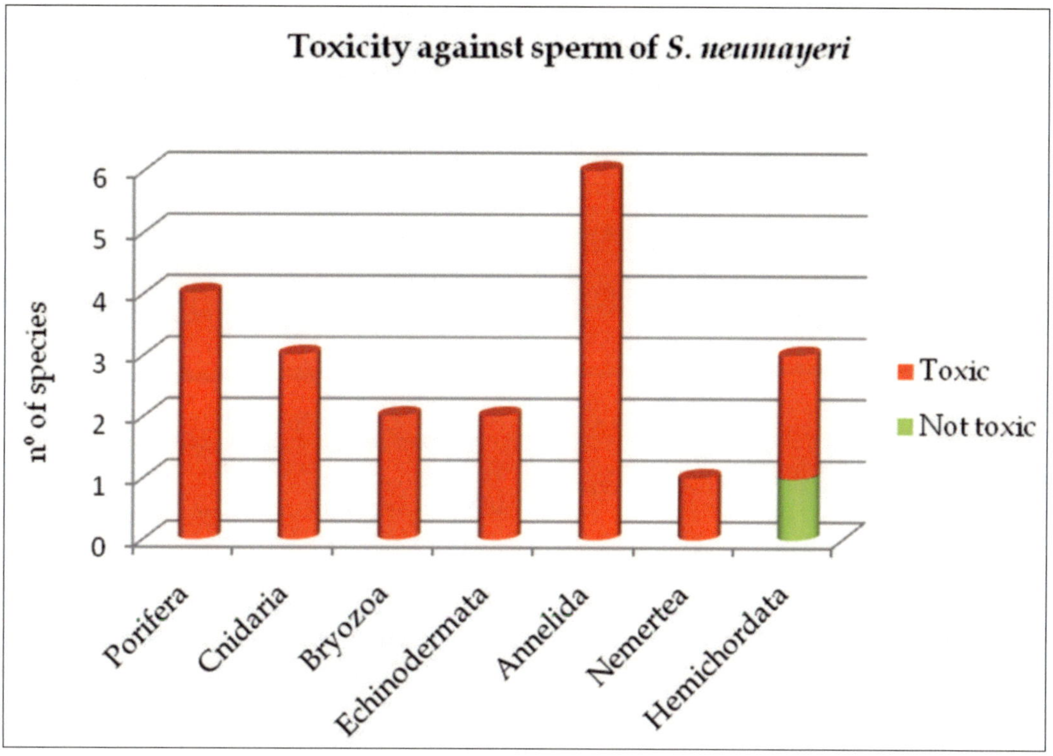

Fig. 7. Toxicity against sperm of *Sterechinus neumayeri* in different taxa.

3.5 Antifouling activity

We evaluated 130 extracts (70 species) from Porifera (22), Cnidaria (7), Tunicata (4), Bryozoa (14), Echinodermata (3), Annelida (6), Nemertea (1), Hemichordata (4), Algae (8) and others (1). A total of 28 extracts (24) were active (30.8%) from Porifera (5), Cnidaria (3) Tunicata (1), Bryozoa (3), Echinodermata(2), Annelida (3), Hemichordata (1) and Algae (2) (Fig. 8). This means all these active extracts produced zones of growth inhibition larger than 2 mm.

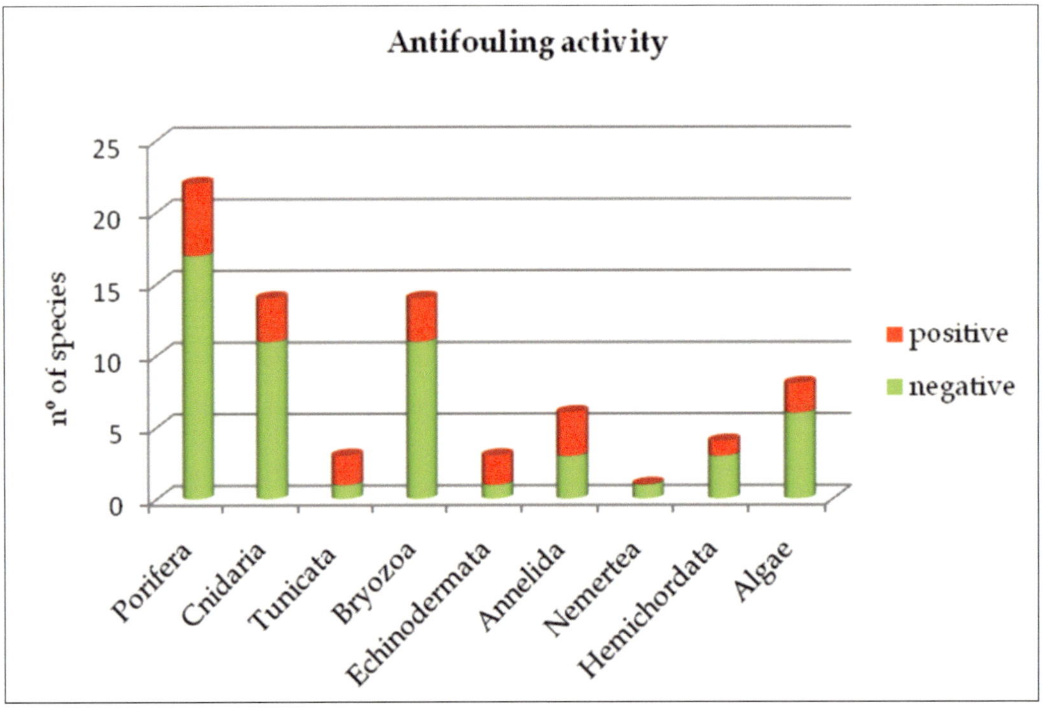

Fig. 8. Antifouling activity in the different taxa.

4. Discussion and conclusions

The Antarctic benthos appears to be greatly regulated by chemical interactions, mainly interfering with competence and predation (Fig. 9). Predation seems to be largely driven by the omnivorous sea star *O. validus,* known to have a noteworthy and extensive diet (McClintock et al., 2010). According to the high predation pressure described for this asteroid (Dayton et al., 1974), our results showed that repellence to avoid sea star predation is present in almost all the zoological groups of Antarctic invertebrates. This is demonstrated by the fact that more than 50% of the studied species of the main taxonomical groups exhibited significant deterrent activities. In agreement with these findings, previous experiments already demonstrated strong feeding deterrency towards this sea star in the opisthobrach molluscs *Austrodoris kerguelenensis* (Gavagnin et al., 2000; Iken et al., 2002) and *Bathydoris hodgsoni* (Avila et al., 2000). In the present survey, the tunicates exhibited the highest repellent activity (73%). Not surprisingly, the tunicate *Aplidium falklandicum* was recently found to possess particular alkaloid metabolites, the meridianins (A-G), responsible for this deterrent activity when tested isolated (Núñez-Pons et al., 2010). The phylum Porifera resulted to be also a quite active group (51%) and, in fact, other studies sustain this strong feeding deterrence reported for Antarctic sponges (21 species active out of the 27 species tested; Peters et al., 2009). Our study also found that 60% of the echinoderm samples were unsuitable for *O. validus,* although this species is known to feed on another seastars, namely *Acodontaster conspicuus* (Dayton et al., 1974). Cnidarians have already demonstrated the presence of chemical defenses, like the gorgonian coral *Ainigmaptilon antarcticus* (Iken & Baker 2003), and our results support this with more than 50% of the studied species (58%)

being active. The bryozoans displayed a similar deterrence as the cnidarians (58%), and also the polychaetes (57%), although this group was much less represented in number of samples tested. In fact, bryozoans have also been reported to be part of the diet of *O. validus* (Dayton et al., 1974). Finally, the algae seem also a potential food for this asteroid. Dearborn (1977) found diatoms, as well as red algae, in the stomach contents of *O. validus*. However the lower activity (14% of species) found in our study for this group may indicate a carnivorous preference of the star, despite being described as an opportunistic omnivorous consumer. All these results support the idea that many species from most of the phyla of Antarctic marine benthic invertebrates studied contain chemical defences against this voracious generalist sea star.

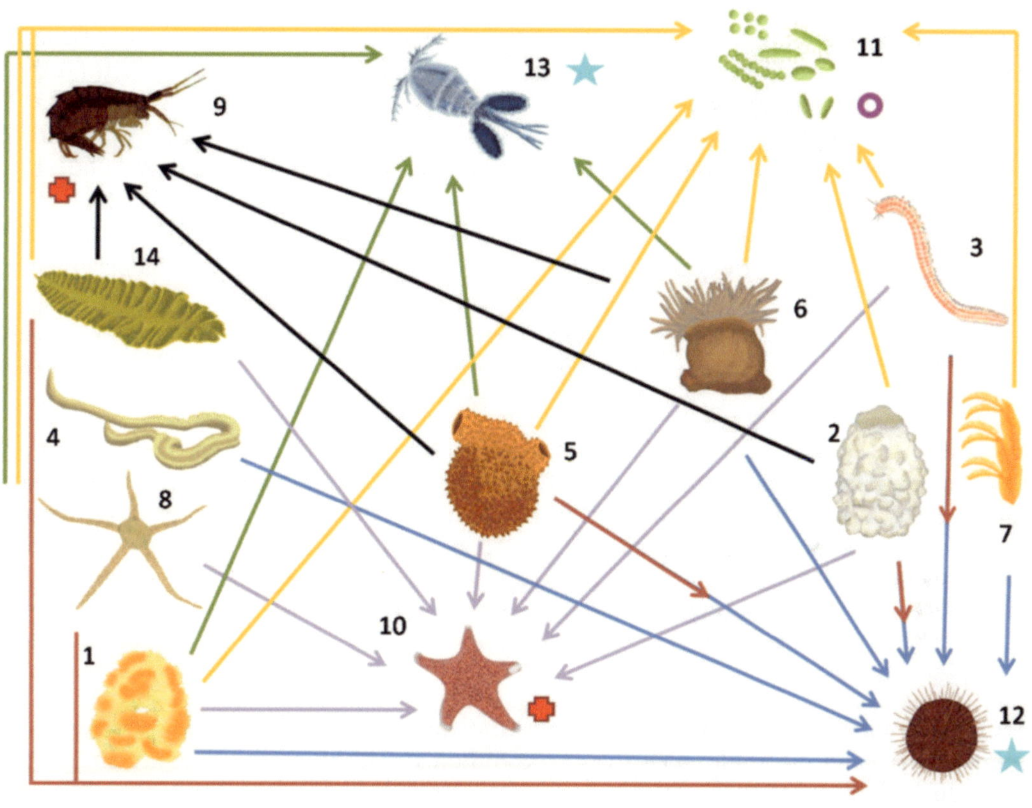

Fig. 9. Diagram of the proposed model of chemical ecology interactions according to the results obtained in our experiments, where: 1. Bryozoa, 2. Porifera, 3. Annelida, 4. Nemertea, 5. Tunicata, 6. Cnidaria, 7. Hemichordata, 8. Echinodermata; 9. *C. femoratus*, 10. *O. validus*, 11. Bacteria, 12. *S. neumayeri*, 13. Copepoda. 14. Algae. Symbols: Star: toxic activity; Cross: repellency activity; Circle: antifouling activity; Red: cytotoxicity activity; Blue: sperm toxicity. Each group is connected by arrows with the organisms used for each *in vivo* experiment.

Feeding preferences tested towards the amphipod *C. femoratus* revealed a high repellent activity in both benthic invertebrates and algae. Among the four majorly represented groups, the cnidarians displayed the highest incidence of feeding deterrent activities (85%), followed by the macroalgae (75%), the tunicates (66%), and finally the sponges (46%). Previous experiments with different species of Antarctic gammarid amphipods suggested that many macroalgae had feeding deterrent properties (Huang et al., 2006). It is worth to note that the amphipod used here, *C. femoratus,* has never been used previously as putative consumer in feeding assays. However, its voracious scavenger-omnivorous habits turn this mesograzer into a very suitable organism to test chemical defense. Actually, Bregazzi (1972) reported a large variety of food items (from algae to copepods and euphausiid larvae) in the stomach contents of this species. The impressive unpalatable activities recorded in our samples could be explained by the fact that, as many benthic amphipods, *C. femoratus* may use sessile organisms (mainly algae and sponges, but also others) both as host (biosubstrata) as well as a potential prey. This exerts a localized, constant pressure, which could be more intense than that caused by wandering mobile predators, such as sea stars or fish (Toth et al., 2007). Another species used as a model in previous Antarctic bioassays is the sympatric omnivorous amphipod *Gondogeneia antarctica*. This species, however, has repeatedly shown the problem of finding artificial foods too attractive, causing a phagostimulatory effect on the crustacean, and hence making the results obtained hard to interpret (Amsler et al., 2005, 2009a and b; Iken et al., 2009; Koplovitz, et al., 2009). Instead, the amphipod *C. femoratus* seems to possess a quite discriminatory potential to detect unpalatabilities, as observed in our results. Only bryozoans and echinoderms did not seem to produce a repellent effect against the mesograzer, but very few species were tested compared to other groups and, therefore, more samples have to be assayed before establishing any general conclusion for these groups.

Regarding the toxic activity against the copepod *M. gerlachei,* all groups except Hemichordata had some active species. Echinoderms showed, surprisingly, the highest activity (100%). Why this happens remains unknown at the moment, since *M. gerlachei* is omnivorous, feeding on phytoplankton, copepod eggs and small metazoans (Metz & Schnack-Schiel, 1995). Toxicity against the copepod could be a very useful mechanism to avoid the competition for phytoplankton food in benthic filter feeders, such as tunicates, sponges and bryozoans, and this is probably what our results reflect. No comparable results are currently available in the literature.

Few experiments have tested the activity from extracts of Antarctic benthic invertebrates against sperm and early life stages of the common Antarctic echinoid *S. neumayeri* (e.g. McClintock et al., 1990, 1992; Heine et al., 1991; Slattery et al., 1995) and, therefore, our contribution may give a wider idea of these cytotoxic mechanisms. In our tests, the sperm toxicity against *S. neumayeri* was evident in all species (100%) except in the group of Hemichordata (66% of species), similarly to what happened with the copepod test. Cytotoxicity against the development of this sea urchin was 100% in almost all groups, except sponges (44%) and bryozoans (0%), although we tested more samples of sponges than other groups. Pearse & Giese (1966) reported that *S. neumayeri* fed mostly on diatoms. Some studies have reported that the grazing by this sea urchin could be the responsible for significant mortality of settling larvae and juveniles of benthic invertebrates (Bowden, 2005; Bowden et al., 2006). Moreover, the settlement of pelagic larvae on or near their prey is frequent in marine predator invertebrates (Pawlik, 1992). Therefore, the presence of cytotoxic compounds found in this study for so many species may play an important role as a mechanism of defence/competence, reducing the recruitment of this sea urchin and, consequently, the grazing pressure and the colonization of the surface (e.g. McClintock et al., 1990; Bowden et al., 2006).

Different marine organisms such as bacteria, algae and invertebrates colonize submerged surfaces. The bacteria are metabolically versatile organisms capable of colonizing multiple surfaces, so it is not surprising that most organisms and, especially, the filter-feeders that are likely to be in trouble if the fouling is intensive, produce defenses against bacterial colonization. In the past, only a few studies were carried out to test the antifouling activity of Antarctic invertebrates against sympatric bacteria. Peters and colleagues (2010) tested the antifouling activity of extracts from Antarctic demosponges isolating bacteria from the surface of them and the majority of extracts did not display an inhibition activity. The reason could be that these bacteria were resistant to the substances from these sponges and, therefore, they were growing on their surface. Contrary to these results, the antifouling activity found in our tests is quite apparent in representatives of most taxa, including sponges (22%). In our case, bacteria were isolated from the water and we tested different species. This could be the cause of a higher activity in our results. Moreover, echinoderms (66%) and polychaetes (50%) were the most actives, followed by the cnidarians (42%). This is in agreement with previous results reporting antimicrobial activity in the soft corals *Alcyonium paessleri* and *Gersemia antarctica* (Slattery et al., 1995) . Tunicates, pterobranchs and algae had the same percentage of active species (25%). This is quite high if compared to the antimicrobial activity detected in only one (*Distaplia colligans*) out of 14 tunicate species tested previously (Koplovitz et al., 2011). The fact that the species tested are different could be the reason for these discrepancies. Compared with other taxa, bryozoans exhibited less activity (21%) and nemertins did not show any activity, although the number of species tested in these groups was too low to allow further considerations.

In summary, our results show that many different benthic organisms showed different strategies of defense, protection and/or competition (Fig. 9). However, it is important to emphasize that, not all phyla were equally tested, and therefore conclusions have to be considered cautiously. In many cases we found different defensive mechanisms in the same organism. This was previously reported for the Antarctic soft corals *Alcyonium paessleri* and *Gersemia antarctica,* which possess compounds with feeding-deterrence, antifouling and toxicity properties (e.g. Slattery & McClintock, 1995, 1997; Slattery et al., 1995). In our case, many species of Porifera and Cnidaria were active in all experiments (frequently over 40% or more of the tested species) and most Tunicata species quite as well (over 50% of species). In contrast, the Antarctic tunicates of the genus *Aplidium* displayed notable repellent activity in the feeding experiments (Núñez-Pons et al., 2010), but not in those of antibacterial nor antifouling.

Not surprisingly, other taxonomical groups were active in just one or a few tests, thus indicating the presence of one or only a few defensive lines (Fig. 9). For example, the phylum Bryozoa exhibited more activity in the experiment of feeding-deterrence against *O. validus* (58% of species) in relation to the antifouling experiment (21%) and no activity was found in the cytotoxicity experiment. The reason of these differences may be the presence in species of this phylum of physical defenses, such as avicularia, used for different roles, such as the prevention of the settlement of epibionts or larvae (e.g. Harmer, 1909). Moreover, in some groups, such as Annelida, only some species were active, and possibly, they have other types of defences no tested here, or physical protection strategies, such as living in their own bio-constructed tubes.

To our knowledge, this is the first ecological model proposed for describing the interactions in the Antarctic marine benthos, considering a wide array of possible chemical ecology relationships. We believe that these interactions are mainly generated to prevent the strong pressure of competition for space and/or food, predation and fouling to which Antarctic organisms are exposed. This general model shows an amazingly complex network of interactions between Antarctic organisms (Fig. 9). Further studies with larger number of samples are needed to complete and enrich this model and to bring some light to the existing gaps of knowledge. Nowadays, the research in marine chemical ecology in Antarctica continues to grow and new data will help to further advance in our knowledge on the role of chemical compounds in the Antarctic benthos. In order to successfully accomplish this task, the close collaboration among ecologists, chemists and microbiologists is essential. Also, further studies, such as those regarding antifouling and cytotoxicity activities, are needed to determine the ecological relevance of these mechanisms in Antarctic environments. Moreover, the bulk of the research in chemical ecology has been done on the phylum Porifera (Paul et al., 2011) compared to the few studies carried out in other phyla, such as Bryozoa, Annelida, Nemertea and Hemichordata, which are quite understudied. To fully understand this "chemical network", we will expand our studies to more types of experiments and more organisms during the development of our current project, ACTIQUIM-II.

5. Acknowledgements

We thank the editors for inviting us to publish our studies in this book. Thanks are also due to W. Arntz and the R/V Polarstern crew for allowing us to participate in the ANTXV/3 and the ANT XXI/2 cruises, as well as the Bentart and the BIO-Hespérides teams during the ECOQUIM cruise. Thanks are due to the taxonomists helping in the identification of some samples: P. Ríos (Porifera), A. Bosch, N. Campanyà and J. Moles (Echinodermata), M. Valera (Tunicata), A. Gómez and M.A. Ribera (Algae). We would like to thank as well the Unidad de Tecnología Marina (UTM) and the crew of Las Palmas vessel for all their logistic support. Special thanks are also given to the "Gabriel de Castilla BAE" crew for their help during the ACTIQUIM-1 and -2 Antarctic expeditions. Funding was provided by the Ministry of Science and Innovation of Spain through the ECOQUIM and ACTIQUIM Projects (REN2003-00545, REN2002-12006E/ANT, CGL2004-03356/ANT and GCL2007-65453/ANT).

6. References

Amsler, C.D., Amsler, M.O., McClintock, J.B. & Baker, B.J. (2009a). Filamentous algal endophytes in macrophytic Antarctic algae: prevalence in hosts and palatability to mesoherbivores. *Phycologia,* Vol.48, pp. 324-334, ISSN 0031-8884

Amsler, M.O., McClintock, J.B., Amsler, C.D. Angus, R.A. & Baker, B.J. (2009b). An evaluation of sponge-associated amphipods from the Antarctic Peninsula. *Antarctic Science*, Vol.21, pp. 579–589, ISSN 0954-1020

Amsler, C.D., Iken, K., McClintock, J.B., Amsler, M.O., Peters, K.J., Hubbard, J.M., Furrow, F.B. & Baker, B.J. (2005) Comprehensive evaluation of the palatability and chemical defenses of subtidal macroalgae from the Antarctic Peninsula. *Marine Ecology Progress Series*, Vol.294, pp. 141–159, ISSN 0171-8630

Amsler, C.D., Iken, K., McClintock, J.B. & Baker. B.J. (2001). Secondary metabolites from Antarctic marine organisms and their ecological implications. In: *Marine Chemical Ecology*, J.B. McClintock & B.J. Baker, pp. 263-296, CRC Press, ISBN 978-0-8493-9064-7, Boca Ratón, Florida.

Ankisetty, S., Nandiraju, S., Win, H., Park, Y.C., Amsler, C.D., McClintock, J.B., Baker, J.A., Diyabalanage, T.K., Pasaribu, A., Singh, M.P., Maiese, W.M., Walsh, R.D., Zaworotko, M.J. & Baker, B.J. (2004). Chemical investigation of predator-deterred macroalgae from the Antarctic peninsula. *Journal of Natural Products* Vol. 67, pp. 1295-1302, ISSN 1520-6025

Antonov, A.S., Avilov, S.A., Kalinovsky, A.I., Anastyuk, S.D., Dmitrenok, P.S., Evtushenko, E.V., Kalinin, V.I., Smirnov, A.V., Taboada, S., Ballesteros, M., Avila, C., & Stonik, V.A. (2008). Triterpene glycosides from Antarctic sea cucumbers. 1. Structure of Liouvillosides A1, A2, A3, B1, and B2 from the sea cucumber *Staurocucumis liouvillei*: new procedure for separation of highly polar glycoside fractions and taxonomic revision. *Journal of Natural Products*, Vol.71, pp.1677-1685, ISSN 1520-6025

Antonov, A.S., Avilov, S.A., Kalinovsky, A.I., Anastyuk, S.D., Dmitrenok, P.S., Evtushenko, E.V., Kalinin, V.I., Smirnov, A.V., Taboada, S., Bosch, A., Avila, C., & Stonik, V.A. (2009). Triterpene glycosides from Antarctic sea cucumbers. 2. Structure of Achlioniceosides A(1), A(2), and A(3) from the sea cucumber *Achlionice violaecuspidata* (=*Rhipidothuria racowitzai*). *Journal of Natural Products*, Vol.72, No.1, pp.33-38, ISSN 0163-3864

Antonov, A.S., Avilov, S.A., Kalinovsky, A.I., Dmitrenok, P.S., Kalinin, V.I., Taboada, S., Ballesteros, M. & Avila, C. (2011). Triterpene glycosides from Antarctic sea cucumbers III. Structures of liouvillosides A(4) and A(5), two minor disulphated tetraosides containing 3-O-methylquinovose as terminal monosaccharide units from the sea cucumber *Staurocucumis liouvillei* (Vaney). *Natural Product Research*, Vol.25, No.14, pp. 1324-33, ISSN 1478-6427

Avila, C. (1995). Natural products of opisthobranch molluscs: A biological review. *Oceanography and Marine Biology: An Annual Review*, Vol.33, pp. 487-559, ISSN 0078-3218

Avila, C., Iken, K., Fontana, A. & Cimino, G. (2000). Chemical ecology of the Antarctic nudibranch *Bathydoris hodgsoni* Eliot, 1907: defensive role and origin of its natural products. *Journal of Experimental Marine Biology and Ecology*, Vol.252, pp. 27-44, ISSN 0022-0981

Avila, C., Taboada, S. & Núñez-Pons, L. (2008). Antarctic marine chemical ecology: what is next? *Marine Ecology*, Vol.29, No.1, pp. 1-71, ISSN 0173-9565

Ballesteros, M. & Avila, C. (2006). A new Tritoniid species (Mollusca: Opisthobranchia) from Bouvet Island. *Polar Biology*, Vol.29, pp. 128-136, ISSN 0722-4060

Bhakuni, D.S. (1998). Some aspects of bioactive marine natural products. *Journal of the Indian Chemical Society*, Vol.75, pp. 191-205, ISSN 0019-4522

Blunt, J.W., Copp, B.R., Hu, W-P., Munro, M.H.G., Northcote, P.T. & Prinsep, M.R. (2009). Marine Natural Products. *Natural Product Reports*, Vol.26, pp. 170-244, ISSN 0265-0568

Blunt, J.W., Copp, B.R., Munro, M.H., Northcote, P.T. & Prinsep, M.R. (2011). Marine Natural Products. *Natural Product Reports*, Vol.28, No.2, pp. 196–268, ISSN 0265-0568

Bosch, I., Beauchamp, K.A., Steele, M.E. & Pearse, J.S. (1987). Development, metamorphosis and seasonal abundance of embryos and larvae of the Antarctic sea urchin *Sterechinus neumayeri*. *The Biological Bulletin*, Vol.173, pp. 126-135, ISSN 0006-3185

Bowden, D.A. (2005). Seasonality of recruitment in Antarctic sessile marine benthos. *Marine Ecology Progress Series*, Vol.297, pp. 101–118, ISSN 1616-1599

Bowden D.A., Clarke, A., Peck, L.S. & Barnes, D.K.A. (2006). Antarctic sessile marine benthos: colonisation and growth on artificial substrata over 3 years. *Marine Ecology Progress Series*, Vol.316, pp. 1–16, ISSN 0171-8630

Bregazzi, P. (1972). Life cycle and seasonal movements of *Cheirimedon femoratus* (Pfeffer) and *Tryphosella kergueleni* (Miers) (Crustacea: Amphipoda). *British Antarctic Survey Bulletin*, Vol.30, pp. 1–34. ISSN 0007-0262

Brey, T., Pearse J.S., Basch, L., McClintock, J.B. & Slattery, M. (1995). Growth and production of *Sterechinus neumayeri* (Echinoidea: Echinodermata) in McMurdo Sound, Antarctica. *Marine Biology*, Vol.124, pp. 279-292, ISSN 0025-3162

Carbone, M., Núñez-Pons, L., Castelluccio, F., Avila, C. & Gavagnin, M. (2009). Illudalane sesquiterpenoids of the alcyopterosin series from the Antarctic marine soft coral *Alcyonium grandis*. *Journal of Natural Products*, Vol.72, No.7, pp. 1357-60, ISSN 1520-6025

Dayton, P.K., Robilliard, G.A., Paine, R.T. and Dayton L.B. (1974). Biological accommodation in the benthic community at McMurdo Sound, Antarctica. *Ecological Monographs*, Vol.44, No.1, pp. 105-128, ISSN 0012-9615

Dearborn, J. H. (1977). Food and feeding characteristics of antarctic asteroids and ophiuroids. In: *Adaptations within antarctic ecosystems*, G. A. Llano , (Ed.), pp. 293-326, Gulf Publishing Co, ISBN 978-087-2010-00-0, Houston.

De Marino, S., Iorizzi, M. & Zollo, F. (1997). Isolation, structure elucidation and biological activity of the steroids glycosides and polyhydroxysteroids from the antarctic starfish *Acodontaster conspicuus*. *Journal of Natural Products*, Vol.60, pp. 959-966, ISSN 0163-3864

Figuerola, B., Ballesteros, M., Avila, C. (2011). Description of a new species of *Reteporella* (Bryozoa Phidoloporidae) from the Weddell Sea (Antarctica) and the possible functional morphology of avicularia. *Acta Zoologica* (doi: 10.1111/j.1463-6395.2011.00531.x).

Gavagnin, M., Fontana, A., Ciavatta, M.L. & Cimino, G. (2000). Chemical studies on Antarctic nudibranch molluscs. *Italian Journal of Zoology,* Vol.67, No.1, pp. 101–109, ISSN 1748-5851

Harmer, S.F. (1909). Presidential address, Report, 78th meeting of the British. *Association for the Advancement for Science*, 1908, pp. 715-731

Heine, J.N., McClintock, J.B., Slattery, M., & Weston J. (1991). Energetic composition, biomass, and chemical defense in the common antarctic nemertean *Parborlasia corrugatus. Journal of Experimental Marine Biology and Ecology,* Vol.153, pp. 15-25 ISSN 0022-0981

Hoyer, K., Karsten, U. &Wiencke, C. (2002). Induction of sunscreen compounds in Antarctic macroalgae by different radiation conditions. *Marine Biology,* Vol.141, pp. 619–627, ISSN 0025-3162

Huang, Y.M., Amsler, M.O., McClintock, J.B., Amsler, C.D. & Baker, B.J. (2007). Patterns of gammaridean amphipod abundance and species composition associated with dominant subtidal macroalgae from the western Antarctic Peninsula. *Polar Biology,* Vol.30, pp. 1417–30, ISSN 0722-4060

Huang, Y.M., McClintock, J.B., Amsler, C.D., Peters, K.J. & Baker, B.J. (2006). Feeding rates of common Antarctic gammarid amphipods on ecologically important sympatric macroalgae. *Journal of Experimental Marine Biology and Ecology,* Vol.329, pp. 55–65, ISSN 0022-0981

Iken, K., Amsler, C.D., Amsler, M.O., McClintock, J.B. & Baker, B.J. (2009). Field studies on deterrent properties of phlorotannins in Antarctic brown algae. *Botanica Marina,* Vol.52, pp. 547–557, ISSN 0006-8055

Iken, K., Avila, C., Fontana, A. & Gavagnin, M. (2002). Chemical ecology and origin of defensive compounds in the Antarctic nudibranch *Austrodoris kerguelenensis* (Opisthobranchia: Gastropoda). *Marine Biology,* Vol.141, pp. 101-109, ISSN 0025-3162

Iken, K. & Baker, B.J. (2003).Ainigmaptilone A: A new sesquiterpene from the Antarctic gorgonian coral *Ainigmaptilon antarcticus* (Octocorallia, Gorgonacea). *Journal of Natural Products*, Vol. 66, pp. 888-890, ISSN 0163-3864

Jayatilake, G.S., Thornton, M.P., Leonard, A.C., Grimwade, J.E. & Baker, B.J. (1996). Metabolites from an Antarctic sponge-associated bacterium, *Pseudomonas aeruginosa. Journal of Natural Products,* Vol.59, No.3, pp. 293-296, ISSN 0163-3864

Kelly, S.R., Jensen, P.R., Henkel, T.P., Fenical, W. & Pawlik, J.R. (2003). Effects of Caribbean sponge extracts on bacterial attachment. *Aquatic Microbial Ecology,* Vol.31, pp. 175-182, ISSN 0948-3055

King, A. & LaCasella, E.L. (2003). Seasonal variations in abundance, diel vertical migration, and population structure of *Metridia gerlachei* at Port Foster, Deception Island, Antarctica. *Deep Sea Research Part II: Topical Studies in Oceanography,* Vol. 50, pp. 1753-1763, ISSN 0967-0645.

Koplovitz, G., McClintock, J.B, Amsler, C.D. & Baker, B.J. (2009). Palatability and chemical anti-predatory defenses in common ascidians from the Antarctic peninsula. *Aquatic Biology,* Vol.7, pp. 81-92, ISSN 1864-7790

Koplovitz, G., McClintock, J.B, Amsler, C.D. & Baker, B.J. (2011). A comprehensive evaluation of the potential chemical defenses of Antarctic ascidians against sympatric fouling microorganisms. *Marine Biology*, pp. 1-11, ISSN 0025-3162

Lebar, M.F., Heimbegner, J.L. & Baker, B.J. (2007). Cold-water marine natural products. *Natural Products Reports*, Vol.24, No.4, pp. 774-797, ISSN 0265-0568

Mahon, A.R., Amsler, C.D, McClintock, J.B., Amsler , M.O. & Baker, B.J. (2003). Tissue-specific palatability and chemical defenses against macropredators and pathogens in the common articulate brachiopod *Liothyrella uva* from the Antarctic Peninsula. *Journal of Experimental Marine Biology and Ecology*, Vol.290, No.2, pp. 197-210, ISSN 0022-0981

Metz, C. & Schnack-Schiel, S.B. (1995). Observations on carnivorous feeding in Antarctic copepods. *Marine Ecology Progress Series*, Vol.129, pp. 71–75, ISSN 0171-8630

McClintock, J.B. (1994). Trophic biology of Antarctic echinoderms. *Marine Ecology Progress Series*, Vol.111, pp. 191–202, ISSN 0171-8630

McClintock, J.B., Amsler, C.D. & Baker, B.J. (2010). Overview of the chemical ecology of benthic marine invertebrates along the Western Antarctic Peninsula. *Integrative and Comparative Biology*, Vol.50, pp. 967-980, ISSN 1540-7063

McClintock, J.B., Heine, J., Slattery, M. & Weston, J. (1990). Chemical bioactivity in common shallow-water Antarctic marine invertebrates. *Antarctic Journal of the United States*, Vol.25, No.5, pp. 204–206, ISSN 0003-5335

McClintock, J.B., Heine, J., Slattery, M. & Weston, J. (1991). Biochemical and energetic composition, population biology and chemical defense of the Antarctic ascidian *Cnemidocarpa verrucosa* Lesson. *Journal of Experimental Marine Biology and Ecology*, Vol.147, pp. 163-175, ISSN 0022-0981

McClintock, J.B., Slattery, M., Heine, J. & Weston, J. (1992). Chemical defense, biochemical composition and energy content of three shallow-water Antarctic gastropods. *Polar Biology*, Vol.11, pp. 623-629, ISSN 0722-4060

Munro, M.H.G., Blunt, J.W., Dumdei, E.J., Hickford, S.J.H., Lill, R.E., Li, S., Battershill, C.N. & Duckworth, A.R. (1999). The discovery and development of marine compounds with pharmaceutical potential. *Journal of Biotechnology*, Vol.70, pp. 15-25, ISSN 0168-1656

Núñez-Pons, L., Forestieri, R., Nieto, R. M., Varela, M., Nappo, M., Rodríguez, J., Jiménez, C., Castelluccio, F., Carbone, M., Ramos-Espla, A., Gavagnin, M. & Avila, C. (2010). Chemical defenses of tunicates of the genus *Aplidium* from the Weddell Sea (Antarctica). *Polar Biology*, Vol.33, No.10, pp. 1319, ISSN 0722-4060

Orejas, C., Gili, J.M., Arntz, W.E., Ros, J.D., López, P.J., Teixido, N., Filipe, P. (2000). Benthic suspension feeders, key players in Antarctic marine ecosystems? *Contributions to Science*, Vol.1, No.3, pp. 299–311, ISSN 1575-6343

Paul, V.J. (1992). *Ecological roles of marine natural products,* pp. 245, ISBN 978-0-8014-2727-5, Comstock, Ithaca, NY, USA. pp.

Paul, V.J., Puglisi, M.P. & Ritson-Williams, R. (2006). Marine chemical ecology. *Natural Product Reports*, Vol.23, No.2, pp. 153-180, ISSN 0265-0568

Paul, V.J., Ritson-Williams, R. & Sharp, K. (2011). Marine chemical ecology in benthic environments. *Natural Product Reports*, Vol.28, pp. 345-387, ISSN 0265-0568

Pawlik, J.R. (1992). Chemical Ecology of the Settlement of Benthic Marine Invertebrates. *Oceanography and Marine Biology: An Annual Review*, Vol.30, pp. 273-335, ISSN 0078-3218

Pearse, J. S. & Giese, A. C. (1966). Food, reproduction and organic constitution of the common antarctic echinoid *Sterechinus neumayeri* (Meissner). *The Biological Bulletin*, Vol.130, pp. 387-401, ISSN 0006-3185

Peters, K.J., Amsler, C.D., McClintock, J.B. & Baker, B.J. (2010). Potential chemical defenses of Antarctic sponges against sympatric microorganisms. *Polar Biology*, Vol.33, pp. 649–58, ISSN 0722-4060

Peters, K.J., Amsler, C.D., McClintock, J.B., Rob, W.M. van Soest & Baker, B.J. (2009). Palatability and chemical defenses of sponges from the western Antarctic Peninsula. *Marine Ecology Progress Series*,, Vol. *385*, pp. 77-85, ISSN 0171-8630

Reyes, F., Fernandez, R., Rodriguez, A., Francesch, A., Taboada, S., Avila, C., & Cuevas, C. (2008). Aplicyanins A-F, new cytotoxic bromoindole derivatives from the marine tunicate *Aplidium cyaneum*. *Tetrahedron*, Vol. 64, No. 22, pp. 5119-5123, ISSN 0040-4020

Ríos, P. & Cristobo, J. (2006). A new species of *Biemna* (Porifera: Poecilosclerida) from Antarctica: *Biemna strongylota*. *Journal of the Marine Biological association of the United Kingdom*, Vol. 86, pp. 949-955, ISSN 0025-3154

Rittschof, D. (2001). Natural Product Antifoulants and Coatings Development. In: *Marine Chemical Ecology*. J. McClintock & P. Baker, (Ed.), pp. 543-557, ISBN 978-0-8493-9064-7, CRC Press, New York

Sivaperumal, P., Ananthan G. & Mohamed Hussain, S. (2010). Exploration of antibacterial effects on the crude extract of marine ascidian *Aplidium multiplicatum* against clinical isolates. *International Journal of Medicine and Medical Sciences*, Vol. 2, No. 12, pp. 382 - 386, ISSN 2006-9723

Slattery, M., Hamann, M.T., McClintock, J.B., Perry, T.L., Puglisi, M.P. & Yoshida , W.Y. (1997). Ecological roles for water-borne metabolites from Antarctic soft corals. *Marine Ecology Progress Series*, Vol. 161, pp. 133–144, ISSN 0171-8630

Slattery, M. & McClintock, J.B. (1995). Population structure and feeding deterrence in three shallow-water Antarctic soft corals. *Marine Biology*, Vol. 122, pp. 461–470, ISSN 0025-3162

Slattery, M., McClintock, J.B. & Heine, J.N. (1995). Chemical defenses in Antarctic soft corals: evidence for antifouling compounds. *Journal of Experimental Marine Biology and Ecology*, Vol.190, pp. 61–77, ISSN 0022-0981

Sokal, R.R. & Rohlf, F.J. (1995). Biometry: the principles and practice of statistics in biological research. Pp. 937, ISBN 0-7167-0663-6, Freeman, W. H. and Co. New York

Taboada, S., García-Fernández, L.F., Bueno, S., Vazquez, J., Cuevas, C. & Avila. C. (2010). Antitumoral activity in Antarctic and Sub-Antarctic benthic organisms. *Antarctic Science*, Vol.22, No.5, pp. 449-507, ISSN 0954-1020

Toth, G.B., Karlsson, M. & Pavia, H., 2007. Mesoherbivores reduce net growth and induce chemical resistance in natural seaweed populations. *Oecologia* Vol.152, pp. 245-255, ISSN 0029-8549

Volpi Ghirardini, A., Arizzi Novelli, A., Losso & C., Ghetti. (2005). Sperm cell and embryo toxicity tests using the sea urchin *Paracentrotus lividus* (Lmk). In: *Techniques in aquatic toxicology*, G. Ostrander, (Ed.), pp. 147-168, Vol.2, No.8, CRC press, ISBN 978-1-5667-0149-5, University of Hawaii, Honolulu, USA

6

An Interdisciplinary Erosion Mitigation Approach for Coral Reef Protection – A Case Study from the Eastern Caribbean

Carlos E. Ramos-Scharrón[1,2],
Juan M. Amador[3] and Edwin A. Hernández-Delgado[4,5]
[1]*Island Resources Foundation,*
[2]*Department of Geography & the Environment, the University of Texas-Austin,*
[3]*Greg L. Morris Engineering COOP,*
[4]*Center for Applied Tropical Ecology and Conservation,*
University of Puerto Rico-Río Piedras,
[5]*Caribbean Coral Reefs Institute,*
University of Puerto Rico-Mayagüez,
[1]*US Virgin Islands*
[2]*USA*
[3,4,5]*Puerto Rico*

1. Introduction

1.1 Project background

Although evidence suggests that the onset of worldwide coral reef degradation dates back to centuries ago, it is unequivocal that human impacts are to some degree responsible for their current frail condition (Jackson, 1997, 2001; Pandolfi et al., 2003). Reefs of the Caribbean have not escaped this global trend as studies have noted an unprecedented increase in the spatial and temporal scale of coral reef turnover events, apparently rooted in the accelerating pace of regional-level ecological change (Aronson et al., 2002). Recent impacts appear to have changed coral community structure in ways not observed in the region over the last 220,000 years (Pandolfi and Jackson, 2006). The generalized decline in coral cover observed throughout the Caribbean has been associated with the heightened prevalence of both regional pressures (e.g., warmer sea surface temperatures, bleaching, and higher incidence of disease) and local stressors (e.g., non-point sources of pollution, fishing, etc.) (Rodríguez, 1981; Gardner et al., 2003, Hawkins and Roberts, 2004; Pandolfi et al., 2005; Miller et al., 2006, 2009). In addition, climate change projections suggest a more challenging future for Caribbean coral reef ecosystems (Hoegh-Guldberg, 1999; Hoegh-Guldberg et al., 2007; Buddemeier et al., 2008, 2010; Hernández-Pacheco et al., 2011).

Coral reefs in the Commonwealth of Puerto Rico (PR) are among the most highly threatened reefs of the entire Caribbean as a consequence of the combined effects of climate change, coral bleaching, increased incidence of disease, overfishing, and the delivery of inland pollutants (Burke and Maidens, 2004; Hernández-Delgado, 2005; Ballantine et al., 2008; García-Sais et al., 2008; Larsen and Webb, 2009; Hernández-Delgado and Sandoz-Vera, 2011). Excess delivery of land-based contaminants into the marine environment of PR cannot be considered an exclusive present-day phenomenon (Goenaga and Cintrón, 1979; Goenaga, 1991; Hernández-Delgado, 2000, 2005). Water quality is inevitably related to land use, and generally it is inversely correlated with economic development, population density, land use patterns, and other socioeconomic indicators (Biagi, 1965; Restrepo and Syvistski, 2006; Oliver et al., 2011). Therefore, the deterioration of coastal water quality in PR likely began in the mid-1800's when an island-wide wave of deforestation cleared the way for timber extraction, cattle grazing, and mass production of agricultural goods (Birdsey and Weaver, 1987). Change of sovereignty at the turn of the 20th century from Spanish colonial rule to U.S. control favored the extensive use of coastal lowlands for sugar cane production under a progressively mechanized and more centralized system (Labadie-Eurite, 1949; Dietz, 1986) and this resulted in its own new suite of water pollutants (Biagi, 1968). Assisted by lax enforcement of environmental safeguards (Concepción, 1988; Berman-Santana, 1996) socioeconomic and political development in PR following Second World War (WWII) explicitly encouraged a move towards industrialization at the expense of agricultural production (Dietz, 1986). Even though implementation of this new economic model allowed

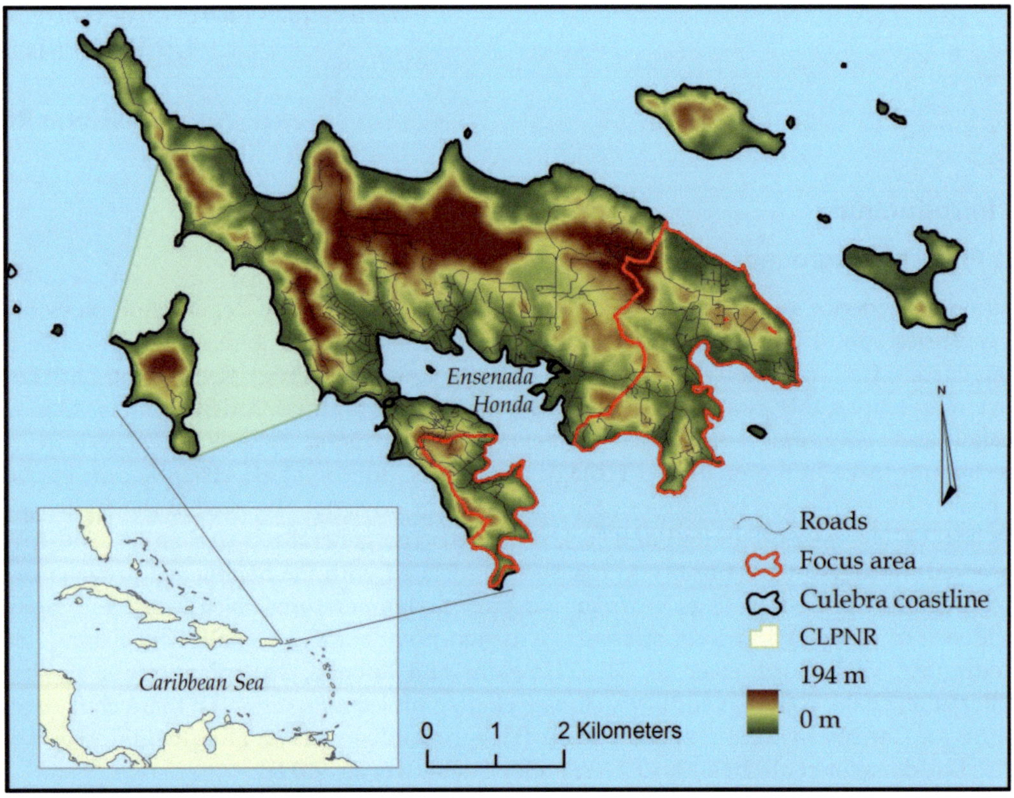

Fig. 1. Map of Isla de Culebra displaying its general location with respect to the Caribbean Region, its road network, the focus watersheds for this article, the Canal Luis Peña Marine Reserve (CLPNR), and the island's topography.

for the recuperation of an island-wide forest cover (Rudel et al., 2000; Grau et al., 2003; Valdés-Pizzini et al., 2011), it also introduced its own new set of water quality issues (Hunter and Arbona, 1995) that have established a legacy of documented stress and detrimental effects on coral reef communities in various parts of the Puerto Rican archipelago (e.g., Loya, 1976; Goenaga and Cintrón, 1979; Goenaga, 1991; Hernández-Delgado, 2000, 2005; Morelock et al., 2001; Larsen and Webb, 2009; Hernández-Delgado et al., 2010, Hernández-Delgado and Sandoz-Vera, 2011).

The island of Culebra (Figure 1) supports coral reef ecosystems characteristic of northeastern Caribbean marine biodiversity (Hernández-Delgado, 2000, 2005; Hernández-Delgado et al., 2000; García et al., 2003; Hernández-Delgado and Rosado-Matías, 2003), and they represent highly valuable sources of fishing, tourism and recreational activities (Estudios Técnicos, 2007; Webler and Jakubowski, 2011). Culebra also supports the first no-take natural reserve established in PR, the Canal Luis Peña Natural Reserve (CLPNR) (Pagán-Villegas et al., 1999), and houses various academic and community-based coral reef and reef fisheries management conservation efforts (Hernández-Delgado et al., 2011). Culebra's nearshore coral reefs have also been described as some of the most exceptional in PR (Hernández Delgado, 2000; Simonsen, 2000), and this is presumably due in part to historic low doses of terrestrial sediment inputs originating from the relatively small watersheds combined with a sub-tropical dry climatic setting. Long-term monitoring within CLPNR has shown an alarming 50-80% decline in percent live coral cover since 1997 (Hernández-Delgado, 2010). Indirect evidence suggests that the decline in percent living coral cover may be associated with increased sedimentation resulting from recent land development, deforestation, and lack of mandatory erosion controls (Hernández-Delgado, 2004; Hernández-Delgado et al., 2006) (Figure 2), in combination with fishing pressure, climate change-related sea surface warming, massive coral bleaching, and post-bleaching mortality events (Hernández-Pacheco et al., 2011), as it has been documented elsewhere (Miller et al., 2006, 2009).

Fig. 2. (Left) Picture of a plume entering Bahía Mosquito by sediment produced from a single, unpaved road segment (photo courtesy of M.A. Lucking-CORALations). (Right) Picture of an unpaved road segment that typifies the road network in Culebra.

Focus groups executing U.S. Coral Reef Task Force mandates have identified Culebra as a top priority site in PR needing a Local Action Strategy (LAS) plan. Two of the most important goals of the LAS efforts are to implement land-use planning at the watershed scale to minimize water quality impacts to the coral reef ecosystem, and to control and reduce pollutant transport to the marine environment. The LAS plan in Culebra is in part required to address the continuous decline in coral reef conditions by reducing the risks posed by the recent acceleration in land development rates (Commonwealth of PR and NOAA, 2010).

1.2 Objectives

The lack of a scientifically-based methodology to guide watershed management strategies is partly to blame for deficient to non-existent erosion control activities on Culebra and on most islands of the Caribbean. The main objective of this article is to describe an innovative framework by which technical knowledge gathered by marine ecologists, watershed scientists, and civil engineers can be best employed in the development of an erosion mitigation strategy. The approach proposed here is intended to explicitly define the principles behind the development of such interdisciplinary strategies and to maximize their benefits. Although the goal of erosion control is simply to alleviate the pressures associated with just one of the many sources of stress affecting coral reefs, the general framework described here could be emulated to address other land-based, non-point pollution sources affecting coral reef systems in Culebra and elsewhere. Isla de Culebra serves as the focus of our efforts because of its imminent need for the implementation of such types of mitigation efforts and to take advantage of previously-existing coral reef databases and watershed assessments (Hernández-Delgado 2000; Hernández-Delgado et al., 2000, 2006; Hernández-Delgado and Rosado-Matías, 2003; Ramos-Scharrón, 2009).

2. Site description

2.1 Natural environment

At 26.6 km^2 and located roughly 28 km east of mainland PR, Isla de Culebra (Latitude: 18.2; Longitude: -65.3) is the second smallest and easternmost municipality (i.e., township) comprising the Commonwealth of PR (Figure 1). Isla de Culebra consists of an irregularly shaped and roughly 10.5 km by 8.5 km main landmass (hereafter referred to as Culebra) and 20 cays. Culebra is an emergent part of the Puerto Rico-Virgin Islands microplate, a broad and tectonically active deformation zone defining the boundary of the Caribbean and Atlantic plates (Masson and Scanlon, 1991). The dominant lithology dates to the Cretaceous Period and is composed of surface volcanics (andesites) and shallow intrusives (Meyerhoff, 1927). Although the maximum elevation in Culebra is just shy of 200 m, Culebra's topography is hilly and characterized by abrupt slopes of up to 36 degrees near the ridge tops, interrupted by flat alluvial deposits and coastal wetlands. The dominant soil type in Culebra is the generally shallow (40-65 cm thick), well-drained, and moderately permeable Descalabrado clay-loam series also found in semi-arid areas of the US Virgin Islands (USVI) and southwestern PR (Beinroth et al., 2003). The annual rainfall rate in Culebra is close to 990 mm per year and the average temperature is about 26–27° C (PR-EQB, 1970; USACE, 1995). Hence, the island displays a sub-tropical dry type of vegetation typical of low altitude tropical forests with high evapotranspiration but low annual rainfall rates (Ewel and Whitmore, 1973). Watersheds are small with none exceeding more than 2-3 km^2 and are drained by poorly defined, intermittent streams.

Culebra is surrounded by a large system of fringing reefs, rocky bottoms, hard grounds and mid-shelf reefs which are representative of the northeastern Caribbean (Hernández-Delgado, 2005). Windward side reefs generally have extensive structural development, including some spur and groove systems. However, many still display significant physical destruction associated to the impacts of Hurricanes David and Frederic in 1979, Hugo in 1989, Marilyn in 1995, and Georges in 1998 (Garrison et al., 2005). Some extensively developed linear fringing reef structures include some of our study sites (i.e., Ensenada Malena, Cayo Dákity, Ensenada Almodóvar, Puerto Del Manglar, and Playa Larga). There are also extensive systems of discontinuous fringing and patch reefs, like those in the Punta Soldado and Playa Zoní areas. There is also an extensive system of mid-shelf reefs from about 1 to 5 miles off the eastern and southeastern coasts of the island, as well as an extensive and poorly studied system of mesophotic coral reef communities extending from Culebra to the east towards St. Thomas, U.S. Virgin Islands.

2.2 Land use history

Although the number and significance of archaeological findings in Culebra have yet to match the magnificence or relevance of those uncovered in the nearby Isla de Vieques and mainland PR, evidence still suggests that the island was transiently inhabited by various groups of Amerindians during pre-colonial times (Hernández-Delgado et al. 2003). No evidence exists of any permanent human presence in Culebra until 1880 when a Spanish decree promoted settlement and habilitation for agricultural production and cattle grazing. In 1901, only three years following the transition of the entire Puerto Rican territory from the Spanish Crown to the United States, the US Navy began to establish a presence on the island. Paradoxically, while areas of Culebra came to house the first wildlife refuge of the entire Insular Caribbean (established in 1909), other nearby areas became live ammunition training grounds for the US Navy's Atlantic Fleet. During and following the WWII access to the island became severely restricted and this is presumed to have had long-term repercussions on the island as it severed Culebra from the new economic plan being propelled over the rest of PR (Estudios Técnicos Inc., 2004). Some sense of civilian normalcy was finally achieved in Culebra following the ouster of the US Navy in 1978.

Land development on Culebra occurred at a rather slow rate over the initial years following 1978 but it has experienced an accelerated pace since the late 1990's. Contemporary land development practices on Culebra generally consists of vegetation removal, combined with ground leveling and compaction associated to construction (e.g., individual home sites) and opening of low-standard, steep roads that tend to remain unpaved and exposed to erosion over relatively long periods (Figure 2). Land disturbance is achieved by heavy machinery and generally lacks construction-phase mandatory erosion control practices. Most activities fail to meet stormwater design standards theoretically required by US and PR Commonwealth regulations. The accelerated pace of development in Culebra, accompanied by an unwillingness to comply with or enforce environmental regulations by both the private and public sectors has led to increases in soil erosion and sediment delivery rates to the marine environment which have been implicated in documented coral reef decline (Hernández-Delgado et al., 2006). We believe that Culebra is currently exhibiting its highest ever sediment yield levels as a result of its recent and ongoing construction activities, as it has been suggested elsewhere (Wolman, 1967; MacDonald et al., 1997; Brooks et al., 2007; Ryan et al., 2008). Limited background information currently exists on the type of land development on a subtropical dry climatic setting such as in Culebra, but data from similar sites in La Parguera (southwestern PR) and St. John (USVI) suggest that disturbed hillslopes can erode at rates that are ten to up to four-orders of magnitude higher than undisturbed, densely-vegetated surfaces (Ramos-Scharrón and MacDonald, 2005, 2007a; Ramos-Scharrón, 2010), and that current watershed-scale sediment yields into coastal waters are upwards to ten times higher than under undisturbed conditions (Ramos-Scharrón and MacDonald, 2007b).

3. Methods

The new assessment framework presented here follows a multi-step approach (Figure 3). The first step requires collecting the basic information to describe coral reef abundance and condition, estimating watershed-scale sediment loading rates, and evaluating the feasibility of on-site erosion control measure installation. The second step is meant to formalize an approach to select the watersheds and associated marine habitats that merit a preferred status for the implementation of erosion control activities [Section 3.4.1]. Evaluation of need to mitigate erosion is gaged based on three main considerations: (1) *resource abundance*- the amount of surface prone for coral reef growth and/or the abundance of particular coral species of concern [Section 3.1.1]; (2) *resource condition*- the observed condition of the coral

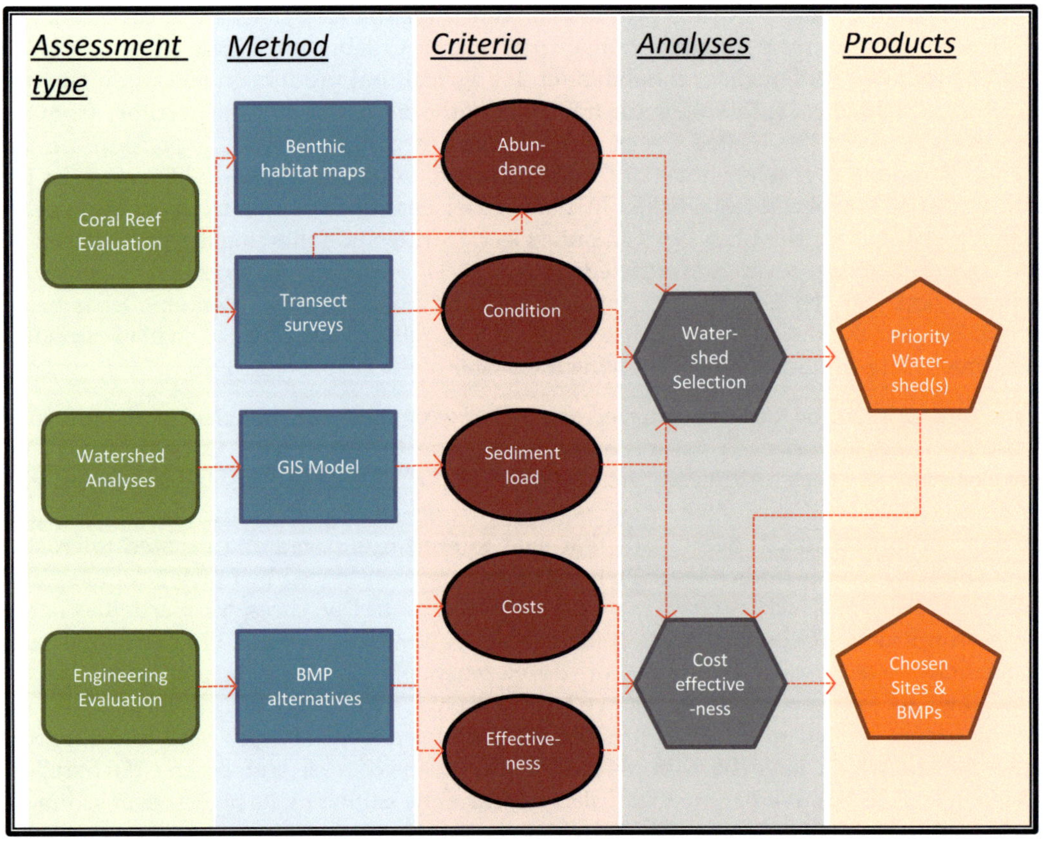

Fig. 3. Flowchart displaying the general scope of the proposed, interdisciplinary erosion mitigation strategy described here.

reef ecosystem [Section 3.1.2]; and (3) *stress level*- sedimentation stress defined by annual sediment yields [Section 3.2]. The final step focuses on the selected watershed(s) and aids in choosing the specific sites (i.e., sediment sources) and methods to be implemented within the priority areas by invoking a sediment reduction cost-effectiveness analysis [Section 3.4.2]. The goal of this final analysis is to minimize the costs of BMP implementation [Section 3.3] while maximizing the reductions in sediment delivery.

Financial constrains will always pose limits to the level and extensiveness of erosion control programs. The guidelines presented here describe a conceptual framework to optimize the use of multi-disciplinary information concerning coral reef abundance and condition, sediment loading rates, and engineering considerations in defining priorities and solutions to tackle erosion control goals. The framework presented here has the potential for becoming incorporated in watershed management plans in other areas of PR and elsewhere throughout the Caribbean.

3.1 Coral reef assessment

3.1.1 Abundance of coral reef habitats

An analysis on the abundance of nearshore submerged zones prone for coral reef establishment and growth was based on its areal coverage, and relied upon pre-existing benthic habitat maps derived from aerial photo-interpretation of images taken in 1999 (NOAA-NOS, 2001). The NOAA-NOS map is publicly available as an ArcGIS shapefile and contains polygons organized following a hierarchical collapsing classification scheme. Classes ranged from eight generalized categories to twenty-six sub-classes that describe in increasing detail each habitat type (Kendall et al., 2001). Since our main intent was to quantify the area prone for coral reef growth, we opted to rely on the third tier of this classification scheme (database attribute: '*HABITAT*') that delineates areas as '*coral reef and colonized hardbottom*', '*submerged aquatic vegetation*' (SAV), '*unconsolidated sediments*', and '*others*'. The original polygon geometry of the NOAA-NOS map was simplified by dissolving polygons based on the '*HABITAT*' attribute.

The premier intention of the abundance analyses was to quantify the areal coverage of coral reef areas in the proximity of each of the eight watersheds of interest. Therefore, our procedure had to assign a submerged area to each of the watersheds. The submerged area for each location is roughly based on a 450 m buffer extending from the coastline defining the downslope end of all watersheds. The edges of each of the areas were manually edited to avoid overlaps whenever conflicts existed between two adjacent watersheds. Each of the newly created submerged area polygons were used to clip or extract the simplified benthic habitat map. The surface area for the clipped and simplified benthic habitat maps was calculated in hectares and summarized to provide a total area for the four '*HABITAT*' categories within each of the submerged areas.

3.1.2 Structure and condition

Coral reef communities adjacent to the terrestrial study sites were assessed in 2007 using digital video imaging and six replicate 25 m-long point-intercept line transects per site, with intersects at 0.5 m intervals (n=50 points/transect). This protocol represents a slight modification from Rogers et al. (1994). Data was obtained at depths typically ranging from 3 to 7 m. Deeper reef zones were not assessed and consequently not included in the analysis. Data used for this study included percent cover of benthic components: live coral, macroalgae, algal turf, crustose coralline algae (CCA), and cyanobacteria. The data obtained also allowed calculations of live coral to algal ratios (i.e., '*coral:algal*') for all algal functional groups, as well as coral to cyanobaterial ratios (i.e., '*coral:cyanobacterial*'). Ratios are used to further describe the condition of the reef ecosystem. Addressing spatial patterns in algal communities is of utmost importance since they often respond relatively fast to runoff, eutrophication and sedimentation disturbance. Observed percent living coral cover values could be considered as either a measure of abundance or descriptive of existing conditions. The same could be said for the percent relative cover of four highly sensitive Scleractinian coral species: Elkhorn coral (*Acropora palmata*), Staghorn coral (*A. cervicornis*), Columnar star coral (*Montastraea annularis*), and Laminar star coral (*M. faveolata*).

3.2 Watershed assessment

The unpaved road network in Culebra is considered to be the island's most important anthropogenic source of sediment (Ramos-Scharrón, 2009; Figure 2). Therefore, our erosion and sediment yield assessments relied upon application of the STJ-EROS model (Ramos-Scharrón, 2004; Ramos-Scharrón and MacDonald, 2007b) as it provides a modeling structure that estimates the annual sediment contribution from both natural sources of sediment and unpaved road networks. STJ-EROS is a Geographical Information System (GIS) model that uses empirical sediment production functions (Ramos-Scharrón, 2004; Ramos-Scharrón and MacDonald, 2005, 2007a) and sediment delivery ratios to estimate sediment yields into coastal waters. STJ-EROS estimates erosion rates from both natural (i.e., streambanks, treethrow, and undisturbed hillslopes) and anthropogenic sources of sediment (i.e., unpaved roads) based on empirical equations developed from data collected on St. John (U.S. Virgin Islands-USVI), an island with a similar physical setting as Culebra's. In STJ-EROS, the estimated sediment delivery rate from the terrestrial to the marine environment is controlled by user-defined sediment delivery ratios (SDRs), where SDR is the ratio of sediment delivered to the gross erosion occurring within the basin (Walling, 1983). STJ-EROS allows users to choose SDR values for areas having different delivery potentials defined by qualitative observations on the location of coastal wetlands relative to the stream network. Areas draining through wetlands before delivering runoff into coastal waters received a SDR value of 25%, while those directly draining into coastlines without an intervening wetland were assigned a SDR of 75% (Ramos-Scharrón, 2009). A 1000 mm per year rainfall value was consistently used for all eight watersheds.

Field surveys consisted in the preparation of input geo-databases required by STJ-EROS. Field reconnaissance determined that the algorithms for two of the natural sources of sediment included in STJ-EROS did not apply to the conditions in Culebra. Treethrow, or the generation of sediment by the uprooting of wind-thrown trees, was not considered relevant in Culebra given the generally low-lying, dry-forest type of vegetation of the island which contrasts with the wetter and taller, treethrow-prone trees found on higher elevation portions of St. John (Reilly, 1991). In addition, none of the watersheds of interest contained well-defined stream channel features from which streambank erosion could be expected. Therefore, the application of STJ-EROS in Culebra assumed that surface erosion from currently undisturbed hillslopes is the only natural source of sediment of significance on the island.

Unpaved roads on Culebra were found to be similar to those from which the STJ-EROS road erosion algorithms were developed in terms of substrate, road prism geometry, and range of slopes. STJ-EROS requires building up a geographical database that involves mapping of individual road segments and their associated drainage points (i.e., culverts, water bars, etc.). Individual road segments were spatially delimited by changes in surface type (i.e., paved versus unpaved and assumed grading frequency) and they consist of road sections with flow patterns uninterrupted by drainage structures. Geographic and attribute data was generated by a combination of on-screen digitizing using an ortho-corrected, full-color, 1-m resolution, 2004 aerial image, in combination with field mapping using a Geographical Positioning Unit (GPS) and field sketches. Field sketches contained information related to surfacing (i.e., paved or unpaved), road segment geometry (i.e., length and width of sub-segments) measured with a tape measure, slope in percent measured with a hand-held clinometer, and a categorical description related to the frequency of road grading (i.e., graded, ungraded, or abandoned- See definitions in Ramos-Scharrón and MacDonald, 2005). In the absence of precise time since construction or information on the frequency of grading, assigning a road segment to a category was based on a qualitative assessment of the road surface texture and vegetation cover. A road was identified as graded if its surface was dominated by a fine granular texture, limited exposure of large rock fragments, and low vegetation cover. Ungraded roads were those still actively travelled but with an abundance of coarse fragments (i.e., armored surface) and low to moderate vegetation cover density. Abandoned roads were those that exhibited an armored surface and a high abundance of vegetation due to scarce or no traffic.

The eight watersheds of interest covered a total area of 6.8 km^2 or about a quarter of the total landmass of Culebra. Individual watersheds ranged in drainage areas from 9.1 ha at Punta Soldado (PSO) to 274 ha at Puerto Del Manglar (PDM) (Table 1). The proportion of area defined as having a high potential for sediment delivery varied widely from a maximum of 99% at PSO and Ensenada Fulladosa (EFU) to a minimum of 16 - 17% at the larger watersheds of PDM and Bahía Mosquito (BMO) that contain extensive wetland areas. The eight watersheds contain almost 35 km of roads out of which 24 km remain unpaved for an overall unpaved road density of 3.5 km km^{-2}. About 24 km of roads were field-surveyed in 2008 within the eight watersheds. Data for road segments for which no access was allowed was estimated based on aerial image interpretation and GIS analyses. Roads that were not surveyed were assigned a standard 4.0 m width and assumed to be ungraded. Drainage points along these roads were presumed to be located at topographical depressions; slope was calculated using the digital elevation model by taking the elevation difference between the top and lower ends of the road segment and dividing it by the length of the segment. Runoff from all roads is delivered off the road network at 160 drainage points.

STJ-EROS calculates total sediment production (i.e., total erosion) from all individual road segments within a watershed as well as their estimated annual sediment yield contribution to coastal waters (i.e., sediment delivery in tons per year). Model results are of three main sorts. First, the model provides an estimate of the total amount of sediment delivered to the marine environment every year from any given watershed. Second, the model may isolate the contribution from each sediment source type (i.e., undisturbed hillslopes, and graded, ungraded, or abandoned unpaved roads), thus allowing us to rank sources according to their net contribution. Finally, the GIS nature of the model allows it to spatially display the

Watershed	Drainage area (ha)	Proportion of Area with High SDR (%)	Proportion of Area with Moderate SDR (%)	Total roads (km)	Unpaved roads (km)	Number of drainage points
Punta Soldado (PSO)	9.1	99%	1%	0.09	0.09	0
Ensenada Malena (EMA)	18	60%	40%	0.52	0.52	7
Cayo Dakiti (CDA)	44	41%	59%	1	0.31	6
Bahia Mosquito (BMO)	132	17%	83%	7.4	5.3	34
Ensenada Fulladosa (EFU)	70	99%	1%	6.9	4.5	42
Puerto Del Manglar (PDM)	274	16%	84%	12.8	9.7	49
Playa Larga (PLA)	52	78%	22%	2.1	1.8	8
Playa Zoni (PZO)	81	28%	72%	3.9	1.9	14
Total	*680*			*34.7*	*24.1*	*160*

Table 1. Summary description of the eight study watersheds including their drainage areas, the proportion of each watershed contained within the two sediment delivery ratio (SDR) categories, the total length of roads and unpaved roads, and the number of road drainage points.

magnitude of sediment contributed by individual road segments or the amount of sediment being delivered from the road network to the marine environment through each individual drainage structure.

3.3 Best Management Practices

Best Management Practices (BMPs) related to soil erosion refer to "… a variety of site planning, design, and construction activities to minimize the production and transport of sediments" (Anderson, 1994). BMPs may refer to precautions taken during the planning and construction stages of new roads that help locate, align, and define their geometry by not only considering their capital costs but also taking into account potential contamination of downstream water bodies. It is our impression from the current state of the road network in Culebra that, with only few exceptions, no erosion control considerations are contemplated when roads are being planned and laid out. Since the intention of this article is to provide guidance on mitigating already existing problems, readers are referred elsewhere for a comprehensive discussion on forest road construction guidelines (e.g., B.C. Ministry of Forests, 2002).

Road erosion mitigation BMPs are of three main types. First, are those methods that improve the resistance to erosion processes by preventing the direct contact of rain and runoff with the soil surface (Type I). These include different methods to promote re-vegetation, use of gravel for added protection (Ziegler and Sutherland, 2006), and paving a surface with concrete. The second type of BMPs is meant to minimize the amount of flow on the unpaved road surface and thus reduce its erosive energy. This is mostly achieved by preventing flow concentration with a variety of stormwater drainage structures including side-ditches, rolling dips, water bars, and culverts, among others (Ramos-Scharrón, in press). The third type of BMP attempts to capture as much sediment as possible while runoff is transported through or discharged from the road prism (Type III). These BMPs reduce flow velocity and thus promote settling of sediment, and include methods such as hay bales, sediment traps, check dams, and settling ponds (Anderson, 1994). Road drainage improvements (Type II) may also be viewed as attempts to reduce the downstream transport of eroded sediment as adequate placement of road drainage structures also promotes reduced connectivity with downslope water bodies (Megahan and Ketcheson, 1996; Croke et al., 2005).

The specific approach to identifying erosion and sediment control BMPs in Culebra is framed by three general limitations including: (a) an already existing and thus mostly immovable road network layout; (b) a characteristically rugged topography, and (c) a lack of locally available specialized materials and equipment that significantly increases costs and therefore reduces the number of BMPs that would otherwise be considered feasible. The following list of BMPs consists of those methods that are being given further consideration for application in Culebra as they are deemed implementable from a technical and economical point of view:

i. *Inside ditch-* An upslope or inside ditch running along the length of the road reduces erosion by providing a surface specifically prepared to handle the runoff generated by the road travelway (Types I and II).
ii. *Vegetated ditch-* Allowing or providing for vegetative cover within ditches stabilizes the ditch surface, reduces flow energy and enhances suspended sediment deposition (Types I and II).
iii. *Check dams-* When installed along ditches, check dams reduce flow energy and thus reduce the potential for erosion. Dams also allow sufficient space for sediment deposition and can be constructed of locally available materials such as rocks, logs, or properly treated native soil (Types II and III).
iv. *Rolling dips-* These stormwater handling structures consist of a reverse grade depression aligned diagonally to the general trend of the road. A mound of soil running parallel to the downslope side of the dip serves to increase their runoff handling capacity. Rolling dips are used to divert water from the unpaved surface into the ditch or out of the road prism and therefore reduce flow concentration and shorten downstream delivery (Types II and III).
v. *Paved gutter-* These play a similar role to rolling dips in that they divert water into a ditch or out of erodible road surfaces. The only difference is that these are covered by pavement and are therefore more costly (Types I, II, and III).
vi. *Energy dissipaters-* These are installed at discharge points on the downslope end of the road prism. They usually consist of riprap, baffled concrete structures or small catchment basins. Implementation costs are low to high, depending on the type. For the purpose at hand we have considered riprap as the preferred energy dissipating BMP with a moderate implementation cost (Types I, II, and III).
vii. *Wire-mesh pavement-* This is a measure that makes the surface impervious and prevents contact between in-situ soil particles with rainfall and runoff. Implementation costs are very high. (Type I).

BMP selection in most cases is site specific and a combination of these individual BMPs is usually the most effective alternative. Therefore, we developed three general road designs or treatments, each incorporating a different sub-set of BMPs (Table 2). The first type maintains an unpaved road segment, but enhances stormwater management by constructing a properly vegetated ditch with check dams, unpaved rolling dips, and energy dissipaters every 30 m ($325 per linear meter of road). The second type exactly matches the first type description but relies on a paved gutter to channel water out of the road surface instead of a rolling dip ($350 m^{-1}). The third type refers to a fully-paved road with an adequately vegetated ditch, no check dams, and one paved gutter every 30 m with an accompanying energy dissipater ($600-$650 m^{-1}). Post-implementation erosion rates for the two treatments that maintain an unpaved road surface is estimated to be 30% of pre-treatment rates and this is based on an effectiveness evaluation study conducted for a singular road segment on St. John (USVI) (Ramos-Scharrón, in press). Erosion rates following treatment by paving the entire road travelway is expected to reduce rates to only about 10% of pre-treatment levels. This is based on field data collected from twenty road segments on St. John which found out that only 10% of the sediment exiting a road prism is generated by road cutslopes, while the remaining 90% of sediment is generated from the road travelway (Ramos-Scharrón and MacDonald, 2007a). Since pavement effectively shuts down the entire contribution from the road travelway, we assume that road cutslopes are the sole source of sediment exiting the road prism. It is important to note that our effectiveness evaluations cannot estimate the role played by the check dams or energy dissipaters in reducing sediment production rates.

Treatment name	**Inside ditch**	**Vegetated ditch**	**Check dams**	**Rolling dips**	**Paved gutter**	**Energy dissipater**	**Wire-mesh pavement**	**Costs (U.S. $ per m)**	**Post-treatment erosion rates**
Unpaved with rolling dips	✓	✓	✓	✓		✓		$325	~30%
Unpaved with paved gutter	✓	✓	✓		✓	✓		$350	~30%
Paved with gutter	✓	✓			✓	✓	✓	$600-$650	~10%

Table 2. Summary of the three main treatment types being considered for implementation in Culebra. Each treatment contains a different assortment of BMPs. Costs represent implementation costs in Culebra in U.S. dollars and apply to 2011 prices; efficiency of the different treatment options are based on the road erosion literature.

3.4 Prioritization strategy

Ecological *restoration* implies the manipulation of an ecosystem with the purpose of returning it back to its 'pristine' condition (Bradshaw, 1997). The approach proposed here recognizes the impracticality of attempting to fulfill the goal implied by this strict definition of *restoration*. The inadequacy of setting such rigorous goals is particularly applicable to coral reefs as they represent open systems affected by diverse biotic and abiotic processes, some of which act at spatio-temporal scales that are inalterable by direct human intervention. In addition, it is questionable if truly pristine coral reefs that could represent a genuine 'reference state' still exist in the Caribbean (Jackson, 1997, 2001; Gardner et al., 2003; Hawkins and Roberts, 2004). Therefore, we propose that an achievable goal of erosion control strategies should be simply to mitigate the effects of land erosion and sediment yields into coral reef systems, where *mitigation* "...refers to activities that lessen the degree of damage to an ecosystem..." (Jackson et al., 1995). Hence, erosion mitigation by itself has the singular purpose of reducing sediment delivery to levels that are somewhat closer to background rates. The presently grim condition of most coral reefs in the Caribbean and in Culebra is sufficient to justify curtailing sediment delivery into any reef-bearing water body (Hernández-Delgado, 2010; Hernández-Pacheco et al., 2011). Unfortunately, funding limitations will always restrain the level and extensiveness of erosion control strategies. Therefore, a scientifically-sound process by which priority areas for erosion control are selected based on local observations, needs, and availability of funds is critical to maximize the benefits of the effort, as well as to ensure goals and expectations are clearly acknowledged, and that activities are in agreement with the intended results.

The analyses presented here are limited in two important ways. First is that the coral reef condition assessments are based on a single, one-time observation. Therefore, the procedure is blind to trends in coral conditions that could serve in making more sound judgments when setting priorities for erosion control efforts. Second is that the erosion analyses explicitly lacks the capacity to understand sediment dynamics and effects once delivered to the marine environment. We acknowledge that an annual estimate of sediment delivery is too coarse to provide the temporal resolution needed to follow a process-based examination of the role of sediments on reefs. Nevertheless, erosion mitigation as defined above does not require *a priori* diagnosis of sedimentation as the cause of any coral deterioration. By reducing sediment loads we can expect to alleviate the light transmissivity limitations, high nutrient concentrations, abrasion, and direct sedimentation stresses on coral reefs that ensue increases in land erosion (Fabricius, 2005). Diminishing sediment stress should benefit reef ecosystems directly by lessening these effects and indirectly by allowing corals to better cope with other sources of stress (Hoegh-Guldberg et al., 2007).

3.4.1 Watershed selection

The combined watershed and marine habitat evaluation procedure presented here is based on three criteria: (1) abundance of the marine resource (i.e., coral reefs) [Section 3.1.1], (2) marine resource condition [Section 3.1.2], and (3) stress level (i.e., sediment load) [Section 3.2]. This type of multi-parameter evaluation is expected to be applicable to relatively homogeneous areas that might have shared similar marine and terrestrial conditions during their pristine states. A contained area like Culebra provides the perfect scenario for this type of analyses as it holds coral reef ecosystems that are quite compatible in their structure, and where factors that control sediment yields, such as rainfall, soils, relief, and watershed size are comparable among different sites. The areas chosen in Culebra all lie outside of the 'special' marine areas of the island that include the former US NAVY training grounds and the CLPN Reserve.

Each of the three criterion being considered for analyses can be graphically portrayed as the axis of a three-dimensional cube in which sediment stress level is displayed along the x-horizontal axis from low to high (left to right), while resource abundance is graphed in the y-vertical axis from low upwards to high abundance. Meanwhile, resource condition lies along the z-depth axis from good to poor (foreground to background) (Figure 4). The range of parameter values represented by each of the three axes making up the cube should represent the range of values found within the areas of interest. Therefore, the cube provides a conceptual space in which each of the areas is compared in terms of quantity and condition of coral reefs and sediment loading stress against the entire population of sites being considered and not to a theoretical reference state. Low/high and poor/good labels consequently refer to relative conditions within the context of the area of interest.

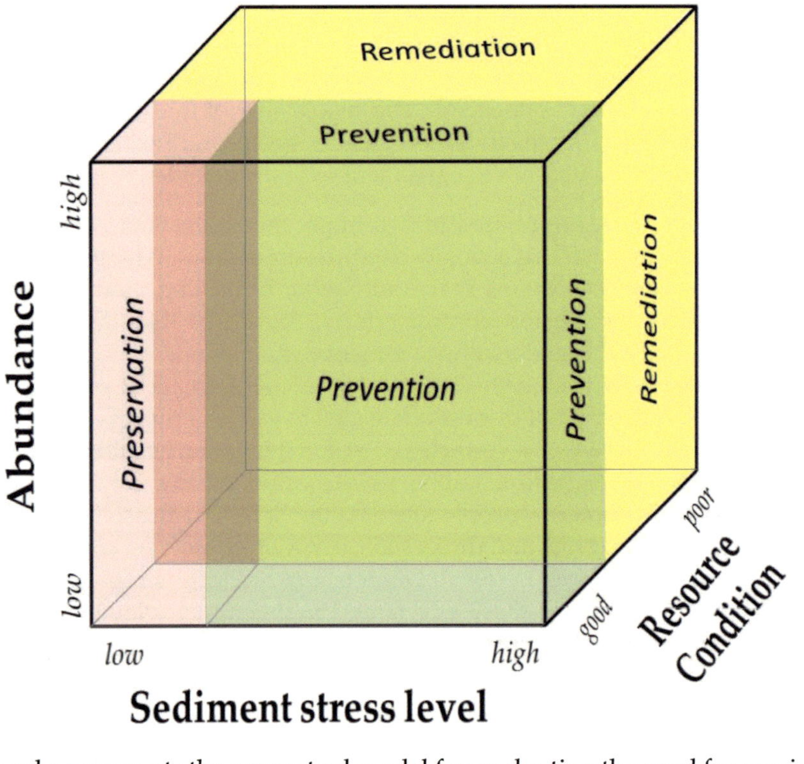

Fig. 4. This cube represents the conceptual model for evaluating the need for erosion mitigation in individual watersheds based on a set of multi-disciplinary parameters. The model is based on (a) abundance of marine areas prone for coral reef growth, (b) an assessment of coral reef condition, and (c) the estimated level of sediment stress originating from the most proximate watershed. Justification for controlling erosion varies according the particular values defined for the marine resource of interest and the watershed directly draining runoff and sediment into it.

The cube may also serve as a graphical aid to guide or understand the management tendencies of conceding some areas with a priority status ranking in relation to others. As mentioned above, all erosion mitigation strategies on watersheds contributing sediments into reef-bearing waters of the Caribbean is likely justifiable but the resource conservation goal of implementing such strategies at each site depends on the existing conditions. Hence, the cube also serves to map the justification or goal of erosion control activities being implemented. Implementation of erosion mitigation strategies for coral reef protection can be justified on the basis of three main motives:

i. *Preservation*- Within the context of the framework being presented here, preservation implies that current coral reef and sediment loading conditions remain close to a 'pristine' or 'desired' state. That is, coral reefs are in good condition and sediment loading rates are low. Any mitigation activities taken in these areas are meant to maintain conditions in their current state.
ii. *Prevention*- This term implies safeguarding or taking actions that anticipate an imminent or potential problem. Erosion control efforts in areas characterized by moderate to good coral reef conditions and moderate to high sediment loads could be considered as preventive measures intended to avoid further coral reef deterioration.
iii. *Remediation*- *Remediation* implies reducing the level of stress to an ecosystem without any concern to an ultimate goal (Bradshaw, 1997; Clark, 1997). In this context remediation is very similar to the meaning attached to rehabilitation in which it implies activities meant to improve conditions on a coral reef with a presently degraded state with an emphasis on the process and not on the end point (Bradshaw, 2002).

Within the context of Figure 4 remediation applies to areas with moderate to poor coral reef conditions regardless of sediment stress levels. Remediation is a term we prefer over *restoration* as it lacks the intention of attempting to reverse conditions back to a pristine state. The approach presented here finds unnecessary to establish a diagnostic cause and effect relationship between high sediment loads and poor coral reef conditions. It simply acknowledges that any increase in sediment yield rates above background levels is potentially harmful to corals, and therefore any reductions in sediment delivery rates will be of benefit to reef ecosystems. Erosion control in the spirit of preservation, prevention, or remediation would then be assumed as a way to alleviate a source of stress related to sediments with the intention of improving the chances at handling other stressors. In other words, erosion mitigation can then be viewed as an attempt to "... restore self-healing processes in an ecosystem that will lead to balance once more." (Jackson et al., 1995). In addition, erosion control would also aid in establishing adequate conditions for enhancing the success of other management measures such as coral farming and transplanting.

Given the limitations imposed by the current state of coral reef science and the diversity of real-world scenarios we believe it to be unbeneficial to attempt to provide a simple generic formula to rank watersheds in terms of need or expected optimization of results. The final decision on prioritization must be left to stakeholders, managers, scientists, and engineers knowledgeable of the local conditions. Other criteria, such as physical connectivity between watersheds and marine resources outside the immediate receiving bays, the presence of areas or species with a special conservation designation, temporal trends in coral abundance or condition, and recent land development activities must also be considered. Nevertheless, the framework might be able to offer some explicit prioritization guidance in two special scenarios. One scenario is when two or more sites display similar coral and sediment load conditions, in which case priority could be awarded to areas with a higher abundance of reefs. A second scenario is one in which multiple sites possess similar reef abundance and coral conditions, in which case priority could be conceded to areas exhibiting the highest sediment loads.

3.4.2 Site and BMP selection, cost-effectiveness analyses

For the purposes of erosion mitigation, cost-effectiveness may be defined by the total amount of funds spent installing BMPs relative to the amount of sediment that will no longer reach coastal waters (i.e., sediment 'savings') as a result of their implementation. Therefore, cost-effectiveness for the case in Culebra will be described in terms U.S. Dollars spent on BMPs per ton of sediment 'saved' (\$ ton^{-1}). Costs of the implementation of the unpaved road BMPs being considered in Culebra and their expected reductions in sediment production have been discussed above [Section 3.3]. Savings in the amount of sediment that would not be reaching the marine environment due to BMP installation are based in relation to STJ-EROS results and standard effectiveness measures. Cost-effectiveness evaluation was executed for only one of the eight study watersheds and consisted in the evaluation of per unit ton costs for treating road segments within the chosen watershed. Analyses were based on the three treatment options described in Table 2. Individual segments were ranked according to their pre-treatment sediment contribution and the cumulative implementation costs and sediment savings were calculated based on the incremental costs and 'savings' based on this ranking.

The cost-effectiveness analysis described here does not prescribe a given level of sediment reduction; it only attempts to maximize effectiveness given a total amount of funds available for mitigation. Coral reef science currently lacks the type of process-based analyses capabilities to define adequate loading levels. Levels that would suit a particular case might not be proper for another location due to differences in coral reef structure, species composition, or oceanographic conditions, to name a few. As previously described, our approach is based on the principle that any attempt to bring sediment loading levels closer to background rates is beneficial to coral reefs.

4. Results, discussion and recommendations

4.1 Coral reef abundance

The GIS procedure used to describe the benthic habitats directly linked to each of the eight study watersheds led to the characterization of 705 ha of submerged areas (Box 1). Almost 40% or 281 ha was identified as coral reef and colonized hardbottom, while 46% (325 ha) was SAV, and 14% (99 ha) was unconsolidated sediments, algal plains or other delineations. The watersheds directly associated with the largest coral reef and colonized hardbottom areas were Playa Larga (PLA; 88 ha) and Puerto del Manglar (PDM; 58 ha). Meanwhile, Ensenada Fulladosa (EFU) contains no corals and is dominated by SAV, mostly extensive seagrass beds largely composed by Turtle grass (*Thalassia testudinum*) and manatee grass (*Syringodium filiforme*) (Hernández-Delgado et al., 2003).

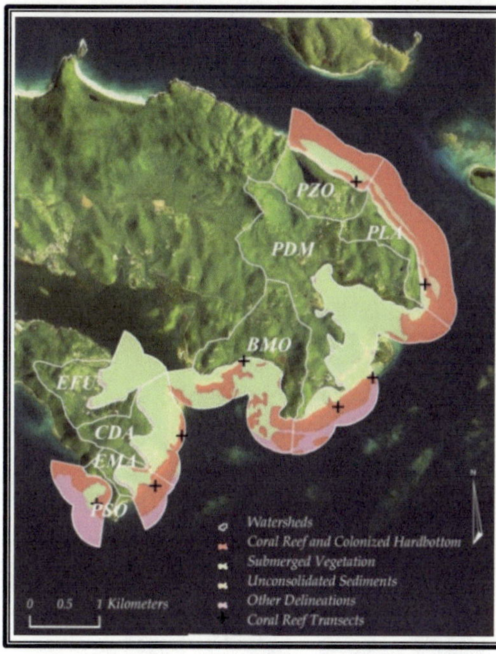

Watershed Name	Total submerged area (ha)	Coral Reef & Colonized Hardbottom (ha)	Submerged Vegetation (ha)	Unconsolidated sediments & other (ha)
Punta Soldado (PSO)	57	21	6.4	29
Ensenada Malena (EMA)	38	18	10.5	8.6
Cayo Dakiti (CDA)	71	14	53	4.1
Bahia Mosquito (BMO)	115	38	72	4.9
Ensenada Fulladosa (EFU)	52	–	49	2.8
Puerto Del Manglar (PDM)	201	58	113	30
Playa Larga (PLA)	106	88	7.9	9.9
Playa Zoni (PZO)	66	43	13	9.6
Total	705	281	325	99

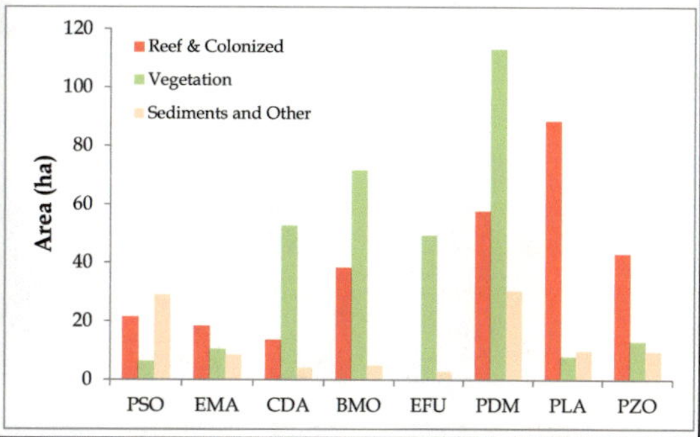

Box 1. Map of focus study area showing the code name and location of the eight study watersheds and the spatial distribution of benthic habitats directly associated to them. The table and figure summarize the quantitative benthic habitat abundance information obtained from the GIS analyses.

4.2 Coral reef condition

Benthic habitat characterization on coral reef communities adjacent to the study sites shows the unequivocal signs of long-term ecological decline. Overall, percent living coral cover averaged 7.8% across all study sites and was highest at Playa Larga (PLA) and Playa Zoní (PZO), with 12% and 10%, respectively (Figure 5a). The lowest percent living coral cover was observed at Bahía Mosquito (BMO) and at Cayo Dákity (CDA), with 4% and 3%, respectively Columnar star coral (*Montastraea annularis*) had a mean 1.4% relative cover across all sites, with a maximum value of 2.7% at PLA and a minimum value of 0.2% at BMO and CDA. Laminar star coral (*M. faveolata*) had a mean relative cover of 0.24% across all sites, with a maximum value of 0.6% at Ensenada Almodóvar within Puerto Del Manglar (PDM-1). Star coral was absent from surveyed transects at BMO and CDA. Threatened Elkhorn coral (*Acropora palmata*) had a mean relative cover of 0.1% across all sites, with a maximum value of 0.3% at PLA. Elkhorn coral was absent from surveyed transects at Punta Soldado (PSO), Las Pelás (PDM-2), and CDA. Also, threatened Staghorn coral (*A. cervicornis*) had a mean 0.5% relative cover across all sites, with a maximum value of 3.6% at PLA. It was absent from surveyed transects at Ensenada Malena (EMA), CDA, BMO, and PZO. It should be noted that numerous areas were covered by dead colonies in standing position of each species at each site, particularly of *Montastraea* spp. and of *A. palmata*. *Montastraea annularis* and *M. faveolata* are dominant components of many coral reefs across the region, even reaching percent relative cover values of 40 to 50% of the coralline fauna at many sites, but have showed significant recent declines as a result of sediment-laden and nutrient-loaded runoff pulses, in combination with climate-related impacts (Hernández-Delgado, 2010; Hernández-Pacheco et al., 2011).

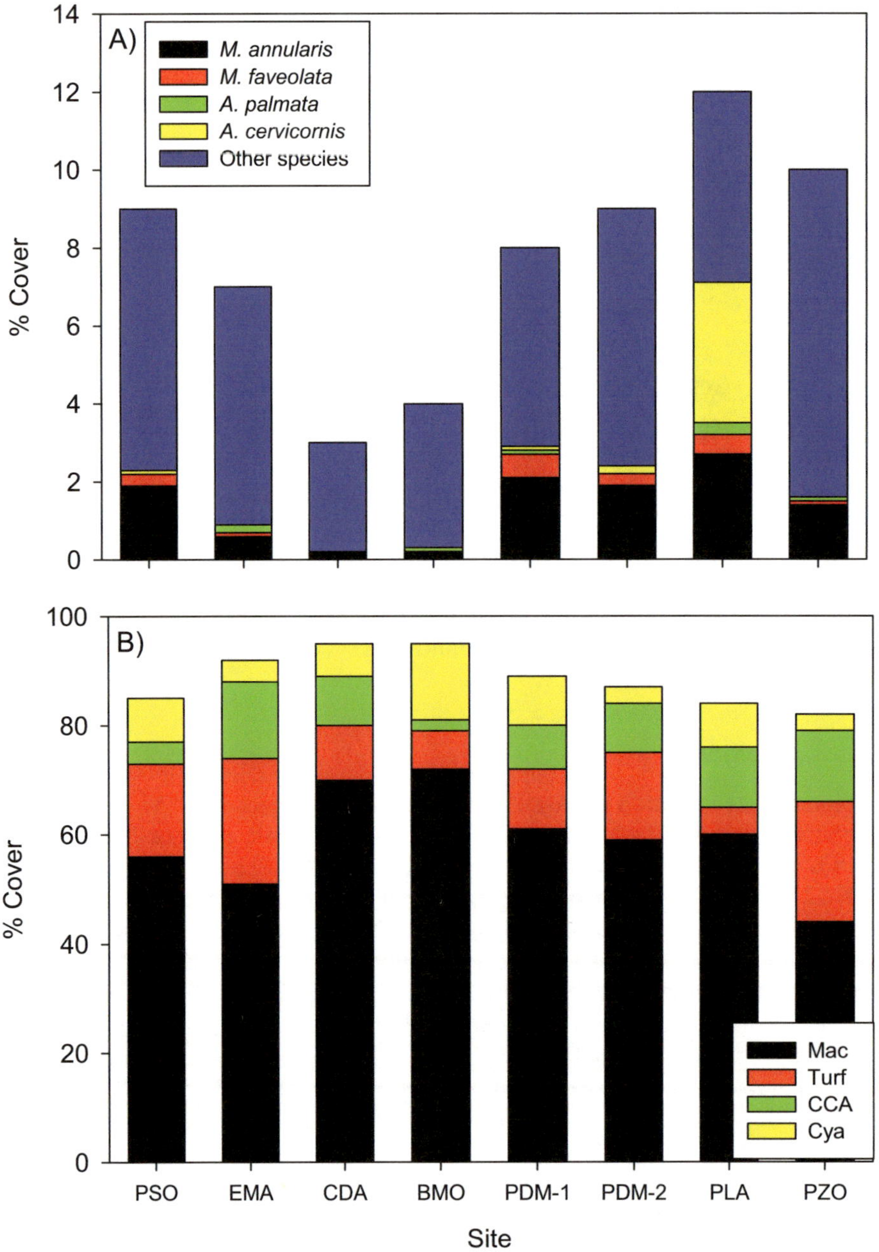

Fig. 5. Mean coral reef benthic parameter values for the eight study sites in Culebra. From top: A) Percent coral cover of four of the most sensitive Scleractinian coral species (*M. annularis*, *M. faveolata*, *A. palmata*, *A. cervicornis*); B) Percent cover of the four most important algal functional groups: macroalgae (Mac), turf, crustose coralline algae (CCA), and cyanobacteria (Cya).

Benthic habitats were largely dominated by non-reef building taxa, mostly algae (Figure 5b). Macroalgae averaged 59% across all sites and was particularly dominant on coral reefs adjacent to watersheds with higher sediment delivery rates like BMO (72%). Unpalatable brown algae *Dyctiota* spp. and *Lobophora variegata* were dominant across all sites, with other

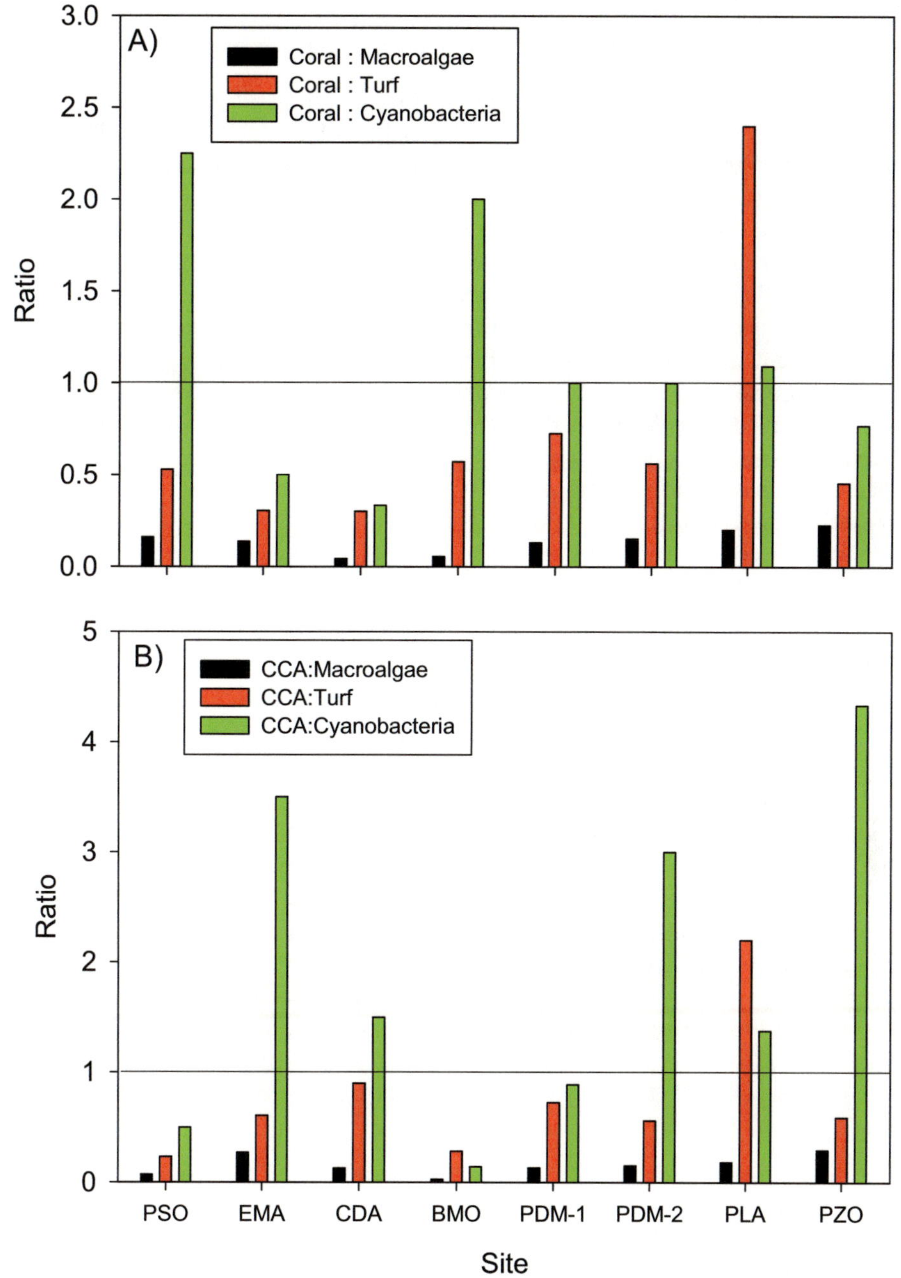

Fig. 6. '*Coral : algal*' ratios (Figure 6a) and '*CCA : algal*' ratios (Figure 6b) across the eight study sites in Culebra.

red and green macroalgae that were locally abundant at CDA and BMO. Lowest macroalgal cover was documented at EMA (51%) and PZO (44%). Algal turf was the second abundant algal functional group (14%), with higher values at EMA (23%) and PZO (22%), and lower values at PLA (5%) and BMO (7%). Crustose coralline algae (CCA) had a 9% mean cover across all sites, with maximum values at EMA (14%) and PZO (13%). The minimum value was observed at BMO (2%). Finally, cyanobacteria showed a nearly 7% cover across sites, with the highest value at BMO (14%), and the lowest at PDM and PZO (3%).

The '*coral : macroalgae*' ratio was highest at PZO (0.23) indicating a higher abundance of coral relative to other sites but still representing a macroalgae-dominated system. This ratio had its lowest value at CDA (0.04) (Figure 6a). The '*coral : turf*' ratio was highest at PLA (2.40), and lowest at CDA (0.30), while the '*coral : cyanobacteria*' ratio was highest at PSO (2.25), and lowest at CDA (0.33). The '*CCA : macroalgae*' ratio was highest at PZO (0.30), and lowest at BMO (0.03) (Figure 6b). The '*CCA : turf*' ratio was highest at PLA (2.20), and lowest at BMO (0.29), while the '*CCA : cyanobacteria*' ratio was highest at PSO (4.33), and lowest at BMO (0.14). Low '*coral : macroalgae*' and '*CCA : macroalgae*' ratios dominant across most sites, as well as the consistent abundant presence of cyanobacteria, strongly suggest that coral reef benthic communities across most sediment- and nutrient-impacted sites are being dominated by non-reef building taxa. Macroalgae, cyanobacteria, and other non-reef building taxa are known to be principal components of highly disturbed reefs, including those impacted by recurrent nutrient pulses (Cloern, 2001), sewage (Pastorok and Bilyard, 1985), low herbivory due to long-term fishing impacts (Bellwood et al., 2004; Hawkins and Roberts, 2004), or a combination of these (Littler et al., 2006a,b; Hernández-Delgado, 2010; Hernández-Delgado et al., 2010).

4.3 Watershed assessment

According to STJ-EROS, natural sources of sediment from within the eight study watersheds contribute 2.6 tons of sediment every year to the coastal waters of Culebra (Box 2). This estimate translates into an area-normalized yield rate of 0.40 tons km^{-2} yr^{-1}, which is an order of magnitude lower than the 2.6 - 6.7 tons km^{-2} yr^{-1} estimated for three watersheds on the island of St. John, U.S.V.I (Ramos-Scharrón, 2004). The difference between these rates is due to the lack of any sediment contribution from treethrow and streambank erosion in Culebra, two important sources of sediment dominating sediment delivery rates under natural conditions in St. John (Ramos-Scharrón and MacDonald, 2007a). STJ-EROS estimated that the total contribution from the unpaved road network in the eight study areas in Culebra is 347 tons per year or 133 times higher than background rates, and that the sediment gets distributed by a total of 160 road drainage points spread throughout the entire area (Box 2). Current sediment yield rates including contributions from both undisturbed hillslopes and the unpaved road network from all eight watersheds are estimated at 37.3 t yr^{-1} (5.6 t km^{-2} yr^{-1}).

STJ-EROS estimated very variable sediment yields for individual watersheds (Box 2). On one extreme, PSO represents an area lacking direct anthropogenic impacts in that it contains no road drainage points and where the entire 0.06 t y^{-1} contribution is solely derived from undisturbed hillslopes. Similarly, the watershed directly fronting CDA represents a barely impacted area with sediment yield rates only slightly above undisturbed conditions due to the reduced length of unpaved roads (Table 1). EMA, BMO, PLA, and PZO represent intermediate disturbance conditions with a more highly significant presence of unpaved roads leading to sediment yield rates ranging from 10.7 to 46.4 tons yr^{-1}. Meanwhile, EFU and PDM represent areas with extremely high sediment delivery rates of 112 and 154 tons yr^{-1}, respectively. The high delivery rates for EFU and PDM are due to the presence of a dense unpaved steep road network and road conditions that are prone to high road sediment production rates (i.e., steeper slopes and abundance of frequently graded roads). In addition, for the particular case of EFU the high delivery rates are also due to the lack of a prominent wetland buffer area that could promote the settling of sediment before it enters the bay. Although EFU does not support coral reef ecosystems, it did have a direct and rapid oceanographic connectivity with CDA, and with BMO in a lesser degree, particularly during ebbing tides. Normalized sediment yield rates for individual watersheds ranged between 0.68 and 220 tons km^{-2} yr^{-1}, which expand beyond the 8–46 tons km^{-2} yr^{-1} rates estimated for three watersheds on St. John (Ramos-Scharrón and MacDonald, 2007b). While the lower rates in Culebra represent rates equal to background conditions, the upper range of these rates represent delivery rates that are up to 320 times higher than background levels.

Watershed	Background sediment yield (t yr^{-1})	Unpaved road sediment yield (t yr^{-1})	Current/ background sediment yields	Normalized sediment yields (tons km^{-2} yr^{-1})
PSO	0.06	0.0	1	0.68
EMA	0.09	10.8	122	61
CDA	0.18	0.36	3	1.2
BMO	0.40	46.4	117	35
EFU	0.48	154	323	220
PDM	0.82	117	144	43
PLA	0.30	10.7	37	21
PZO	0.29	12.7	45	16
Total	*2.6*	*352*		

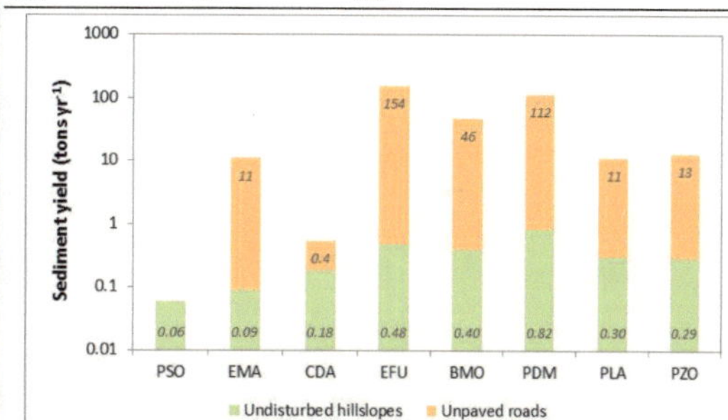

Box 2. Map presents one option for geographically displaying the STJ-EROS model results. Points in the map represent the annual amount of unpaved road sediment reaching a particular road drainage structure within the eight study watersheds. The table and figure summarize the estimated sediment yield rates related to both natural undisturbed hillslopes and the unpaved road network according to STJ-EROS.

4.4 Watershed-marine habitat selection

Watershed size and sediment delivery potential, as well as the length and characteristics of the unpaved road network have a direct influence on anthropogenic-driven sediment yield, which in turn impact adjacent coral reef ecosystems. Larger watersheds having a

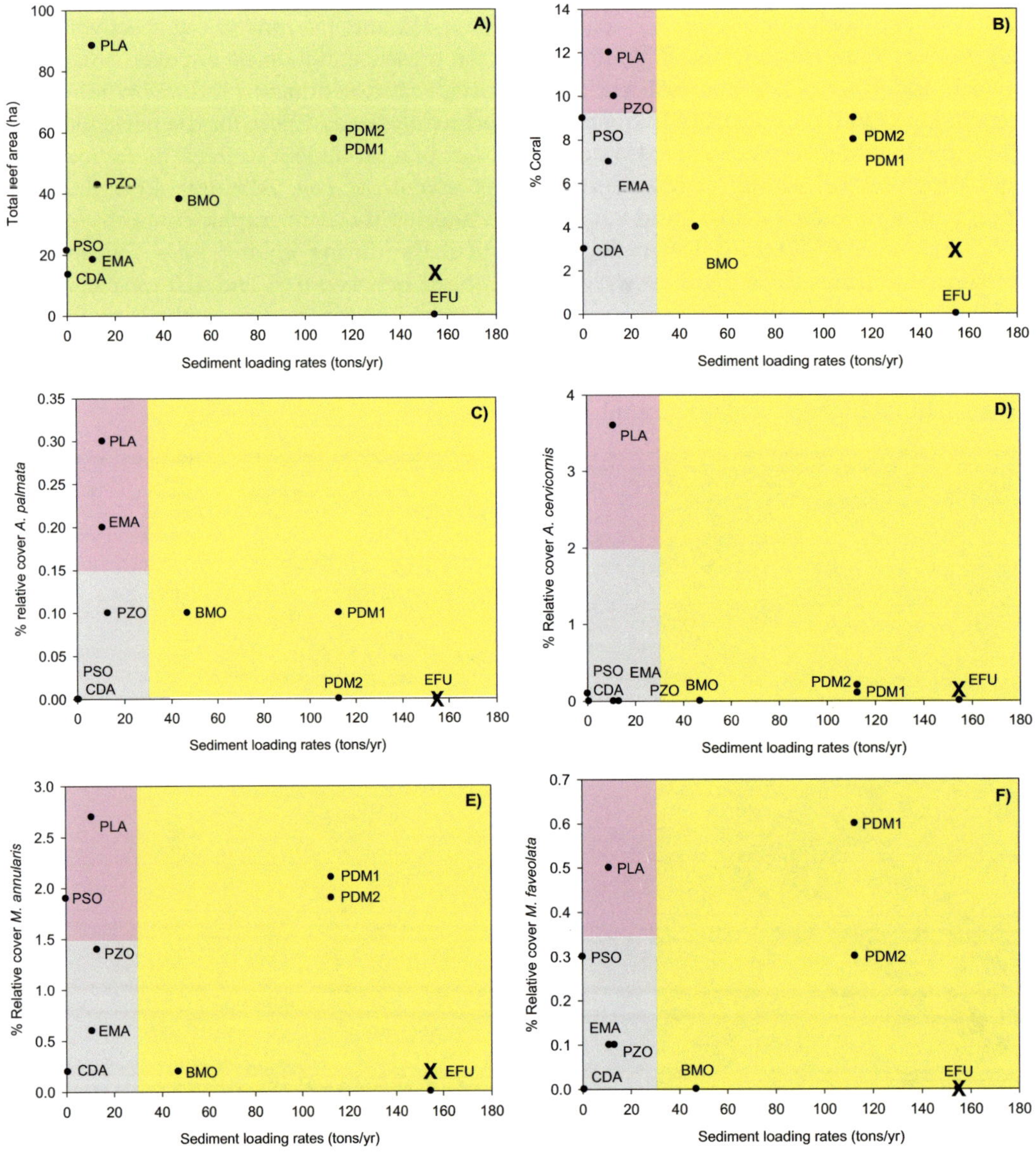

Fig. 7. Relationship between sediment loading rates and several benthic parameters across impacted coral reefs in Culebra: A) Total reef area; B) Percent living coral cover; C) Percent relative cover of Staghorn coral (*Acropora cervicornis*); D) Percent relative cover of Elkhorn coral (*A. palmata*); E) Percent relative cover of Columnar star coral (*Montastraea annularis*); and F) Percent relative cover of Laminar star coral (*M. faveolata*). Colors represent the justification for erosion control actions as follows: Yellow= remediation; Gray= prevention; and Pink= preservation as indicated in Figure 4. Point 'X' denotes a more realistic condition for CDA due to its down current oceanographic connectivity with EFU and the rest of Ensenada Honda (Figure 1).

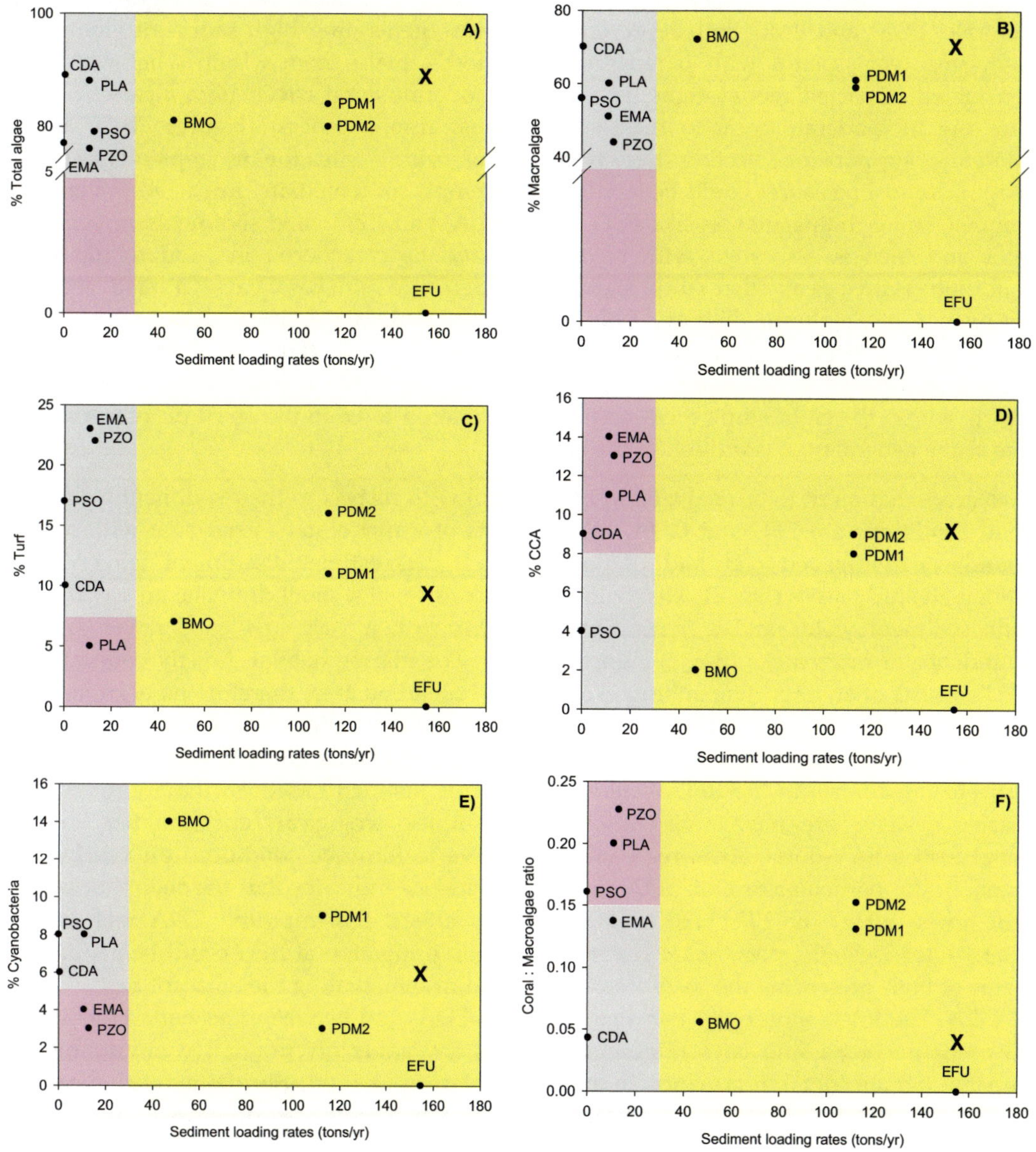

Fig. 8. Relationship between sediment loading rates and several benthic parameters across impacted coral reefs in Culebra: A) Percent total algal cover; B) Percent macroalgal cover; C) Percent algal turf cover; D) Percent crustose coralline algae (CCA) cover; E) Percent cyanobacterial cover; and F) *Coral : Macroalgae* ratio. Colors represent the justification for erosion control actions as follows: Yellow= remediation; Gray= prevention; and Pink= preservation as indicated in Figure 4. Point 'X' denotes a more realistic condition for CDA due to its down current oceanographic connectivity with EFU and the rest of Ensenada Honda (Figure 1).

higher density of steep unpaved and graded roads showed the highest sediment delivery rates to adjacent waters, often impacting large coral reef areas (Box 2; Figure 7a). For example, PDM and BMO both represent large areas generating high sediment yields and associated reef systems with parameters that mostly place them within what would be considered impacted reef systems (i.e., low to moderate coral cover, high algal coverage, and low to moderate coral to macroalgae ratios, among others) (Figures 7b-7f, 8a-8f). Therefore, support for selecting these two areas as priority sites for the implementation of erosion control measures could be justified as attempts to 'remediate' impacted systems. In contrast, small drainage areas like PSO, EMA, PLA, and PZO yield sediments at very low rates and each is associated with reefs that could be considered in good to moderate condition relative to all other study sites (i.e., moderate to high coral cover, a relatively high abundance of Staghorn, Elkhorn, Columnar, and Laminar star coral, low to high algal coverage, and high coral to macroalgal ratio) (Figures 7b-7f, 8a-8f). Support for erosion control on these watersheds could be catalogued as a 'preventive' effort. No erosion control efforts within the eight study areas could be considered to be in the spirit of 'preservation' due to the general poor condition of the reefs.

Two areas that merit to be analyzed in more detail with respect to their sediment loads and coral conditions are EFU and CDA. EFU consists of a moderately-sized area with a high abundance of unpaved roads and it represents the highest estimated sediment delivery rates among all study areas (Box 2). Meanwhile, CDA consists of a small drainage area with very little sediment yields and a marine environment with a very low coral cover and an abundance of macroalgae (Box 2, Figures 7b, 8b). The marine habitat directly connected to EFU consists of an important submerged aquatic vegetation area, therefore no argument for erosion control could be justified based on a strict interpretation of our scheme that only considers the abundance and condition of the immediately adjacent reef systems. We argue that erosion control in the EFU watershed could be justified based on the argument that marine systems are interconnected through complex ecological functionalities so that benefits to a SAV-dominated area could also serve to improve conditions on nearby reef areas. In the particular case of EFU, anecdotal evidence indicates that the sediment plume that flows out of the EFU marine area directly affects the impacted CDA reef system (Hernández-Delgado, pers. obs.). Therefore, erosion mitigation at EFU could be justified in terms of both preserving the SVA area at EFU and in remediating the adjacent reef systems at CDA. The interconnectivity between EFU and CDA, and between Ensenada Honda and all of its encompassing bays (Figure 1), signals the value of cumulative environmental impacts and anecdotal information in making final decisions for prioritizing erosion control efforts and the potential for incorporating other factors such as oceanographic current patterns in our analyses.

4.5 Site and BMP selection

The PDM watershed was chosen as the target area for conducting cost-effectiveness analyses because of its high sediment yield rates, its relatively extensive unpaved road network, and the poor to moderate condition of its adjacent marine resources. PDM contains a total of 9.4 km of unpaved roads, sub-divided into 104 individual road segments which in total deliver 112 tons of sediment every year into the receiving coastal waters (Box 2). The average road segment has a length of 90 m and a slope of 7% with individual values ranging between 12 –

390 m and from 0% to 25%, respectively. Twenty-seven road segments individually contribute more than 0.82 tons yr^{-1}, which is the estimated background sediment yield level for this watershed (Box 3). Although these road segments represent only 36% of the total unpaved road network and approximately 0.6% of the entire watershed surface area (~ 1.6 ha), they are responsible for 86% of its sediment yield. Three segments encompassing 0.74 km of roads individually contribute an excess of 10 tons of sediment per year (road segment id's 1-3 in Box 3) and together yield 52 tons yr^{-1} or 44% of the annual sediment load. The spatial distribution and delivery rates from individual road segments found throughout PDM reminds us that sediment pollution in this and most watersheds has a true non-point source nature but that particular road segments outweigh their counterparts in their relative contribution to watershed-scale sediment yields.

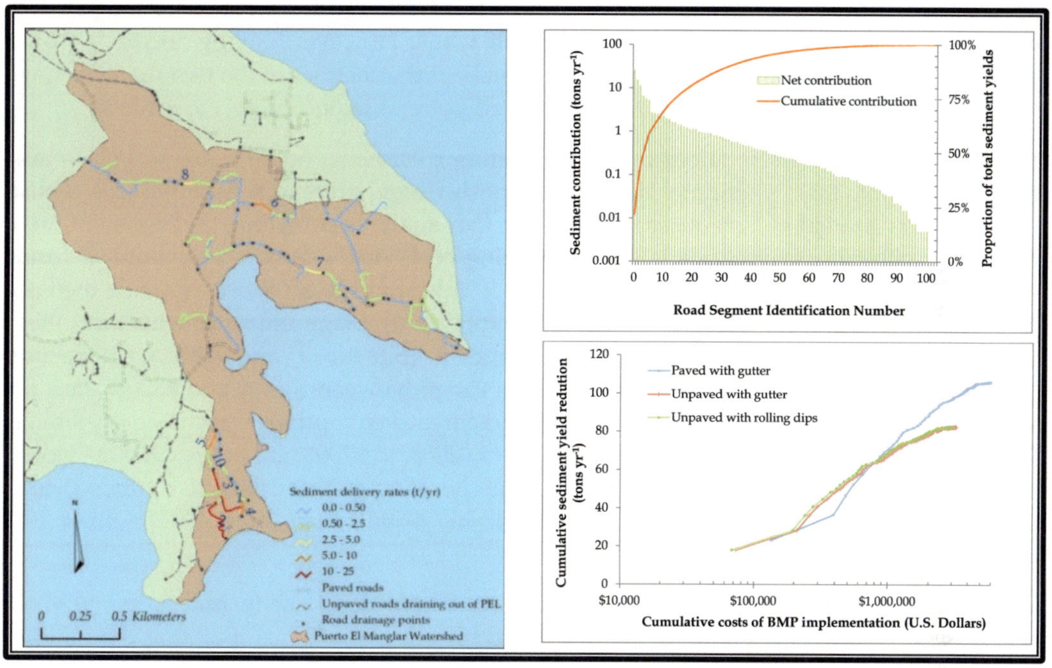

Box 3. Map contains another possibility for displaying the results of the STJ-EROS model by using a color-coded scheme to represent the amount of annual sediment contribution from individual unpaved road segments in the Puerto del Manglar (PDM) watershed. Numbers in the map represent the top-ten ranked road segments based on their individual sediment contribution estimates. Top-right graph displays the annual sediment contribution from each of the 104 unpaved road segments of the PDM and the cumulative proportion of the total estimated sediment yield. Bottom-right graph displays the relationship between cumulative implementation costs and cumulative reductions in sediment yields for the three treatment options described in Table 2. Cumulative costs and savings are consecutively added based on the sediment load rankings displayed on the top graph.

Paving all roads within PDM would reduce sediment yields by 106 tons yr^{-1} according to our estimates (Box 3). These reductions would imply a post-treatment sediment yield rate of 11 tons yr^{-1}, or roughly 10% of pre-treatment levels (117 tons yr^{-1}). These delivery rates would still be 13 times higher than background. Funds required to achieve this goal would amount to $6.1M for an overall cost-effectiveness measure of $57.7K per ton reduced. The remaining discrepancy between post-treatment sediment delivery rates with background load levels and the costs required for achieving those levels highlight the unfeasibility of attempting to fully restore conditions to pre-disturbance rates. Road drainage improvements accomplished by placing rolling dips or paved gutters every 30 m on all road segments in PDM would reduce sediment yields by 82 tons yr^{-1} (Box 3). Post-treatment delivery rates would be 35 tons yr^{-1}, or roughly 30% of pre-treatment levels. Costs related to the installation of the rolling dips and paved gutter treatments with their accompanying sediment check dams and energy dissipaters on all roads would cost $3.0M and $3.3M, respectively. The overall cost-effectiveness measure would be $37.1K per ton for the rolling dips method and $40.0K ton^{-1} for paved gutters, or 64% - 69% more cost-effective (i.e., less expensive per unit ton reduced) than paving all roads. However, paving all roads would save the marine environment an additional 24 tons yr^{-1} that neither of the two road drainage improvement methods (i.e., rolling dips and paved gutters) would be able to achieve.

In reality, the high costs required for implementing treatments on all roads in PDM make this an unfeasible task. Therefore, devising a prioritization strategy is essential to establish price tags for different sediment reduction goals. Our analyses show that the best solution in terms of maximizing reductions while minimizing costs depends on the amount of funds available for treatment implementation (Box 3). If only roughly $70K are available then the only feasible options are the two treatments involving drainage improvements, and these funds would only properly address one road segment (Site No. 1) and achieve a reduction of approximately 18 tons yr^{-1}. If available funds range between $130K and $200K then the reductions in sediment yields achieved by the three treatment options would be very similar (25–28 tons yr^{-1}), but if available funds range from $200K to $700K greater reductions would be achieved by road drainage improvements than by paving. Nevertheless, if funds exceed $700K then road paving becomes a more favorable option than either of the other two treatment options.

It is important to note that the analysis presented here does not include other possible treatment scenarios. One possibility would be to further explore manipulating the ranking of individual road segments to attempt to further maximize the cost-effectiveness measure. Manipulations of site priority rankings for the PDM watershed did not display much difference to the one based simply on sediment yield contributions shown in Box 3, but this does not appear to be the case for some of the other seven watersheds studied. In addition, the analysis shown here does not explore applying a mix of the three treatments options and this might provide another alternative that generates more cost-effective results.

Although no spatial information was used to establish the priority ranking of the road segments, this kind of information should also be considered when making decisions. Roads with an obvious direct connectivity with the marine environment should be contemplated as high priority candidates. In the case of the PDM watershed, many of the top ranked sites based simply on annual sediment contribution (e.g., sites 1-5 and 10 in Box 3) not only show up as contributing large amounts of sediment but are also located in close proximity to the marine environment and are likely contributing sediment very effectively into coastal waters.

5. Conclusions

High sediment delivery rates on highly erodible, anthropogenic-disturbed soils can have significant long-term deleterious impacts on coral reef biodiversity, sustainability, productivity, resilience, and on its ecosystem services, which could in turn affect its socio-economic value and benefits to island communities. Reducing sediment loads into coastal marine ecosystems is one feasible mitigation activity by which humans can help alleviate a key stressor affecting coral reefs worldwide. The costs of implementing BMPs will always pose a limit to the type and extensiveness of erosion control efforts. Therefore, selecting priority areas, targeting the most relevant sediment sources, and choosing adequate BMPs to optimize efforts are critical steps in the development of effective erosion control plans. In this chapter we have presented an interdisciplinary approach to erosion mitigation that weighs information resulting from coral reef assessments, watershed analyses, and engineering considerations. The general framework described could also be used to help devise mitigation strategies for other non-point sources of pollution that also affect reefs and its associated ecosystems (i.e., seagrass communities, mangroves, estuarine systems). The addition of long-term coral reef community dynamics data as well as environmental parameter information (i.e., turbidity, high resolution sediment loading rates, sediment composition analysis, oceanographic currents, etc.) may further enhance the capacities of our proposed framework.

The interdisciplinary approach presented here was applied within the context of Isla de Culebra, part of the Puerto Rican archipelago in the Eastern Caribbean. The strategy serves in part to choose priority target watersheds for erosion control on the basis of the intentions of the mitigation efforts. Here we recognize that, depending on coral reef condition and sediment load rates, erosion mitigation efforts may have three different motivations: (1) to preserve reefs that are still in a good condition; (2) to prevent further damage to reefs that have a good to moderate condition but are being influenced by inland sediment sources; and (3) to remediate conditions for deteriorated reefs receiving high sediment loads. The approach also includes a cost-effectiveness analyses that aids in choosing specific sites and erosion control methods to maximize the net reductions in sediment loads while minimizing costs. Application of this cost-effectiveness analysis to one watershed in Culebra suggests that the choice of most effective erosion control method varies according to the amount of funds available for implementation. However, it is important to emphasize the need to strictly implement existing erosion-sedimentation regulations. Controlling the current rampant deforestation trends is crucial if further degradation of marine habitats in Culebra and throughout the rest of the Caribbean is to be prevented. The combination of climate change-related impacts and the cumulative degradation associated to localized anthropogenic factors, including negligent land use practices, may cause further irreparable coral reef decline if local stressor factors are not effectively managed and mitigated.

6. Acknowledgements

Funding for this project was provided through a contract with the Coastal Zone Management Program of the Puerto Rico Department of Natural and Environmental Resources (Federal Grant Number NA08NOS4190468). Partial funding to E.A. Hernández-Delgado was provided by the Caribbean Coral Reefs Institute-University of Puerto Rico at Mayagüez, and by the National Science Foundation through grant NSF HRD 0734826 to the

Center for Applied Tropical Ecology and Conservation at UPR Río Piedras. We kindly acknowledge the support provided by Mr. Ernesto Diaz and Mr. Raúl Santini (PR-DNER), and Mr. Samuel E. Suleimán-Ramos (Sociedad Ambiente Marino).

7. References

Anderson, D.M. (1994). Guidelines for Sediment Control Practices in the Insular Caribbean. CEP Technical Report No. 32, United Nations Environment Programme-Caribbean Environment Programme, 58 p.

Aronson, R.B., Mcintyre, I.G., Precht, W.F., T.J.T. Murdoch, & Wapnick, C.M. (2002). The expanding scale of species turnover events on coral reefs in Belize. *Ecological Monographs*, Vol. 72, No. 2, pp. 233-249.

Ballantine, D.L., Appeldoorn, R.S., Yoshioka, P., Weil, E., Armstrong, R., García, J.R., Otero, E., Pagán, F., Sherman, C., Hernández-Delgado, E.A., Bruckner, A., & Lilyestrom, C. (2008). Biology and ecology of Puerto Rican coral reefs. In, B.M. Riegl & R.E. Dodge (eds.), *Coral Reefs of the World, Vol. I. Coral Reefs of the USA*, Springer-Science + Business Media B.V., pp. 375-406.

Beinroth, F.H., Engel, R.J., Lugo, J.L., Santiago, C.L., Ríos, S., & Brannon, G.R. (2003). Updated taxonomic classification of the soils of Puerto Rico, 2002. Agricultural Experiment Station, Bulletin No. 303, University of Puerto Rico-Mayaguez, 73 p.

Bellwood, D.R., Hughes, T.P., Folke, C., & Nyström. (2004). Confronting the coral reef crisis. *Nature*, Vol. 429, pp. 827-833.

Berman-Santana, D. (1996). Kicking Off the Bootstraps- Environment, Development, and Community Power in Puerto Rico, The University of Arizona Press, ISBN 0-8165-1590-5, Tucson, AZ.

Biagi, N. (1965). Puerto Rico's water pollution image. *Water Pollution Federation Journal*, Vol. 37, No. 3, pp. 381-391.

Biagi, N. (1968). The sugar industry in Puerto Rico and its relation to the industrial waste problem. *Water Pollution Federation Journal*, Vol. 40, No. 8, Part I, pp. 1423-1433.

Birdsey, R.A. & Weaver, P.L. (1987). Forest Area Trends in Puerto Rico. USDA-US Forest Service, Research Note SO-331, New Orleans, LA, pp.

Bradshaw, A.D. (1997). What do we mean by restoration?. In: Restoration Ecology and Sustainable Development, K.M. Urbanska, N.R. Webb, & P.J. Edwards (eds.), Cambridge University Press, Cambridge, UK, pp. 8-14.

Bradshaw, A.D. (2002). Introduction and philosophy. In: Handbook of Ecological Restoration- Volume 1: Principles of Restoration, Perrow, M.R, & Davy, A.J. (eds.), Cambridge University Press, Cambridge, UK, pp. 3-9.

British Columbia Ministry of Forests. (2002). Forest road engineering guidebook. B.C. Ministry of Forests, Forest Practices Code of British Columbia Guidebook, Victoria, BC, 208 p.

Brooks, G.R., Devine, B., Larson, R.A., Rood, B.P. (2008). Sedimentary development of Coral Bay, St. John, USVI: A shift from natural to anthropogenic influences. *Caribbean Journal of Science*, Vol. 43, No. 2, pp. 226-243.

Buddemeier, R.W., Jokiel, P.L., Zimmerman, K.M., Lane, D.R., Carey, J.M., Bohling, G.C., & Martinich, J.A. (2008) A modeling tool to evaluate regional coral reef responses to changes in climate and ocean chemistry. *Limnology and Oceanography: Methods*, Vol. 6, pp. 395-411.

Buddemeier, R.W., Lane, D.R., & Martinich, J.A. (2010). Modeling regional coral reef responses to global warming and changes in ocean chemistry: Caribbean case study. *Climatic Change*, DOI10.1007/s10584-011-0022-z, pp. 1-23

Burke, L. & Maidens, J. (2004). *Reefs at Risk in the Caribbean*, World Resources Institute, Washington, DC, 5 p.

Clark, M.J. (1997). Ecological restoration - the magnitude of the challenge: an outsider's view. In: Restoration Ecology and Sustainable Development, Urbanska, K.M., Webb, N.R., & Edwards, P.J. (eds.), Cambridge University Press, Cambridge, UK, pp. 353-377.

Cloern, J.E. (2001). Our evolving conceptual model of the coastal eutrophication problem. *Marine Ecology Progress Series*, Vol. 210, pp. 223-253.

Commonwealth of Puerto Rico & NOAA. (2010). Puerto Rico's Coral Reef Management Priorities. NOAA, Silver Springs, MD. 40 p.

Concepción, C.M. (1988). El conflict ambiental y su potencial hacia un desarrollo alternative: el caso de Puerto Rico. *Ambiente y Desarrollo*, Vol. IV, No. 1 y 2, pp. 125-135.

Croke, J., Mockler, S., Fogarty P., & Takken, I. (2005). Sediment concentration changes in runoff pathways from a forest road network and the resultant spatial pattern of catchment connectivity. *Geomorphology*, Vol. 68, pp. 257-268.

Dietz, J.L. (1986). Economic History of Puerto Rico: Institutional Change and Capitalist Development, Princeton University Press, ISBN 0-691-07716-9, Princeton, NJ, 337 p.

Estudios Técnicos Inc. (2004). Plan Maestro para el Desarrollo Sustentable de Culebra- Parte I: Análisis de Situación. Unpublished Report to Grupo Interagencial Vieques y Culebra, 187 p.

Estudios Técnicos Inc. (2007). Valoración económica de los arrecifes de coral y ambientes asociados en el este de Puerto Rico: Fajardo, Arrecifes de La Cordillera, Vieques y Culebra. Report submitted to the Department of Natural and Environmental Resources, San Juan, PR., pp. 1-99, + App.

Ewel, J.J. & Whitmore, J.L. (1973). The ecological life zones of Puerto Rico and the U.S. Virgin Islands. US Forest Service Research Paper ITF-18, Río Piedras, PR, 72 p.

Fabricius, K.E. (2005). Effects of terrestrial runoff on the ecology of corals and coral reefs: review and synthesis. *Marine Pollution Bulletin*, Vol. 50, pp. 125-146.

García-Sais, J.; Appeldoorn, R.; Battista, T.; Bauer, L.; Bruckner, A.; Caldow C.; Carruba, L.; Corredor, J.; Díaz, E.; Lylyestrom, C.; García-Moliner, G.; Hernández-Delgado, E.; Menza, C.; Morell, J.; Pait, A.; Sabater, J.; Weil, E.; Williams, E. & Williams, S. (2008). The State of Coral Reef Ecosystems of Puerto Rico. In, J.E. Waddell & A.M. Clarke (eds.), *The State of Coral Reef Ecosystems of the United States and Pacific Freely Associated States: 2008*. NOAA Technical Memorandum NOS NCCOS 73. NOAA/NCCOS Center for Coastal Monitoring and Assessment's Biogeography Team, Silver Spring, MD., pp. 75-116.

Gardner, T.A.; Côté, I.M.; Gill, J.A.; Grant, A. & Watkinson, A.R. (2003). Long-term region-wide declines in Caribbean corals. *Science*, Vol. 301, pp. 958-960.

Garrison, V.H., Shinn E.A., Miller, J., Carlo, M., Rodríguez, R.W., & Koltes, K. (2005). Benthic cover on coral reefs of Isla de Culebra (Puerto Rico) 1991-1998 and a comparison of assessment techniques. US Geological Survey, Open-File Report 2005-1398.

Goenaga, C. (1991). The state of Puerto Rican corals: An aid to managers. Technical report submitted to the Caribbean Fishery Management Council, San Juan, PR., pp. 1-71.

Goenaga, C., & Cintrón, G. (1979). Inventory of the Puerto Rican coral reefs. Report submitted to the Department of the Department of Natural Resources, San Juan, PR., pp. 1-190.

Grau, H.R.; Aide, T.M.; Zimmerman, J.K.; Thomlinson, J.R.; Helmer, E. & Zou, X. (2003). The ecological consequences of socioeconomic and land-use changes in postagriculture Puerto Rico. *Bioscience*, Vol. 53, No. 12, pp. 1159-1168.

Hawkins, J.P., & Roberts, C.M. (2004). Effects of artisanal fishing on Caribbean coral reefs. *Conservation Biology*, Vol. 18, No. 1, pp. 215-226.

Hernández-Delgado, E.A. (2000). Effects of anthropogenic stress gradients in the structure of coral reef epibenthic and fish communities. Ph.D. Dissertation, Department of Biology, University of Puerto Rico, Río Piedras, P.R., pp. 1- 330.

Hernández-Delgado, E.A. (2004). Análisis del estado de los recursos y de la situación ambiental actual de la Reserva Natural del Canal Luis Peña, Culebra, P.R. Technical report submitted to the Culebra Conservation and Development Authority, Culebra, PR., pp. 1-133.

Hernández-Delgado, E.A. (2005). Historia natural, caracterización, distribución y estado actual de los arrecifes de coral Puerto Rico. *In*, R.L. Joglar (Ed.), *Biodiversidad de Puerto Rico: Vertebrados Terrestres y Ecosistemas. Serie Historia Natural.* Editorial Instituto de Cultura Puertorriqueña, San Juan, PR. pp. 281-356.

Hernández-Delgado, E.A. (2010). Thirteen years of climate-related non-linear disturbance and coral reef ecological collapse in Culebra Island, Puerto Rico: A preliminary analysis. In, E.A. Hernández-Delgado (ed.), Puerto Rico Coral Reef Long-Term Ecological Monitoring Program, CCRI-Phase III and Phase IV (2008-2010) Final Report. Caribbean Coral Reef Institute, Univ. Puerto Rico, Mayagüez, PR. pp. I.1-I.62.

Hernández-Delgado, E.A., Alicea-Rodríguez, L., Toledo-Hernández, C.G. & Sabat, A.M. (2000). Baseline characterization of coral reef epibenthic and fish communities within the proposed Culebra Island Marine Fishery Reserve, Puerto Rico. *Proceedings of the Gulf and Caribbean Fisheries Institute*, Vol. 51, pp. 537-556.

Hernández-Delgado, E.A., & Rosado-Matías, B.J. (2003). Suplemento técnico al Plan de Manejo para la Reserva Natural del Canal Luis Peña, Culebra, Puerto Rico. II. Inventario biológico. Technical Report submitted to the Coastal Zone Management Program, Department of Natural and Environmental Resources. San Juan, PR., pp. 1-60.

Hernández-Delgado, E.A., Lucking, M.A., & González, R.L. (2003). Ecological impacts of private peer structures and operation on critical seagrass communities in Fulladosa Cove, Culebra, Puerto Rico. Tech. Rept. submitted to the US Corps of Engineers, San Juan, P.R., pp. 1- 61.

Hernández-Delgado, E.A., Vázquez, M.T., Díaz, A., Rentas, X., López, J.C., Cortés, K., Rainford, A.F., Colón, D., López, A., Rodríguez, A., Hernández, G., Navarro, I., Gómez, J., González, M., & Hernández, S. (2003). Suplemento técnico al Plan de Manejo de la Reserva Natural del Canal Luis Peña, Culebra, Puerto Rico. III. Usos históricos, actuales y valor arqueológico del Canal Luis Peña. Technical Report submitted to the Coastal Zone Management Program, Department of Natural and Environmental Resources. San Juan, PR., pp. 1-45.

Hernández-Delgado, E.A., Rosado-Matías, B.J., & Sabat, A.M. (2006). Management failures and coral decline threatens fish functional groups recovery patterns in the Luis Peña Channel No-Take Natural Reserve, Culebra Island, PR. *Proceedings of the Gulf and Caribbean Fisheries Institute*, Vol. 51, pp. 577-605.

Hernández-Delgado, E.A.; Sandoz, B.; Bonkosky, M.; Norat-Ramírez, J. & Mattei, H. (2010). Impacts of non-point source sewage pollution on Elkhorn coral, *Acropora palmata* (Lamarck), assemblages of the southwestern Puerto Rico shelf. *Proceedings of the 11th International Coral Reef Symposium*, pp. 747-751.

Hernández-Delgado, E.A., & Sandoz-Vera, B. (2011). Impactos antropogénicos en los arrecifes de coral. In, J. Seguinot-Barbosa (ed.), *Islas en Extinción: Impactos Ambientales en las Islas de Puerto Rico*. Ediciones SM, Cataño, pp. 62-72.

Hernández-Delgado, E.A., Suleimán, S., Olivo, I., Fonseca, J., & Lucking, M.A.. (2011). Alternativas de baja tecnología para la rehabilitación de los arrecifes de coral. In, J. Seguinot-Barbosa (ed.), *Islas en Extinción: Impactos Ambientales en las Islas de Puerto Rico*. Ediciones SM, Cataño, PR, pp. 178-186.

Hernández-Pacheco, R., Hernández-Delgado, E.A. & Sabat, A.M. (2011). Demographics of bleaching in the Caribbean reef-building coral *Montastraea annularis*. *Ecosphere*, Vol. 2, No. 1:art9. 1-13. doi:10.1890/ES10-00065.

Hoegh-Guldberg, O. (1999). Climate change, coral bleaching and the future of the world's coral reefs. *Marine and Freshwater Resources*, Vol. 50, pp. 839-866.

Hoegh-Guldberg, O., Mumby, P.J., Hooten, A.J., Steneck, R.S., Greenfield, P., Gomez, E., Harvell, C.D., Sale, P.F., Edwards, A.J., Caldeira, K., Knowlton, N., Eakin, C.M., Iglesias-Prieto, R., Muthiga, N., Bradbury, R.H., Dubi, A. & Hatziolos, M.E. (2007). Coral reefs under rapid climate change and ocean acidification. *Science*, Vol. 318, pp. 1737-1742.

Hunter, J.M. & Arbona, S.I. (1995). Paradise lost: An introduction to the geography of water pollution in Puerto Rico. *Social Science and Medicine*, Vol. 40, No. 10, pp. 1331-1355.

Jackson, J.B.C. (1997). Reefs since Columbus. *Coral Reefs*, Vol. 16, Suppl., pp. S23-S32.

Jackson, J.B.C. (2001). What was natural in the coastal oceans. *Proceedings of the Natural Academy of Sciences*, Vol. 98, No. 10, pp. 5411-5418.

Jackson, L.L., Lopoukhine, N., & Hillyard, D. (1995). Ecological restoration: A definition and comments. Restoration Ecology, Vol. 3, No. 2, pp. 71-75.

Kendall, M.S., Monaco, M.E., Buja, K.R., Christensen, J.D., Kruer, C.R., Finkbeiner, M., & Warner, R.A. (2001). Methods used to map the benthic habitats of Puerto Rico and the U.S. Virgin Islands. URL: http://ccma.nos.noaa.gov/products/biogeography/usvi_pr_mapping/manual.pdf

Labadie-Eurite, J. (1949). La Mecanización Agrícola en Puerto Rico. Departamento de Instrucción, San Juan, PR, 31 p.

Larsen, M.C., & Webb, R.M.T. (2009). Potential effects of runoff, fluvial sediments, and nutrient discharges on the coral reefs of Puerto Rico. *Journal of Coastal Research*, Vol. 25, No. 1, pp. 189-208.

Littler, M.M., Littler, D.S., Lapointe, B.E., & Barile, P.J. (2006). Toxic cyanobacteria (blue-green algae) associated with groundwater conduits in the Bahamas. *Coral Reefs* DOI 10.1007/s00338-005-0010-8, pp. 1-2.

Littler, M.M., Littler, D.S., & Brooks, B.L. (2006). Harmful algae on tropical coral reefs: Bottom-up eutrophication and top-down herbivory. *Harmful Algae*, Vol. 5, pp. 565-585.

Loya, Y. (1976). Effects of water turbidity and sedimentation on the community structure of Puerto Rican corals. *Bulletin of Marine Science*, Vol. 26, No. 4, pp. 450-466.

MacDonald, L.H., Anderson, D.M., & Dietrich, W.E. (1997). Paradise threatened: Land use and erosion on St. John, US Virgin Islands. *Environmental Management*, Vol. 21, No. 6, pp. 851-863.

Masson, D.G. & Scanlon, K.M. (1991). The neotectonic setting of Puerto Rico. *Geological Society of America Bulletin*, Vol. 103, No. 1, pp. 144-154.

Megahan, W.F., & Ketcheson, G.L. (1996). Predicting downslope travel of granitic sediments from forest roads in Idaho. *Water Resources Bulletin*, Vol. 32, pp. 371-382.

Meyerhoff, H.A. (1927). Geology of the Virgin Islands, Culebra and Vieques-Physiography. Scientific Survey of Porto Rico and the Virgin Islands, New York Academy of Sciences, Vol. IV, Part II, pp. 145-219.

Miller, J., Waara, R., Muller, E., & Rogers, C. (2006). Coral bleaching and disease combine to cause mortality on reefs in US Virgin Islands. *Coral Reefs*, Vol. 25, pp. 418.

Miller, J., Muller, E., Rogers, C., Waara, R., Atkinson, A., Whelan, K.R.T., Patterson, M., & Witcher, B. (2009). Coral disease following massive bleaching in 2005 causes 60% decline in coral cover on reefs in the U.S. Virgin Islands. *Coral Reefs*, Vol. 28, pp. 925-937.

Morelock, J.; Ramírez, W.R.; Bruckner, A.W. & Carlo, M. (2001). Status of coral reefs southwest Puerto Rico. *Caribbean Journal of Science*, Special Publication No. 4, pp. 1-57.

National Oceanic and Atmospheric Administration-National Ocean Service-National Centers for Coastal Ocean Science Biogeography Program. (2001). Benthic Habitats of Puerto Rico and the U.S. Virgin Islands (CD-ROM). Silver Spring, MD, National Oceanic and Atmospheric Administration.

Oliver, L.M., Lehrter, J.C., & Fisher, W.S. 2011. Relating landscape development intensity to coral reef condition in the watersheds of St. Croix, US Virgin Islands. *Marine Ecology Progress Series*, Vol. 427, pp. 293-302.

Pagán-Villegas, I.M., Hernández-Delgado, E.A., & Vicente, V.P. (1999). Documento de designación de la Reserva Natural del Canal Luis Peña, Departamento de Recursos Naturales y Ambientales, San Juan, P.R., 21 de mayo de 1999.

Pandolfi, J.M.; Bradbury, R.H.; Sala, E.; Hughes, T.P.; Bjorndal K.A.; Cooke, R.G., McArdle D.; McClenachan L.; Newman, M.J.H.; Paredes, G.; Warner, R.R. & Jackson, J.B.C. (2003). Global trajectories of the long-term decline of coral reef ecosystems. *Science*, Vol. 301, pp. 955-958.

Pandolfi, J.M., Jackson, J.B.C., Baron, N., Bradbury, R.H., Guzman, H.M., Hughes, T.P., Kappel, C.V., Micheli, F., Ogden, J.C., Possingham, H.P. & Sala, E. (2005). Are U.S. coral reefs in the slippery slope to slime? *Science*, Vol. 307, pp. 1725-1726.

Pandolfi, J.M. & Jackson, J.B.C. (2006). Ecological persistence interrupted in Caribbean coral reefs. *Ecology Letters*, Vol. 9, pp. 818-826.

Pastorok, R.A., & Bilyard, G.R. (1985). Effect of sewage pollution on coral-reef communities. *Marine Ecology Progress Series*, Vol. 21, pp. 175-189.

Puerto Rico Environmental Quality Board. (1970). An Island in Transition, Culebra 1970- A Staff Report on the Environment to the Governor's Special Committee on Culebra, 106 p.

Ramos-Scharron, C.E. (2004). Measuring and predicting erosion and sediment yields on St. John, US Virgin Islands. PhD Dissertation, Department of Geosciences, Colorado State University, Fort Collins, CO.

Ramos-Scharrón, C.E. (2009). The effects of land development on sediment loading rates into the coastal waters of the Islands of Culebra and Vieques, Puerto Rico. Unpublished report to the Coastal Zone Management Program of the Puerto Rico Department of Natural and Environmental Resources, San Juan, PR, 94 p.

Ramos-Scharrón, C.E. (2010). Sediment production from unpaved roads in a sub-tropical dry setting- Southwestern Puerto Rico. *Catena*, Vol. 82, pp. 146-158.

Ramos-Scharrón, C.E. (in press). Effectiveness of drainage improvements in reducing sediment production rates from and unpaved road. *Journal of Soil and Water Conservation*.

Ramos-Scharrón, C.E. & MacDonald, L.H. (2005). Measurement and prediction of sediment production from unpaved roads, St. John, US Virgin Islands. *Earth Surface Processes and Landforms*, Vol. 30, pp. 1283-1304.

Ramos-Scharrón, C.E. & MacDonald, L.H. (2007a). Measurement and prediction of natural and anthropogenic sediment sources, St. John, U.S. Virgin Islands. *Catena*, Vol. 71, pp. 250-266.

Ramos-Scharrón, C.E. & MacDonald, L.H. (2007b). Development and application of a GIS-based sediment budget model. *Journal of Environmental Management*, Vol. 84, pp. 157-172.

Restrepo, J.D & Syvitski, J.P.M (2006). Assessing the effect of natural controls and land use change on sediment yield in a major Andean river: The Magdalena drainage basin, Colombia. *Ambio*, Vol. 35, No. 2, pp. 65-74.

Reilly, A.E. (1991). The effects of Hurricane Hugo in three tropical forests in the U.S. Virgin Islands. *Biotropica*, Vol. 23, No. 4a, pp. 414-419.

Rodríguez, A. (1981). Marine and coastal environmental stress in the Wider Caribbean Region. *Ambio*, Vol. 10, No. 6, pp. 283-294

Rogers, C.S., Garrison, G, Grober, R., Hillis, Z.M., & Franke, M.A. (1994). Coral Reefs Monitoring Manual for the Caribbean and Western Atlantic. National Park Service, St. John, USVI.

Rudel, T.K.; Perez-Lugo, M. & Zichal, H. (2000). When fields revert to forest: Development and spontaneous reforestation in post-war Puerto Rico. *Professional Geographer*, Vol. 52, No. 3, pp. 386-397.

Ryan, K.E., Walsh, J.P., Corbett, D.R., & Winter, A. (2008). A record of recent change in terrestrial sedimentation in a coral-reef environment, La Parguera, Puerto Rico: A response to coastal development. *Marine Pollution Bulletin*, Vol. 56, pp. 1177-1183.

U.S. Army Corps of Engineers. (1995). Defense environmental restoration program for formerly used defense sites, ordnance and explosive waste, Archives Search Report, Findings for Culebra Island National Wildlife Refuge, Culebra, Puerto Rico, Project Number I02PR006802. Rock Island, Illinois.

Valdés-Pizzini, M., González-Cruz M., & Matínez-Reyes, J.E. (2011). La Transformación del Paisaje Puertorriqueño y la Disciplina del Cuerpo Civil de Conservación, 1933-1942. Centro de Investigaciones Sociales, Universidad de Puerto Rico, San Juan, PR.

Walling, D.E. (1983). The sediment delivery problem. *Journal of Hydrology*, Vol. 65, pp. 209-237.

Webler, T., & Jakubowski, K. (2011). Characterizing harmful behaviors of snorkelers and SCUBA divers to coral reefs in Puerto Rico. Interim Project Report. Preliminary Technical Report submitted to the Department of Natural and Environmental Resources, San Juan, PR, pp. 1-15.

Wolman, M.G. (1967). A cycle of sedimentation and erosion in urban river channels. *Geografiska Annaler, Series A, Physical Geography*, Vol. 49, No. 2/4, pp. 385-395.

Ziegler, A.D, & Sutherland, R.A. (2006). Effectiveness of a coral-derived surfacing material for reducing sediment production on unpaved roads, Schoffield Barracks, Oahu, Hawaii. *Environmental Management*, Vol. 37, No. 1, pp. 98-110.

7

A Re-Visit to the Evolution and Ecophysiology of the Labyrinthulomycetes

Clement K. M. Tsui[1] and Lilian L. P. Vrijmoed[2]
[1]Department of Forest Sciences, The University of British Columbia, Vancouver, BC,
[2]Department of Biology and Chemistry,
City University of Hong Kong, Hong Kong SAR,
[1]Canada
[2]China

1. Introduction

The labyrinthulomycetes (also known as Labyrinthulomycota or Labyrinthulea) are marine heterotrophic fungus-like protists and belong to the eukaryotic Kingdom Stramenopiles (Honda et al., 1999, Tsui et al., 2009). Most labyrinthulomycete species are unicellular, and they are ubiquitous in the ocean, and their occurrence and distribution in water column and sediments have been well documented (Kimura et al., 1999, Naganuma et al., 1998, Raghukumar, 2002). Their main ecological role may be as saprotrophic decomposers, recycling nutrients in marine and coastal ecosystems, by chemical alteration of detritus through extra-cellular enzymes (Raghukumar, 2002, Taoka et al., 2009). Their role in facilitating the settlement of barnacle cyprids has also been demonstrated (Raghukumar et al., 2000).

Labyrinthulomycetes have been studied by mycologists, and two comprehensive reviews were published by Raghhukumar and her co-workers on their ecology (Raghukumar, 2002, Raghukumar & Damare, 2011). In these reviews, the authors dealt mainly with the general ecological role of these organisms in the marine ecosystems; their associations/interactions with living or decaying plant materials, phytoplankton, animals and bacteria, either in sediments or in the oceanic water column. Their role in the marine food web either as "remineralizers" and possible "left-over" scavengers were also discussed.

Though labyrinthulomycetes belong in the Stramenopiles, they evolved a fungus-like, absorptive mode of osmotrophic nutrition by developing rhizoids on detritus. Convergently with true fungi and oomycetes (also in Stramenopiles), some labyrinthulomycetes are pathogenic, causing diseases such as turf grass and eelgrass wasting disease, and the hard clam disease 'QPX', a role discovered only over the last two decades (Bigelow et al., 2005, Craven et al., 2005, Muelstein et al. 1988, Stokes et al., 2002). Many representatives in labyrinthulomycetes accumulate high level of omega-3 long-chain polyunsaturated fatty acids (PUFAs), such as, eicosapentaenoic acid (EPA), docosahexaenoic acid (DHA) and docosapentaenoic acids (DPA) within the cells, thus being an important component in the detrital food web (Findlay et al., 1986, Yongmanitchai & Ward 1989). As a result, a number of species are currently serving as sources of valuable DHA used in dietary supplements and for DHA production in industry (Abril et al., 2000, Sijtsma & de Swaaf, 2004). Recent studies have also revealed their potential in carotenoid and squalene production (Carmona et al., 2003, Jiang et al. 2004), and as aquacultural feeds (Yamasaki et al., 2007).

The labyrinthulomycetes are important in nutrient recycling, and in the food and biotechnology industry. However their ecophysiology and evolution are not well understood. This chapter will bring together the latest information on their evolution, ecology and physiology. We also review some current approach to unravel their evolutionary origins and ecological role in the oceans and mangrove environment, particularly on the thraustochytrids.

2.Techniques for physiological, ecological and evolutionary investigation

2.1 Isolation and cultivation

Representatives of labyrinthulomycetes can be isolated from mangrove leaves, sediment, open water, and from the guts of marine invertebrates. Normally mangrove or marine samples collected are rinsed and directly placed on yeast extract-peptone (YEP) agar (Fan et al., 2002a, 2009). Alternatively samples are collected and placed in the centrifuge tubes/ test tubes containing 10 ml sterilized, full strength artificial/ natural seawater, together with small amount of sterilized pine pollen (approx. 50-100 pollens). The pine pollens are then aseptically placed on GYP agar [glucose 2 g, polypeptone 1 g, yeast extract 0.5 g, chloramphenicol 0.2 g, agar 15 g, seawater 500 ml, D water 500 ml] or YEP agar [yeast extract 1 g, mycological peptone 1 g, agar (technical grade) 15g and 1 L 15‰ artificial seawater] for microscopy and further isolation.

Similarly marine invertebrates are collected, and the diluted gut contents are plated onto various media (Porter, 1990, Tsui et al., 2009). Undiluted coelomic fluid samples can be directly plated onto corresponding media (Porter, 1990, Tsui et al., 2009). Plates were checked every 4 -5 days under a dissecting microscope. Transmission electron microscopy (TEM) can be carried out according to Honda et al. (1998). Colonies exhibiting thraustochytrid-like morphology can be sub-cultured several times until axenic. Thraustochytrid colonies can be maintained in sterile broth too [yeast extract 1g, mycological peptone 1g, glucose 10 g and 1L 15‰ artificial seawater prepared from artificial sea salts (Sigma)].

2.2 Fatty acids analysis

Fatty acid profiles have become important biochemical characters in the delineation of genus, species, and isolates (Fan et al., 2009, Yokoyama et al., 2007a, b). Fatty acids composition are analysed using a modified method of Lepage & Roy (1984). The freeze-dried cells of labyrinthulomycetes are methylated with sulfuric acids in methanol with the addition of an internal standard (e.g. heptadecaenoic acid, C17:0). Then the fatty acid methyl esters (FAMEs) are extracted by water and hexane (1:1). The FAMEs (1μl) in the hexane layer were subjected to gas chromatography equipped with a flame ionization detector (Agilent 6890 GC-FID), and a DB-225 capillary column (30 mm 5 0.25 mm diam). Injector is held at 220°C with initial temperature at 90°C for 3 min then increases from 90°C to 210°C at 20°C/ min. The detector is held at 230°C and helium is used as carrier gas and the column flow rate is 1ml/ min. The amount of DHA is identified and quantified by a comparison of retention time for laboratory standard and internal standard.

2.3 Carotenoid analysis

To characterize the carotenoid pigment composition of the taxa, cells are extracted with chloroform-methanol. The solvent is removed in vacuum to obtain a crude residue of the extract. The dried extraction is dissolved in a small amount of chloroform and applied to the column of silica gel packed by hexane. The fraction is reconstituted with methanol and loaded onto the HPLC instrument, which is capable of detecting UV-visible wavelength carotenoid spectra (Carmona et al., 2003).

2.4 DNA extraction, PCR and sequence analyses

For molecular phylogeny, cells of labyrinthulomycetes on agar or in liquid broth are harvested, and DNA is extracted by commercial kit. Primers of various genes are used to amplify corresponding fragments under the conditions in White et al. (1990) and Tsui et al. (2009). In case of having several fragments after PCR, products corresponding to the expected size are gel-purified and cloned into the vector pCR2.1 using the TOPO TA cloning kit (Invitrogen). Five to ten clones are sequenced using the vector primers and designed internal primers. Sequence data is then aligned with homologous sequences from a representative sampling of eukaryotes from GenBank databases with computer softwares, such as Clustal X (Thompson et al., 1997) or MacClade (Maddison & Maddison, 2000). Alignment data are subjected to various methods of phylogenetic analysis; Maximum Parsimony (MP), Neighbor Joining (NJ) and Maximum-likelihood (ML) using PAUP*4.0 (Swofford, 2003) and Phylip 3.6 (Felsenstein et. al., 2002).

Culture independent methods are getting popular recently for environmental characterization. Clone libraries of SSU rRNA from water and environmental samples facilitate the investigation of natural communities and unknown lineages in various habitats (Massana et al., 2004a, b). Fluorescent *in situ* hybridisation probes (FISH) and quantitative PCR probes have also been developed for detection of thraustochytrids (Takao et al., 2007), and QPX from marine water simultaneously (Liu et al., 2009).

3. Position in the 'tree of life'

Labyrinthulomycetes have been traditionally classified under the Kingdom Fungi based on morphology, as well as their life histories and mode of nutrition. The labyrinthulomycetes presently belong to the Kingdom Stramenopiles, which also accommodate the photosynthetic ochrophytes (brown algae, golden brown algae and diatoms), along with the non-photosynthetic free-living bicoeceans, and oomycetes which are well known as serious plant pathogens (Fig. 1) (Cavalier-Smith, 1998, Keeling et al., 2005, Leipe et al., 1994, Oudot-Le Secq et al., 2006, Tsui et al., 2009). Labyrinthulomycetes share Stramenopile characters in having cell walls of thin scales (Chamberlain & Moss, 1988), tubular mitochondria, and biflagellate zoospores with one smooth flagellum and one bearing tripartite tubular hairs (Patterson, 1989). Together with the alveolate relatives, which include the apicomplexa, ciliates and dinoflagellates, they form the super-kingdom "Chromalveolate" defined firstly in Baldalf et al. (2000).

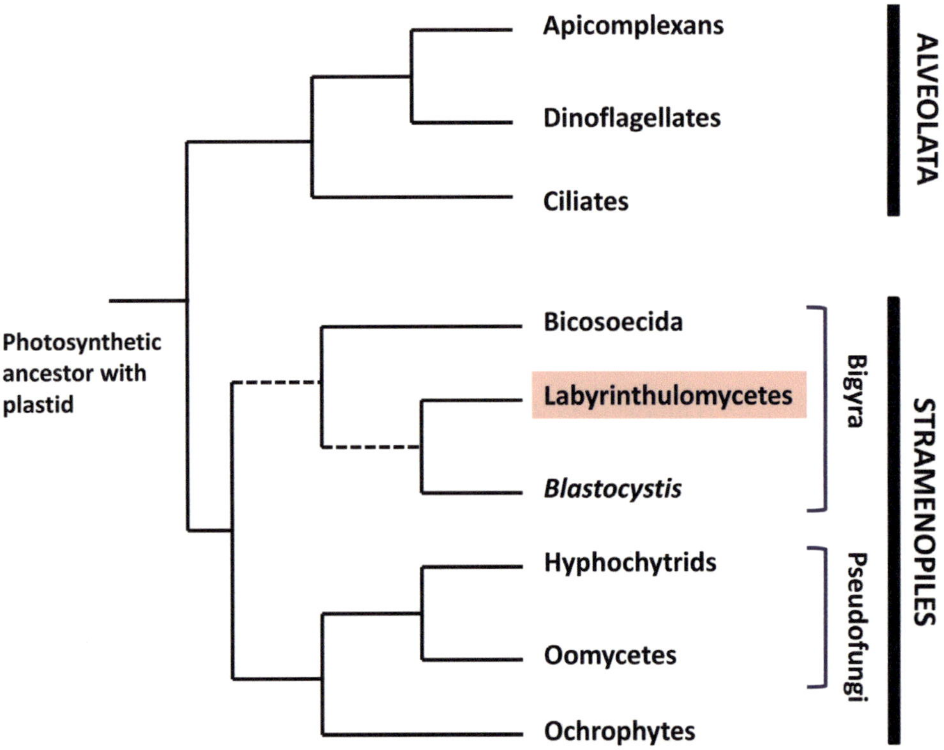

Fig. 1. A simplified phylogenetic tree showing the relationships among Labyrinthulomycetes and other members in Chromalveolate based on Riisberg et al. (2009) (dotted lines indicate unsolved relationship).

The Stramenopiles form a strong, monophyletic group, but the branching order among early-diverging lineages including the heterotrophic labyrinthulomycetes, bicoecida and oomycetes, and the photosynthetic ochrophytes has been difficult to resolve until recently (Cavalier-Smith, 1998, Keeling et al., 2005, Oudot-Le Secq et al., 2006, Tsui et al., 2009). Published phylogenies strongly support the oomycetes and photosynthetic ochrophytes as a monophyletic group (Tsui et al., 2009, Tyler et al., 2006). While the labyrinthulomycetes appeared as the closest relative to the Bicosoecida, and the phylum Bigyra diverged at the earliest bifurcation of ancestral stramenopiles based on three protein coding genes and SSU rRNA (Tsui et al., 2009). However the sister relationship between labyrinthulomycetes and Bicosoecida was not recovered with seven genes phylogenies when additional representatives of Bicosoecida and *Blastocystis* were included (Riisberg et al., 2009). The basal relationships among the labyrinthulomycetes, bicoesida and *Blastocystis* were unsolved and not supported (Riisberg et al., 2009), as previous SSU rDNA phylogenies (Cavalier-Smith et al., 1994, Van de Peer et al., 2000). Those studies either showed that the labyrinthulomycetes as the sister group of the bicoeceans or showed the labyrinthulomycetes, then bicoeceans emerging from successive divergences at the base of the stramenopiles (Cavalier-Smith & Chao 2006, Leipe et al., 1994). In contrast, Oudot-Le Secq et al. (2006)'s analysis of mitochondrial data showed the labyrinthulomycetes and oomycetes forming a monophyletic group.

No matter what is the branching order in the basal heterotrophic stramenopiles, evidence is accumulating that the ancestors of Stramenopiles and "Chromalveolate" were photosynthetic/ phagotrophic algae (mixotrophs) (Cavalier-Smith & Chao 2006, Harper et al., 2005). Therefore photosynthesis had been lost once in the oomycetes and at least once in the common ancestor to the bicoeceans and labyrinthulomycetes (Riisberg et al., 2009, Tsui et al., 2009). Phagotrophy is the main mode of nutrition in the bicoeceans, which feed on bacteria by the invagination of cell membrane (Boenigk & Arndt, 2002). This may be a shared primitive character for the bicoeceans and the labyrinthulomycetes too. In the labyrinthulomycetes lineage, phagotrophy may have preceded the development of an ectoplasm and cell well. In addition to their dominant walled, osmotrophic vegetative stage, labyrinthulomycetes including *Thraustochytrium striatum*, *Aurantiochytrium mangrovei*, *Ulkenia* and *Labyrinthula* sp. can produce a transient phagotrophic amoeboid stage that ingests bacteria through the development of pseudopodia (Raghukumar, 1992). Oomycetes secrete enzymes and absorb dissolved nutrients across a continuous cell wall, while labyrinthulomycetes are believed to secrete enzymes and absorb dissolved nutrients across their wall-less ectoplasm (Moss, 1991), possibly reflecting the convergent origins of osmotrophy in these two groups.

It is well established that the plastids (cyanobacterial origin) of all photosynthetic stramenopiles originated from a common ancestor. So scientists are interested in the process of plastid loss or the lost of plastid function in those non-photosynthetic stramenopiles (Leipe et al., 1996). The identification of an apparently plastid-derived 6-phosphogluconate dehydrogenase gene and genes of algal origin in *Phytophthora infestans* (a non-photosynthetic stramenopiles) supported it has a photosynthetic ancestor (Tyler et al., 2006). The labyrinthulomycetes also have characters that may have originated from ancestral chloroplasts. Many thraustochytrids produce omega-3 PUFA using desaturase and elongase which are usually located in chloroplasts (Sargent et al., 1995). A few members can be phototactic (e.g. *Labyrinthula* sp. (Perkins & Amon, 1969) and *Ulkenia* sp. (Amon & French, 2004)). The eyespot of *Labyrinthula* zoospores (Perkins & Amon, 1969) also resembles eyespots of other stramenopiles and it may mark the remains of an ancestral chloroplast. In the stramenopiles and in dinoflagellates, eyespots are either within the chloroplast (Motomura, 1994), or are believed to be derived from a chloroplast that underwent evolutionary reduction (Dodge, 1984). Eyespots are absent in the basal thraustochytrids and aplanochytrids (Chamberlain & Moss, 1988, Porter, 1990) and the phylogeny suggests that if these were the last remnants of chloroplasts/plastids, they must have undergone multiple, convergent losses in the labyrinthulomycetes.

4. Phylogenetic relationships within the labyrinthulomycetes

The current taxonomic classification of labyrinthulomycetes is based on the framework of Porter (1990) and Dick (2001). They share a morphological synapomorphy in that their cells secrete an 'ectoplasmic' network, a radiating network of cytoplasm bound by a plasma membrane (Perkins, 1972). Cells extrude ectoplasm through an electron opaque organelle at the periphery of the cell body that is variously called a 'bothrosome,' (Porter, 1969) or a 'sagenogenetosome' (Perkins, 1972). The ectoplasmic network appears to help cells adhere to and penetrate substrates, and it secretes the digestive enzymes required to solubilize nutrients that can be absorbed by the cells (Raghukumar, 2002).

Morphologically they are divided into two major lineages - labyrinthulids and thraustochytrids, largely corresponding to the family Labyrinthuaceae and Thraustochytriaceae. The labyrinthulids include the genera *Labyrinthula* and *Aplanochytrium* (Leander & Porter, 2001). In contrast to thraustochytrids, they are commonly recorded from living algae and seagrasses. The cell bodies of *Labyrinthula* are colonial and glide within the shared ectoplasmic net (containing spindle-shaped vegetative cells) that gives them their common name, 'net slime molds.' The vegetative cells multiply by mitotic division and reproduce by forming zoosporangia and biflagellate zoospores. The cell bodies of *Aplanochytrium* species also crawl via ectoplasmic filaments but unlike *Labyrinthula* species, cells are solitary, not colonial and they are not embedded in ectoplasm (Leander et al., 2004). In addition to the difference in the function of their ectoplasmic filaments, *Labyrinthula* species produce biflagellate zoospores with eyespots (Perkins & Amon, 1969) while *Aplanochytrium* species often reproduce by aplanospores rather than by zoospores. For *Aplanochytrium* species that do have zoospores, eyespots have not been reported (Leander et al., 2004, Porter, 1990).

The remaining labyrinthulomycete genera, commonly referred to as the 'thraustochytrids' produce unicellular, non-motile thalli and although they secrete an ectoplasmic network, they do not use the network for mobility as expressed in the labyrinthulids. Thraustochytrids are abundant heterotrophs in marine and mangroves habitats, and there are three major genera according to Porter (1990) – *Thraustochytrium, Schizochytrium,* and *Ulkenia.* The mode of zoospore production is the basis for genus differentiation. The cytoplasmic content of a vegetative cell develops into a zoosporangium, and then divides directly into zoospores in the genus *Thraustochytrium*. The cytoplasm escapes as an amoeboid mass, prior to the zoospore division in *Ulkenia. Schizochytrium* is characterised by the successive bipartition of a vegetative cell, resulting in the formation of the stages called the diad and the tetrad. Eventually the individual cells within a tetrad develop into zoosporangia and zoospores (Porter, 1990). However there is a high level of morphological variability and overlapping among the genera.

Molecular data consistently support the monophyly of the labyrinthulomycetes (Cavalier-Smith et al. 1994, Honda et al., 1999, Leipe et al., 1996). Multi-gene phylogenies divided them into two well-supported clades. Clade I includes only thraustochytrids, while Clade II includes the labyrinthulids, which include both gliding species and colonial species, as well as thraustochytrids (Fig. 2) (Honda et al., 1999, Tsui et al., 2009). So thraustochytrids that are nonmotile in their assimilative phase are paraphyletic. Also the nesting of labyrinthulids (representatives of *Aplanochytrium* and *Labyrinthula*) among thraustochytrids in Clade II suggested that the ectoplasmic trackways that allow gliding movement of *Aplanochytrium* and *Labyrinthula* had their origin in thraustochytrid's ectoplasmic networks used for anchorage and for nutrient absorption but not movement (Fig. 2) (Tsui et al., 2009).

Molecular data support the sister relationship between *Aplanochytrium* and *Labyrinthula* (Fig. 2) (Honda et al., 1999, Tsui et al. 2009, Yokoyama and Honda 2007a), but provide little resolution on the branching order of genera in thraustochytrids *sensu* Porter (1990) and earlier taxonomic treatment. None of the genera *Thraustochytrium, Schizochytrium* and *Ulkenia* were monophyletic, indicating that the morphological characters employed as taxonomic criteria are unreliable (Honda et al., 1999).

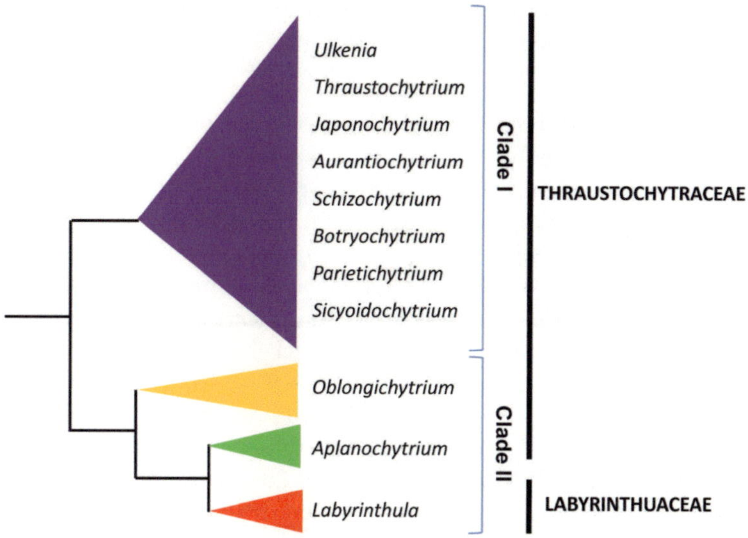

Fig. 2. A schematic summary of the phylogenetic relationships among the genera within the labyrinthulomycetes (modified from Tsui et al., 2009 and Yokoyama et al., 2007b).

Recent studies have delineated the thraustochytrids into multiple monophyletic genera with their morphology, biochemistry, and molecular data. Genera of *Oblongichytrium, Aurantiochytrium, Botryochytrium, Parietichytrium,* and *Sicyoidochytrium* have been erected during the taxonomic revisions of *Schizochytrium* and *Ulkenia* (Fig. 2) (Yokoyama et al., 2007a, b). For example, the genus *Aurantiochytrium* has been erected for a phylogenetic lineage of *Schizochytrium* species that could accumulate DHA for over 30% of the total fatty acids (Yokoyama et al., 2007a). Also the content of C18 and C20 precursor unsaturated fatty acids in *Aurantiochytrium* for DNA synthesis in the elongation/ desaturation pathway were much lower than those in the genera *Thraustochytrium* and *Schizochytrium* (Nagano et al., 2011).

5. Ecophysiology of thraustochytrids

Thraustochytrids, are ubiquitous in oceanic water column (Bahnweg & Sparrow, 1974, Raghukumar, 2002) and they are associated with the wide range of substrata and habitats; e.g. from both fresh and decaying algal surfaces (e.g. in UK waters - Miller & Jones, 1986; in Indian waters - Raghukumar 1986), from decaying leaves of sea grass (e.g. in US waters - Jensen et al., 1998) , from decaying leaves of mangrove plants (e.g. in Hong Kong waters - Fan et al., 2002a) and from invertebrate tissues (e.g. in abalone tissues – Bower, 1987; in clam tissues – Azevedo & Corral 1997). Raghukumar & Damare (2011) gave a short concise chronological account of the development of the research of this group of organisms since their first discovery in US waters in the mid-30s (Sparrow, 1936).

In the past decade, there were two areas of research in thraustochytrids where efforts were concentrated; phylogeny studies based on molecular analysis as described earlier in this chapter and the physiology of production of compounds which have important nutritional values (e.g. fatty acids - Fan et al., 2001, 2007; squalene – Li et al., 2009). In the following paragraphs, a review of some ecophysiological investigations of thraustochytrids isolated from decaying mangrove leaves in subtropical mangroves is presented (Fan et al., 2002a, b, Tsui et al. 2011, Wong et al., 2005).

Thraustochyrids are well adapted to the mangrove environment where salinity and temperature levels fluctuate daily, monthly and seasonally. A series of ecological and physiological investigations have been undertaken on various isolates of thraustochytrids isolated from the subtropical mangroves where salinity levels could vary between 5 and 34 ‰ in summer and winter in Hong Kong respectively (Fan et al., 2002a, b, Tsui et al., 2011, Wong et al. 2005). Some of these species were isolated from low saline waters (ca. 5‰). These isolates were, namely *Schizochytrium* sp. KF1, *Aurantiochytrium mangrovei* KF-2, KF-7 KF-12, *Thraustochytrium striatum* KF-9, and *Ulkenia* KF-13. Their growth response under different salinities (distilled water, 7.5 - 30 ‰), pH (4 - 9) and temperature (15 - 30 °C) levels in yeast extract -peptone-glucose seawater (YPGS) broth were reported (Fan et al. 2002a). In general, all cultures grew equally well in all tested pH levels, and the overall optimal temperature range was at 22 - 25°C between 7.5 and 30‰ salinity levels. *Aurantiochytrium* and *Schizochytrium* isolates produced overall higher dry weight biomass (ca. 150 - 300 mg/50mL) at all tested temperature and salinity levels compared to *Ulkenia* and *Thraustochytrium* isolates. Although each isolate had their own specific, optimal response to varying salinities and temperature levels, the interaction of salinity and temperature affected their growth significantly ($P<0.001$) (Fan et al., 2002a, b).

The zoospore production capacity and their motility profile are also highly influenced by salinity (Tsui et al., 2011). A summary of the zoospore features at various salinities of *Schizochytrium* sp.KF1, *Aurantiochytrium mangrovei* KF-6, *Thraustochytrium striatum* KF-9 and *Ulkenia* KF-13. is shown in Table 1. Zoospores of thraustochytrids were also strongly attracted to the mangrove leaf extracts when comparing to various amino acids and carbohydrates (Fan et al., 2002b). Zoospores of *A. mangrovei* KF-6 showed highest response followed by *Ulkenia* sp. KF-13 whereas those of *T. striatum* KF-9 were very weak, showing almost no differentiation amongst all the test compounds. The summary data shown in Table 1, and the results of the chemotactic response experiment (Table II in Fan et al., 2002b) indicate the overall competitiveness of these strains in the mangrove environment where fluctuating saline waters could be encountered within each tidal cycle throughout the year.

Aurantiochytrium mangrovei was the most abundant thraustochytrid species in the Hong Kong mangroves, followed by *Schizochytrium* spp. and *Ulkenia* spp., whereas *Thraustochytrium* spp. were seldom encountered (Vrijmoed unpublished). The very small number of zoopores being produced by *T. striatum* (Table 1) coupled with the weak chemotactic response to mangrove leaf extracts and nutrients and the overall low biomass produced in batch cultures may explain their low occurrence in spite of the fairly active zoospores after release from the zoosporangium. The average zoospore production capacity of *Ulkenia* sp. was nearly 10-fold of that of *Thraustochytrium* sp. However, its moderate motility and lowest VCL and VSL amongst the test strains lower the chance of the zoospores to locate a substrate for settlement and growth. *A. mangrovei* had the highest growth rate in batch cultures; its zoospores were also most attracted to mangrove leaf extracts and nutrients. Their VSL and VSL were high which compensate the moderate zoospore production amount and the average motility % within a 4h period. These are the probable reasons for their dominance in the subtropical mangroves in Hong Kong (Tsui et al., 2011).

	Schizochytrium sp.KF1	*Aurantiochytrium mangrovei* KF-6	*Thraustochytrium striatum* KF-9	*Ulkenia* KF-13	General Comments
Average Zoospore production[a] (x10^3 mL^{-1})	19.34	29.04	4.88	40.30	zoospore production of all strains suppressed at levels > 15‰
Average motility[b] within a 4h-period (%)	86.6	78.1	91.7	84.3	Motility of all strains remained at 90 to 100% after 2h but were reduced 60 - 90% after 4h.
Average curvilinear velocity (VCL)[b] (µm sec^{-1}) within a 4h-period	89.3	99.2	103.1	71.0	Not applicable
Average straight line velocity (VSL)[c] (µm sec^{-1}) within a 4h-period	60.2	70.6	71.3	35.8	Not applicable

[a] The motility of zoospores was recorded using the image analysis system consisting of a phase contrast microscope with a lens at 20x10 magnification (Olympics BX50 Japan) equipped with a progressive scan charged-coupled device (CCD) camera (Basler Scout, SCA640-70FM, Ahrensburg, Germany).
[b] Zoospores were induced from 2-day old cultures in yeast extract peptone plates flooded separately with distilled water, and artificial seawater at 7.5, 15, 22.5 and 30‰.
[c] VCL – the time average velocity of the zoospore head along its actual trajectory.
[d] VSL- the time average velocity of the zoospore head along the straight line between its first detected position and its last position.

Table 1. A summary of zoospore profile of mangrove thraustochytrids (adapted from Tsui et al., 2011).

The temporal variation of abundance of thraustochytrids in decaying mangrove leaves (*Kandelia obovata*) and sediments were also investigated, and the results indicate that thraustochytrid abundance in decaying leaves were much higher (4.8x103 - 5.6x105 $CFUg^{-1}$ of oven-dried weight of leaves) compared with the levels in surface sediments (1.0x102 - 1.6x103 $CFUg^{-1}$ of oven-dried weight of sediment) (Wong et al., 2005). Thraustochytrids colonies were enumerated by spreading the leaf homogenate and sediment suspension on YEP agar plates incorporated with antibiotics and incubated at 25 °C for two days. This is supported by a similar pattern of thraustochytrid occurrence in the samples, being an average of 85.5% vs. 57.5% in leaves and sediments respectively. However statistical analyses revealed no significant correlations in the occurrence between leaves and sediments, as well as between the samples and the air temperature and water salinities.

Data of several experiments indicate that thraustochytrids provide the necessary long-chain polyunsaturated fatty acids (LCPUFAs) to marine organisms which cannot synthesize them. Mangrove crabs (e.g. *Parasesarma affinis* and *Parasesarma bidens*) which mainly ingest decay leaves (Lee & Kwok, 2002) would be enriched with the LCPUFAs laden in the leaves. Partially digested thraustochytrid cells were also detected amongst diatom skeletons in the gut content of the mudskipper *Boleopthalmus pectinirostris* which are prevalent in the intertidal mangrove shores in Hong Kong (Vrijmoed, unpublished data). Mudskippers sieved sediment to obtain their food. So there is partial evidence on the importance of thraustochytrids in the food web in the mangrove ecosystem.

6. Future research and conclusion

Labyrinthulomycetes occupy an important position in the eukaryote tree of life and they play a critical role in the ecosystems by upgrading the 'nutritional value of detritus' due to their ability to produce LCPUFAs. Although labyrinthulomycetes, specifically the labyrinthulids, are important ecologically, there is no formal estimate to the number of species but many unknown representatives have been described only from sequences in metagenomics studies from marine ecosystems (Massana et al., 2004a, Not et al., 2007). Currently four labyrinthulomycete genomes are being sequenced at Joint Genome Institute. The data will offer genome-scale insight into the physiology of an ecological and biotechnological significant group of organisms. For example, the genome data will provide new information about the genetic basis for the ectoplasmic net development, and virulence to organisms and their evolutionary history. The genome data will also provide specific insight into genetic basis for differences between species that are of ecological and biotechnological relevance. Additionally, the information will make possible further investigations of degrading enzymes of biotechnological interest.

7. Acknowledgements

Drs D Honda and R Yokoyama (Konan University, Japan) are thanked for continued scientific support and discussion. Parts of this chapter are derivatives of article published in Tsui et al. (2009).

8. References

Abril, J. R.; Barclay, W. R. & Abril, P. G. (2000). Safe use of microalgae (DHA GOLD™) in laying hen feed for the production of DHA-enriched eggs. In: *Egg nutrition and biotechnology*, J.S. Sim, S. Nakai, W. Guenter (Eds.), 197-202, ISBN 0851993303, CAB International, Wallingford.

Amon J. P. & French K. H. (2004). Photoresponses of the marine protist *Ulkenia* sp. zoospores to ambient, artificial and bioluminescence light. *Mycologia* 96, 463–469

Azevedo, C.; Corral L. (1997). Some ultrastructural observations of a thraustochytrid (Protoctista, Labyrinthulomycota) from the clam *Ruditapes decussates* (Mollusca, Bivalvia). *Diseases of Aquatic Organisms* 31, 73-78

Bahnweg, G.; Sparrow, F.K., (1974). Occurrence, distribution and kinds of zoosporic fungi in subantarctic and Antarctic waters. Veröff. Inst. Meeresforscg. Bremerh. Supplement 5, 149-157.

Baldauf, S. L.; Roger, A. J.; Wenk-Siefert, I. & Doolittle, W. F. (2000). A kingdom-level phylogeny of eukaryotes based on combined protein data. *Science* 290, 972–977

Bigelow, D. M.; Olsen, M. W. & Gilbertson, R. L. (2005). *Labyrinthula terrestris* sp. nov., a new pathogen of turf grass. *Mycologia* 97, 185-190

Boenigk, J. & Arndt, H. (2002). Bacterivory by heterotrophic flagellates: community structure and feeding strategies. *Antonie van Leewenhoek* 81, 465-480

Bower, S. M. (1987). The life cycle and ultrastructure of a new species of thraustochytrid (Protozoa: Labyrinthomorpha) pathogenic to small abalone. *Aquaculture,* 67, 269-272

Carmona, M. L.; Naganuma, T. & Yamaoka, Y. (2003). Identification by HPLC-MS of Carotenoids of the *Thraustochytrium* CHN-1 Strain Isolated from the Seto Inland Sea. *Bioscience, Biotechnology, and Biochemistry* 67, 884-888

Cavalier-Smith, T. (1998). A revised six-kingdom system of life. *Biological Reviews* 73, 203-266.

Cavalier-Smith, T. & Chao, E. E. Y. (2006). Phylogeny and megasystematics of phagotrophic heterokonts (Kingdom Chromista). *Journal of Molecular Evolution* 62, 388-420

Cavalier-Smith, T.; Allsopp, M. T. E. P. & Chao, E. E. (1994). Thraustochytrids are chromists, not fungi: 18S rDNA signatures of heterokonta. *Philosophical Transactions of the Royal Society B: Biological Sciences* 346, 387-397

Chamberlain, A. H. L. & Moss, S. T. (1988). The thraustochytrids: a protist group with mixed affinities. *BioSystems* 21, 341-349

Craven, K. D.; Peterson, P. D.; Windham, D. E.; Mitchell, T. K. & Martin, S. B. (2005). Molecular identification of the turf grass rapid blight pathogen. *Mycologia* 97, 160-166

Dick, M. W. (2001). *Straminipilous fungi: systematics of the peronosporomycetes, including accounts of the marine straminipilous protists, the plasmodiophorids, and similar organisms.* ISBN 0792367804 Dordrecht, Boston, Kluwer Academic Publishers

Dodge, J. D. (1984). The functional and phylogenetic significance of dinoflagellate eyespots. *BioSystems* 16, 259-267

Fan, K. W.; Chen, F.; Jones, E. B.G.; Vrijmoed, L. L.P. (2001). Eicosapentaenoic and docosahexaenoic acids production by and okara-utlizing potential of thraustochytrids. *Journal of Industrial Microbiology and Biotechnology* 27, 199-202

Fan, K. W.; Vrijmoed, L. L. P. & Jones, E. B.G. (2002a). Physiological Studies of Subtropical Mangrove Thraustochytrids. *Botanica Marina* 45, 50-57

Fan, K. W.; Vrijmoed, L. L.P. & Jones, E. B.G. (2002b). Zoospore chemotaxis of mangrove thraustochytrids from Hong Kong. *Mycologia* 94, 569-578

Fan, K. W.; Jiang, Y.; Faan, Y. W. & Chen, F. (2007). Lipid Characterization of Mangrove Thraustochytrid - *Schizochytrium mangrovei*. *Journal of Agricultural and Food Chemistry* 55, 2906-2910

Fan, K. W.; Jiang, Y.; Ho, L. T. & Chen, F. (2009). Differentiation in fatty acid profiles in pigmented and nonpigmented *Aurantiochytrium* isolated from Hong Kong mangroves. *Journal of Agricultural and Food Chemistry* 57, 6334–6341

Felsenstein, J., (2002). PHYLIP (Phylogeny inference package). Version 3.6a3. Department of Genome Science. University of Washington, Seattle, Washington.

Findlay, R. H.; Fell, J. W.; Coleman, N. K. & Vestal, J. R. (1986). Biochemical indications of the role of fungi and thraustochytrids in mangrove detrial systems. In: *The biology of marine fungi*. S. T. Moss (Ed.), 91–103, ISBN 0521308992, Cambridge University Press, Cambridge, UK

Harper, J. T.; Waanders, E. & Keeling, P. J. (2005). On the monophyly of chromalveolates using a six-protein phylogeny of eukaryotes. *International Journal of Systematic and Evolutionary Microbiology* 55, 487-496

Honda, D.; Yokochi, T.; Nakahara, T.; Erata, M. & Higashihara, T. (1998). *Schizochytrium limacinum* sp. nov., a new thraustochytrid from a mangrove area in the west Pacific Ocean. *Mycological Research* 102, 439–448

Honda, D.; Yokochi, T.; Nakahara, T.; Raghukumar, S.; Nakagiri, A.; Schaumann, K. & Higashimhara, T. (1999). Molecular phylogeny of labyrinthulids and thraustochytrids based on sequencing of 18S ribosomal RNA gene. *Journal of Eukaryotic Microbiology* 46, 637-647.

Jensen, P. R.; Jenkins, K. M.; Porter, D. & Fencial, W. (1998). Evidence that a new antibiotic flavones glycoside chemically defends the sea grass *Thalassia testudinum* against zoosporic fungi. *Applied and Environmental Microbiology* 64, 1490-1496.

Jiang, Y.; Fan, K.W.; Wong, R.T.Y. & Chen, F. (2004). Fatty acid composition and squalene content of the marine microalgae *Schizochytrium mangrovei*. *Journal of Agricultural and Food Chemistry* 52, 1196-1200

Keeling, P.J.; Burger, G.; Durnford, D. G.; Lang, B. F.; Lee, R. W.; Pearlman, R. E.; Roger, A. J. & Gray, M. W. (2005). The tree of eukaryotes. *Trends in Ecology and Evolution* 20, 670-676.

Kimura, H.; Fukuba, T. & Naganuma, T. (1999). Biomass of thraustochytrid protoctists in coastal water. *Marine Ecology Progress Series* 189, 27–33

Leander, C.A. & Porter, D. (2001). The Labyrinthulomycota is composed of three distinct lineages. *Mycologia* 93, 459-464

Leander, C. A.; Porter, D. & Leander, B. S. (2004). Comparative morphology and molecular phylogeny of aplanochytrids (Labyrinthulomycota). *European Journal of Protistology* 40, 317-328

Lee, S. Y. & Kwok, P. W. (2002). The importance of mangrove species association to the population biology of the sesarmine crabs *Parasesarma affinis* and *Parasesarma bidens*. *Wetlands Ecology and Management* 10, 215-226.

Leipe, D. D.; Wainright, P. O.; Gunderson, J. H.; Porter, D.; Patterson, D. J.; Valois, F.; Himmerich, S. & Sogin, M. L. (1994). The stramenopiles from a molecular perspective: 16S-like rDNA sequences from *Labyrinthuloides minuta* and *Cafeteria roengergensis*. *Phycologia* 33, 369-377

Leipe, D. D.; Tong, S. M.; Goggin, C.;L.; Slemenda, S.;B.; Pieniazek, N.;J. & Sogin, M. L. (1996). 16S-like rDNA sequences from *Developayella elegans, Labyrinthuloides halioidis,* and *Proteromonas lacertae* confirm that the stramenopiles are a primary heterotrphic group. *European Journal of Protistology* 32, 449-458

Lepage, G. & Roy, C. C. (1984). Direct trans-esterifi cation of all classes of lipids in a one step reaction. *Journal of Lipid Research* 25, 1391–1396

Li, Q., Chen, Q. G., Fan, K.W., Lu, F. P., Aki, T. & Jiang, Y. (2009). Screening and Characterization of Squalene-Producing Thraustochytrids from Hong Kong Mangroves. *Journal of Agricultural and Food Chemistry* 57, 4267–4272.

Liu, Q.; Allam, B & Collier, J. L. (2009). Quantitative real-time PCR assay for QPX (Thraustochytriidae), a parasite of the hard clam (*Mercenaria mercenaria*). *Applied and Environmental Microbiology* 75, 4913-4918

Maddison, W. P. & Maddison, D.R. (2000). MacClade: analysis of phylogeny and character evolution. Version 4.0. Sinauer, Sunderland, MA.

Massana, R.; Balague, V.; Guillou, L. & Pedrós-Alió, C. (2004a). Picoeukaryote diversity in an oligotrophic coastal site studied by molecular and culturing approaches. *FEMS Microbiology Ecology* 50, 231–243

Massana, R.; Castresana, J.; Balague, V.; Guillou, L. & many others (2004b). Phylogenetic and ecological analysis of novel marine stramenopiles. *Applied and Environmental Microbiology* 70, 3528–3534

Miller, J. D. & Jones, E. B. G. (1983). Observations on the association of thraustochytrid marine fungi with decaying seaweed. *Botanica Marina* 24, 345-351.

Moss, T. S. (1991). Thraustochytrids and other zoosporic marine fungi. In: *The Biology of Free-Living Heterotphic Flagellates. Systematics Association Special Volume No. 45,* D. J. Patterson, J. Larsen (Eds), 415-425, ISBN 978-0-19-857747-8, Clarendon Press, Oxford

Motomura, T. (1994). Electron and immunofluorescence microscopy on the fertilization of *Fucus distichus* (Fucales, Phaeophyceae). *Protoplasma* 178, 97-110

Muelstein, L. K.; Porter, D. & Short, F. T. (1988). *Labyrinthula* sp. A marine slime mold producing the symptoms of wasting disease in eelgrass, *Zostera marina*. *Marine Biology* 99, 465-472

Naganuma, T.; Takasugi, H. & Kimura, H. (1998). Abundance of thraustochytrids in coastal plankton. *Marine Ecology Progress Series* 162, 105–110

Nagano, N.; Sakaguchi, K.; Taoka, Y.; Okita, Y.; Honda, D.; Ito, M. & Hayashi, M. (2011). Detection of genes involved in elongation and desaturation in thraustochytrid marine eukaryotes. *Journal of Oleo Science* 60, 475-481.

Not, F.; Gausling, R.; Azam, F.; Heidelberg, J. F. & Worden, A. Z. (2007). Vertical distribution of picoeukaryotic diversity in the Sargasso Sea. *Environmental Microbiology* 9, 1233–1252

Oudot-Le Secq, M.-P.; Loiseaux-de Goër, S.; Stam, W. T. & Olsen, J.L. (2006). Complete mitochondrial genomes of the three brown algae (Heterokonta: Phaeophyceae)

Dictyota dichotoma, Fucus vesiculosus and *Desmarestia viridis. Current Genetics* 49, 47-58

Patterson, D.J. (1989). Stramenopiles, chromophytes from a protistan perspectives. In: *The chromaphyte algae, problems and perspectives.* J. C. Green, B. S. C.Leadbeater, W. L. Diver, (Eds), 357-379, ISBN 0198577133, Clarendon Press, Oxford

Perkins, F.O. (1972). The ultrastructure of holdfasts, "rhizoids", and slime tracks" in thraustochytriaceous fungi and *Labyrinthula* spp. *Archiv für Mikrobiologie* 84, 95-118

Perkins, F. O., & Amon, J. P. (1969). Zoosporulation in *Labyrinthula* sp.: an electron microscopic study. *Journal of Protozoology* 16, 235-257

Porter, D. (1969). Ultrastructure of *Labyrinthula. Protoplasma* 67, 1-19

Porter, D. (1990). Phylum Labyrinthulomycota. In: *Handbook of Protoctista.* L. Margulis, J. O. Corliss, M. Melkonian, D. J. Chapman, (Eds), 388-398, ISBN 0867200529, Jones and Barlett, Boston

Raghukumar, C. (1986). Thraustochytrid fungi associated with marine algae. *Indian Journal of Marine Science* 15, 121-122.

Raghukumar, S. (1992). Bacterivory: a novel dual role for thraustochytrids in the sea. *Marine Biology* 113, 165-169

Raghukumar, S. (2002). Ecology of the marine protists, the Labyrinthulomycetes (Thraustochytrids and Labyrinthulids). *European Journal of Protistology* 38, 127-145

Raghukumar, S. & Damare, V. S. (2011). Increasing evidence for the important role of Labyrinthulomycetes in marine ecosystems. *Botanica Marina* 54, 3-11.

Raghukumar, S.; Anil, A. C.; Khandeparkar, L. & Patil, J. S. (2000). Thraustochytrid protists as a component of marine microbial films. *Marine Biology* 136, 603-609

Riisberg, I.; Orr, R. J. S.; Kluge, R.; Shalchian-Tabrizi, K.; Bowers, H. A.; Patil, V. & Edvardsen, B., Jakobsen, K.S. (2009). Seven gene phylogeny of heterokonts. *Protist* 160, 191-204

Sargent, J. R.; Bell, M. V. & Henderson, R. J. (1995). Protists as sources of (n-3) polyunsaturated fatty acids for verterate development. In: *Proceedings of the Second European Congress of Protistology.* G. Brugerolle, J.-P. Mignot, (Eds), 55-64, Clermont-Ferrand

Sijtsma, L. & de Swaaf, M.E. (2004). Biological production and applications of the omega-3 polyunsaturated fatty acid, docosahexaenoic acid. *Applied Microbiology and Biotechnology* 64, 146-153

Sparrow, F. K. (1936). Biological observations on the marine fungi of Woods Hole waters. *Biological Bulletin, Marine Biological Laboratory, Woods Hole.* 70, 236-273.

Stokes, N. A.; Ragone Calvo, L. M.; Reece, K. S. & Burreson, E. M. (2002). Molecular diagnostics, field validation, and phylogenetic analysis of Quahog Parasite Unknown (QPX), a pathogen of the hard clam *Mercenaria mercenaria. Diseases of Aquatic Organisms* 52, 233-247

Swofford, D. L. (2003). *PAUP*. Phylogenetic Analysis Using Parsimony (*and Other Methods).* Sinauer Associates, Sunderland, Massachusetts.

Takao, Y.; Tomaru, Y.; Nagasaki, K.; Sasakura, Y.; Yokoyama, R. & Honda, D. (2007). Fluorescence in situ hybridization using 18S rRNA targeted probe for specific detection of thraustochytrids (Labyrinthulomycetes). *Plankton Benthos Research* 2, 91–97

Taoka, Y.; Nagano, N.; Okita, Y.; Izumida, H.; Sugimoto, S. & Hayashi, M. (2009). Extracellular enzymes produced by marine eukaryotes, thraustochytrids. *Bioscience, Biotechnology and Biochemistry* 73, 180–182

Thompson, J. D.; Gibson, T. J.; Plewniak, F.; Jeanmougin, F. & Higgins, D. G. (1997). The Clustal X windows interface: flexible strategies for multiple sequence alignment aided by quality analysis tools. *Nucleic Acids Research* 25, 4876-4882

Tsui, C. K. M.; Marshall, W.; Yokoyama, R.; Honda, D.; Lippmeier, J. C.; Craven, K. D.; Peterson, P. D. & Berbee, M. L. (2009). Labyrinthulomycetes phylogeny and its implication for the evolutionary loss of chloroplasts and gain of ectoplasmic gliding. *Molecular Phylogenetics and Evolution* 50, 129–140

Tsui, C. K. M.; Fan, K. W.; Chow, R. K. K.; Jones, E. B. G. & Vrijmoed, L. L. P. (2011). Zoospore production and motility of mangrove thraustochytrids from Hong Kong under various salinities. *Mycoscience* DOI 10.1007/s10267-011-0127-2.

Tyler, B. M.; Tripathy, S.; Zhang, X.; Dehal, P.; Jiang, R. H. Y.; Aerts, A. & 47 others. (2006). *Phytophthora* Genome Sequences uncover evolutionary origins and mechanisms of pathogenesis. *Science* 313, 1261-1266.

Van de Peer, Y.; Baldauf, S. L.; Doolittle, W. F. & Meyer, A. (2000). An updated and comprehensive rRNA phylogeny of (crown) eukaryotes based on rate-calibrated evolutionary distances. *Journal of Molecular Evolution* 51, 565-576

White, T. J.; Bruns, T.; Lee, S. & Taylor, J. (1990). Amplification and direct sequencing of fungal ribosomal RNA genes for phylogenetics. In: *PCR protocols, A guide to methods and applications.* M. A. Innis, D. H. Gelfand, J. J. Sninsky, T. J. White (Eds), 315-322, ISBN-10 0123721814 Academic Press Inc, San Diego, California

Wong, K. M. M.; Vrijmoed, L. L. P. & Au, W. T. D. (2005). Abundance of thraustochytrids on fallen decaying leaves of Kandelia candel and mangrove sediments in Futian National Nature Reserve, China. *Botanica Marina* 48, 374-378

Yamasaki, T.; Aki, T.; Mori, Y.; Yamamoto, T.; Shinozaki, M.; Kawamoto, S. & Ono, K. (2007). Nutritional enrichment of larval fish feed with thraustochytrid producing polyunsaturated fatty acids and xanthophylls. *Journal of Bioscience and Bioengineering* 104, 200–206

Yokoyama, R. & Honda, D. (2007a). Taxonomic rearrangement of the genus *Schizochytrium sensu lato* based on morphology, chemotaxonomical characteristics and 18S rRNA gene phylogeny (Thraustochytriaceae, Labyrinthulomycetes, stramenopiles): emendation for *Schizochytrium* and erection of *Aurantiochytrium* and *Oblongichytrium* gen. nov. *Mycoscience* 48, 199-211

Yokoyama, R.; Salleh, B. & Honda, D. (2007b). Taxonomic rearrangement of the genus *Ulkenia sensu lato* based on morphology, chemotaxonomical characteristics, and 18S rRNA gene phylogeny (Thraustochytriaceae, Labyrinthulomycetes): emendation for *Ulkenia* and erection of *Botryochytrium, Parietichytrium*, and *Sicyoidochytrium* gen. nov. *Mycoscience* 48, 329–341

Yongmanitchai, W. & Ward, O. P. (1989). Omega-3 fatty acids: alternative source of production. *Process Biochemistry* 24, 117–125

8

Seabed Mapping and Marine Spatial Planning: A Case Study from a Swedish Marine Protected Area

Genoveva Gonzalez-Mirelis, Tomas Lundälv, Lisbeth Jonsson, Per Bergström, Mattias Sköld and Mats Lindegarth
University of Gothenburg
Swedish University of Agricultural Sciences
Sweden

1. Introduction

Knowledge of spatial patterns of fauna and flora is in high demand among the policy-making and management community, not least in areas where the biological value is such that conservation efforts are warranted. This type of information enables assessing the distribution of biodiversity and other resources (e.g. fisheries), monitoring habitat change and defining (scale-specific) representative and unique features, as prescribed for the design of reserve networks and, more generally, the realisation of spatial planning. From the micro-habitat scale to the scale of biogeographic provinces, geospatial ecological data alongside habitat mapping have significantly helped fill this knowledge gap by providing local and regional models that capture the spatial distribution of various user-defined or typology-derived classes (e.g. forest types) which can in turn be used to understand or predict species distributions.

Across the landscape lying beneath the water masses, limited means of access precludes the intensification of survey effort, hampering any broad-scale mapping endeavours. Pioneer benthic ecologist Petersen (1924) expressed this problem by stating that "'botanizing' out at sea is a very expensive affair" (Petersen 1924, p. 688). Embedded in this accurate statement is the notion of the high cost associated with acquiring survey data in the benthos that can compare in quality and quantity to those that plant ecologists have readily at their disposal. In addition, the extensive collection of remotely sensed data, which have boosted habitat mapping on land by virtue of efficiently complementing vegetation survey data, can only be extended seaward as far as the width of the narrow littoral fringe circling the land masses. Mapping of seabed, sublittoral environments was therefore for many years limited to what could be inferred from small numbers of scattered point samples, yielding maps with large gaps of information or restricted to very small scales.

The advent of acoustic technologies (namely sidescan sonar and single- and multibeam echo sounders) rendered it possible to acquire high-resolution, full-coverage imagery of the seafloor over extensive areas, beyond the limit of light penetration. As a result, the geophysical attributes of the terrain that can be derived from its acoustic properties can now form the basis for a classification of, for example, depth and seafloor texture. These advances endowed benthic researchers with the ability to carry out spatially continuous, 'wall-to-wall' mapping, following in the steps of terrestrial remote sensing science. Nonetheless, the challenges of direct observation, though also lessened by the emergence of new technologies (e.g. underwater video camera systems, benthic sleds, etc.) have largely remained, just as they have in other remote areas across the globe.

With ecological data being sparse at best and remotely sensed data bearing mostly an indirect (and not fully disentangled) relationship with biological composition, in the benthos the challenge of mapping the various components of biodiversity (from genes to ecosystems) is largely a methodological one. To inform the classification and mapping of the communities supported by the different habitats, effort is geared toward developing techniques to better integrate point-source field data with remotely sensed data (e.g. Brown et al. 2002; 2004; Hewitt et al. 2004; Holmes et al. 2008; Jordan et al. 2005; Kloser et al. 2001; Kostylev et al. 2001).

In addition to the methodological challenges, the upper part of the continental shelf is considered to be the area likely to benefit the most from marine spatial planning (rather, its users) because it is where the highest biodiversity conservation values and the greatest threats overlap. Multiple economic interests compete for space on the continental shelf and upper slope, including fisheries, causing by far the most widespread impact to the megafauna and the ecosystem that it depends on, and to a lesser extent, oil and gas exploration, shipping, mining, acquaculture and tourism.

In this chapter we review the most widespread approaches to mapping the distribution of benthic fauna, limiting ourselves to the offshore circalittoral zone, because this is where all of the described challenges are being faced simultaneously, including methodological and management-related. Sessile benthic organisms are perceived as particularly useful for habitat characterisation because substrate is critical for their survival and proliferation; being spatially fixed, they also become indicative of environmental conditions of the adjacent seafloor (Kostylev et al., 2001). Therefore, we will focus on the particular mapping 'school' which places the emphasis on obtaining the best possible picture of variation within this subset of benthic biota. As far as mapping is concerned, the epibenthic megafauna is regarded as the vegetation of the benthic landscape.

By means of a case study taken from a Swedish Fjord, recently designated as a multiple-use marine protected area, we will show how benthic biotope mapping provides the most effective means to document and spatially manage seabed-dwelling biodiversity.

2. Selection of an appropriate mapping theme

From single species to broad ecosystems, any level of biological organisation can be described in terms of its spatial distribution patterns and hence be depicted in map form. Irrespective of the particular level of choice, units can also be defined with a varying degree of reference to the properties of the environment associated with them, ranging from these being completely absent (e.g. a given 'assemblage of gorgonian corals') to completely replacing any biological information (e.g. a 'sand bank'). Typically, the higher up in the hierarchy of ecosystems one operates, the more weight is carried by the abiotic component, where classes are described on the basis of their physiography, geology or morphology rather than their biological composition, because the latter is often much harder to summarise than the former. Similarly, the area occupied by a single species is more easily characterised by referring to the presence of that species, rather than attempting to describe the habitat conditions in which the species can be found. This rule of thumb does not always hold up, like in the case of a biogenic reef or other habitat-forming species. This is a direct consequence of the way in which ecosystems are structured. Intermediate to high-level units (i.e. broader categories) respond more strongly to changes in the environment and processes that operate at larger scales, than changes in biology (e.g. the presence of a predator), which affect the lower level classes more ostensibly.

Additionally, compositional turnover responds differently to different habitat gradients depending on the complexity and scale of the units under consideration (e.g. very detailed communities versus broad categories), with patterns that are not fully consistent across environments. Gonzalez-Mirelis et al. (2011) found that when considering biological variation at various levels of detail of taxon groups (admittedly a narrow window, within the gene-to-ecosystem continuum), classes at the coarsest end responded more strongly to substrate than any of the other gradients considered, including depth, while this effect was reversed in finer levels, discriminating regions with more homogeneous characteristics whose boundaries change at smaller spatial scales. However, Bergen et al. (2001) found a pattern where the coarsest divisions of a dendrogram of infaunal communities were strongly associated with depth and the finest divisions were better explained by sediment, with depth no longer significant. A key difference lies in the target of each study, where the former focused on epibenthic fauna of circalittoral environments, whereas the latter looked at infauna and also included infralittoral sites. All the above factors provide for a multitude of methods to classify the environment into units, including hierarchical systems, which can in turn be so with respect to faunal composition, or the relative importance of environmental factors in structuring the ecosystem, providing an ensemble of alternatives to choose from when solving different management problems.

Despite the apparent complexity of the task and lack of universally-applicable definitions of each of the categories potentially involved, in the marine realm, and especially in the applied environmental literature, researchers have gravitated towards one of the following two themes: (1) biotopes (Connor et al., 2004; CORINE, 1991; EUNIS, 2005; HELCOM, 1998) and (2) habitats (Allee et al., 2001; Greene et al., 1999; Valentine et al., 2005). We argue that this is so because they meet two crucial conditions: they are easy to map and they are biologically meaningful.

The two concepts are sometimes synonymised, but there are fundamental differences. The current meaning of 'biotope', which in fact became popular in the marine realm before the terrestrial, combines the "physical environment [...] and its distinctive assemblage of conspicuous species" (Olenin & Ducrotoy 2006, p. 22) where, crucially the concept incorporates geographic location, thus rendering it scale-dependent. 'Habitat' has been defined as a spatially recognisable area where the physical, chemical and biological environment is distinctly different from surrounding environments (Kostylev et al., 2001; Valentine et al., 2005). A review of the literature and available definitions reveals a focus that merely shifts from the biological properties in the case of biotopes to the environmental properties in the case of habitats, and indeed, a biotope can be defined as the sum of community and habitat. Also, the concept of habitat is generally more loose and has been used with a much broader array of meanings than biotopes have.

Both habitats and biotopes require that their boundaries be delineated in order to be fully characterised, so they are inherently 'mapping units' (Foster-Smith et al., 1999). Not only are they mappable, they are so across the scales where management and planning typically occur. Biotopes are biologically meaningful for trivial reasons, but habitats also have been shown to be adequate surrogates for patterns of species richness in marine environments (Ward et al., 1999). It is clear that any ordered system of classes with these properties, whether biotopes, habitats or another aggregation level, can be easily translated into a suitable mapping theme and thus be incorporated in the framework that we review below.

3. Mapping the benthic landscape

The integration of data from multiple surveying techniques, typically including one or more full-coverage layers depicting features visible at medium to large spatial scales (e.g. pinnacles, canyons, etc.), and at least one dataset from some *in situ* benthic survey technique providing insight into small-scale variation occurring mostly at the biological level, nested within the former, has proven the most rewarding technique for mapping extensive areas of the seafloor. Hereforth we will refer to any data obtained via an *in situ* sampling technique (e.g. video, dredge, trawl, etc.) as 'survey data', in contrast to 'remotely-sensed data' (note that we include video and photographic sampling methods within methods labelled as *'in situ'*, contrarily to other authors, on the grounds that the sampling device is located directly at the site that is being sampled, even if the operator of the device is not; we therefore reserve the term 'remote sensing' to refer to hydroacoustic techniques). Survey data delivers spatially-explicit information on the value of ecological variables and/or variables relating to the sediment or bedrock, whether quantitative (e.g. species abundance, granulometry information) or qualitative (e.g. 'presence of sessile invertebrates', 'presence of mobile sediment').

Alternatives to this integrative approach include using either only remotely-sensed data, or only survey data, as a basis for mapping, but both come with significant caveats. Using only hydroacoustic data severely limits the level of detail that can be attained by the map, as well as puts into question its validity as a means for elucidating biodiversity and biological patterns. Calibration is in any case needed for the results to be reliable, so some amount of sediment sampling must always accompany the remote sensing survey. Even in the early days studies would use at least some *in situ* information, if not systematically collected, to support the characterisation of acoustic habitats (Ferns & Hough, 2002; Kendall et al., 2003; McRea et al., 1999).

The highest possible accuracy in delineating faunal boundaries and/or depicting faunal occurrence patterns can only be achieved by obtaining survey data from as much of the area as possible, ideally, the whole of it. This, as Petersen (1924) pointed out, is prohibitively expensive. Riegl et al. (2001) and Norris et al. (1997) obtained highly accurate maps of coral reefs and seagrass beds respectively, but had to face the time and monetary costs of surveying 100% of the study area. This of course may be practicable depending on the accessibility to the area (e.g depth range) and sampling method utilised, or in cases where large budgets are available. Stevens & Connolly (2005) used a more cost-effective method that combined a staggered array of sampled locations and a tessellation technique to draw boundaries around groups of similar stations. Increased cost-effectiveness is achieved by making use of observed spatial autocorrelation patterns to extrapolate beyond the sampled locations and thus fill in the blanks between data points, but the limitations of these approaches, whether budgetary or areal, are clear. Their extreme accuracy and the ability to discover unreported biological features, however, should not be underestimated. The merits of incorporating spatial patterns (e.g. scale of patchiness) into the mapping process, which can only be done in the design of the field sampling surveys, are not exclusive to this way of mapping, as will be discussed below.

The main benefits of data integration techniques stem from an induced ability to gain insight into the empirical relationships between biota occurrence and environmental gradients. The popularity of this approach has exploded in recent years, even producing 'schools', which we review briefly below, focusing on the differences at a very fundamental level and stressing the non-technical issues within each approach.

3.1 Approaches

Two general approaches can be distinguished on the basis of the role that survey data play in the mapping process: a 'top-down approach', where survey data are used merely for ground-truthing purposes and the process is driven by the acoustic patterns; and a 'bottom-up approach', where biotic patterns, as inferred from the survey data, drive the definition and mapping of classes. Even though the final result is equivalent, a thematic map showing the distribution of classes that echo biotic patterns of the seafloor to a greater or lesser degree, the path followed is fundamentally different, and at the most abstract level it can be described as (a) an attempt to find the attributes of polygons of (mostly) known boundaries, in the top-down case, or (b) an attempt to find the boundaries of polygons of (mostly) known attributes, in the bottom-up case. Stressing the importance of whether or not boundaries are known, the former approach is also known as 'supervised' and the latter as 'unsupervised'.

In the top-down approach, first, a classification technique based on the identification of patterns in the remotely-sensed data, usually acoustic imagery, is used to derive homogeneous and distinct regions, often referred to as acoustic habitats. Techniques used range from visual interpretation to highly sophisticated classification algorithms (e.g. Lamarche et al. 2011). These are essentially used as a framework within which reference sites are defined. Samples of *in situ* data are then collected from all the detected regions, or the reference sites, so as to validate the classified habitats (see Brown et al. 2002; 2004; Freitas et al. 2003; Jordan et al. 2005; Kloser et al. 2001; Kostylev et al. 2001), and classes are occasionally merged if they can be proven to have non-distinct faunas.

Remote sensing by hydroacoustics is highly effective in classifying habitats over large areas of seabed. The approach effectively reveals boundaries created by discontinuities in substrate types, which in turn give rise to sharp changes in community composition. But the more gradual changes, which may emerge in response to factors other than substrate, are wholly overlooked. A more serious issue is that assemblages may be identified from the *in situ* data which have no corresponding acoustic class (Freitas et al., 2003). In fact, this is, according to Brown et al. (2005), to be expected. The problem is that the spatial detail of the map is limited by the scale at which the acoustic regions are defined, and although boundaries can be modified to a limited extent on the basis of the biotic patterns (which is possible only at a scale defined by the distance between reference sites, i.e. the same scale at which acoustic habitats are defined), the resolution can only decrease as a result of classes being dropped, but never increase.

Indeed, the question is increasingly being raised as to whether acoustically derived habitats are a good representation of the patterns of variability of epibenthic communities (Eastwood et al., 2006; Hewitt et al., 2004). Stevens & Connolly (2004) concluded that the ability of abiotic surrogates to predict patterns of biological similarity was indeed poor. Parry et al. (2003) detected a nested hierarchy of spatial structure within the megafaunal assemblage of a large, apparently homogeneous, soft-bottom habitat unit. Because the subset of biota of interest is precisely the epibenthic megafauna, dissatisfaction has led to a call for improved mapping methods.

The bottom-up approach emerged in response to this call and Field et al. (1982) summarises it as letting the species tell their story. In this approach, the mapping units are defined on the basis of multivariate species patterns (e.g. peaks of similarity within the continuous gradient of faunal composition, Brown et al. 2002), which are in turn assumed to define sets of distinct environmental factors (Kostylev et al., 2001). Eastwood et al. (2006) compared top-down versus bottom-up approaches to classifying and mapping seabed assemblages and found that when "the seabed comprises relatively homogeneous, unconsolidated sediments and the main driver is the development of the best possible biological assemblage map, then a bottom-up, unsupervised approach is likely to arrive at a set of assemblages that are defined equally well or slightly better compared with a top-down approach" (Eastwood et al. 2006 p.1544).

The evolution of bottom-up methods has largely tracked, whether explicitly or implicitly, that of species (or communities) distribution modelling in its broadest meaning i.e. including concepts such as 'habitat modelling', 'habitat suitability modelling', 'predictive mapping' etc. which in turn focus on obtaining spatial predictions of an ecological phenomenon. This ecological phenomenon can be equated with the concept of mapping unit as used throughout this chapter.

The field of distribution modelling has been largely developed by plant ecologists and vegetation scientists, and it has seen explosive growth in the last decade, evidenced by the steadily increasing rate of published papers using this approach. Benthic ecologists have been able to capitalise on this growth due to (1) the advent of hydroacoustic technology and (2) the mechanistic similarities between conspicuous epibenthic fauna and vegetation. Below we describe this approach as it is applied to the benthos in detail.

3.2 Biotope mapping as distribution modelling of communities

Distribution modelling is by and large an extension of the habitat-association approach to ecology, by which biological populations, whether marine or terrestrial, are seen to distribute themselves in space according to habitat gradients (note that the alternative view would be randomly), leading to community zonation when taken as a whole. In distribution modelling, biota-environment relationships, as derived from a set of surveyed sites, are employed to predict biological properties of the unsurveyed intervening areas on a location-by-location basis, so that 'wall-to-wall' maps of biotic components can be cost-effectively generated, where predicting the presence of a single species or the presence of an entity at a higher hierarchical level, such as a community, are methodologically equivalent. Extensive reviews of community-level modelling can be found in Ferrier et al. (2002) and Ferrier & Guisan (2006).

Modelling the spatial distribution of megabenthos on the basis of empirical relationships between its biological composition and coinciding habitat properties as derived from hydroacoustic remote sensing techniques has been pioneered by Hewitt et al. (2004), Holmes et al. (2008), Buhl-Mortensen, Dolan & Buhl-Mortensen (2009) and Monk et al. (2011). The current trend includes variations on a framework that involves at a minimum: (1) biological (response) data compiled by means of underwater video footage analysis and (2) geophysical (predictor) data collected through echosounder (multibeam or single-beam) or sidescan sonar, from which various proxies can be derived (Buhl-Mortensen, Buhl-Mortensen, Dolan, Dannheim & Kröger, 2009; Holmes et al., 2008; Ierodiaconou et al., 2007; Rattray et al., 2009). Further, Ierodiaconou et al. (2007) were the first to use backscatter (an acoustic property of the seabed obtained as a by-product of multibeam data) in combination with bathymetry to predict dominant biotic categories and, as shown in the recent GeoHab (Marine **Geo**logical and Biological **Hab**itat Mapping) 2011 Conference, this is becoming established as a branch in its own right of predictive mapping of the benthos.

Particularly, benthos distribution modelling has mostly followed the specific approach described by Ferrier & Guisan (2006) as 'classification-then-modelling', where biological survey data are classified into the units to be mapped prior to modelling, so that the relationships revealed by the model are indicative of the collective, aggregated response of all of the species detected during the surveys, at a chosen classification detail (e.g. a similarity cut off level).

The process consists of the following stages:

- Conceptualisation
- Data gathering and choice of data model
    - Field surveys (response data)
    - Preparation of explanatory variables (predictor data)
- Model fitting
- Model evaluation
- Spatial predictions
- Assessment of model applicability

If the sample, obtained from the field surveys, was fully representative of the distribution of the unit(s) to be modelled (e.g. a pre-specified assemblage of species) and if all biotic and abiotic phenomena surrounding and potentially interacting with it were adequately represented by the predictor variables, then the modelled overall ecological response obtained from the model would summarise all systematic variation in its aggregated performance, and the spatial predictions from the model would be in full accordance with its real distribution, except for stochastic variation (R. Halvorsen, pers. comm.). In other words, the only two prerequisite conditions for an adequate model are related to the data model, underlining the fact that the power and reliability of empirical models depend strongly on the data. The choice of options should be guided by properties of the model system (species, study area, sampling design, spatial domain, etc.) and, in equal measure, by the purpose of the mapping.

One of the most consequential choices to be made concerning the data model relates to the 'grain size' of the map or the level of spatial detail to be depicted by the map (and by extension, the amount of detail that will be ignored). Distribution modelling is typically conducted around a grid that is overlaid on the data layers (including survey and remotely-sensed), so that a prediction can be obtained for every grid cell outside of the surveyed ones, where naturally no *in situ* information on the biological composition is available and only environmental data exist, and full-coverage is thus achieved. It is assumed that each sampling unit is wholly contained within a single grid cell and that it is representative of it, and that each cell contains no more than one sampling unit. Ideally, then, either the size of the grid cells or the size of the sampling units (the surveyed sites), bears a relation to patterns of spatial heterogeneity of faunal composition, to ensure that the fraction of variation that is ignored (most notably, within-cell variation) only reflects variation that we are not interested in (e.g. stochastic).

Regarding the specific predictive model, the amount of choice available is overwhelming and only a handful have been tested by the benthic ecology community, where the field is still in its infancy. As a general rule, more complex methods produce models that fit the data more closely. The selection of method can have a profound impact on the reliability of the final outcome (see Elith et al. 2006 for a comprehensive review). Moreover, when the outcome is intended to be used as a form of decision support for conservation it is vital that the strengths and limitations of the method are made explicit.

Arguably, following this approach can be regarded as a formal way of defining biotopes. First, the fact that classes are defined according to multispecies patterns accounts for the biotic aspect of the concept. Second, predicting the presence of classes on the basis of environmental data accounts for the abiotic aspect. And third the spatially-explicit nature of the model resolves the geographic boundaries.

3.3 Predicting the distribution of benthic biotopes: a case study from a Swedish fjord

Off the west coast of Sweden, the first marine national park of the country was designated in 2009 (*Kosterhavet Nationalpark*), conferring the fjord and archipelago of this unique site a new status, not without its responsibilities. Since the approval of the denomination a series of Remotely Operated Vehicle (ROV) surveys have been conducted across the area and underwater video footage amassed, resulting in a comprehensive inventory of epibenthic megafaunal species. In the spirit of taking full advantage of the existing library of video material and in view of a pressing need for further documenting the distribution of biological diversity of (the benthic portion of) the national park, a project was commissioned that would use a predictive mapping approach, using existing ROV data as a basis, to produce a map of benthic biotopes. This case study reports the achievements of that project and is currently under review for publication in a scientific journal. Our aim was to produce a map to support marine spatial planning (a biotope map), subject to being as close to the 'truth' as possible, while using methods as objective, automated and repeatable as possible, as well as using existing data.

3.3.1 Conceptualisation and data model

The first step was to create a grid lattice in a Geographic Information System (GIS) to bin all data so that surveyed sites and unsurveyed areas are all modelled at the same spatial scale and the coverage achieved is 100%. Following recommendations in Gonzalez-Mirelis et al. (2009) based on spatial patterns of epibenthic megafauna, the mesh size (linear scale) chosen for the grid was 15 m.

Species data was obtained from underwater video footage, recorded between 2006 and 2008 by means of a video camera mounted on a Sperre Subfighter 7500DC ROV. ROV navigation data was time-synchronised with the video signal, enabling the reconstruction of the ROV's path for each of 52 survey sites (Figure 1), as well as the formalisation of a function linking the video footage to the path.

The grid was first used to clip the survey tracks into sampling units. Through this procedure, a sufficient (and parsimonious) number of presence/absence datapoints was obtained incurring no extra (monetary) costs.

For a grand total of 417 cells, or sites, equivalent to approximately 70 hours of footage, faunal data was then compiled. Species data comprised all epibenthic megafauna recorded at the relevant clip, including both attached and free-living lifeforms. Calcareous sponges, macrophytes and epibiontic fauna, were excluded from this study. Organisms were identified at least to genus level; otherwise they were not included. Additionally, a number of taxon complexes were used, encompassing those taxa difficult to tell apart sharing similar habitat requirements (e.g. the two species of sponge of the genus *Phakellia* and the species *Axinella*

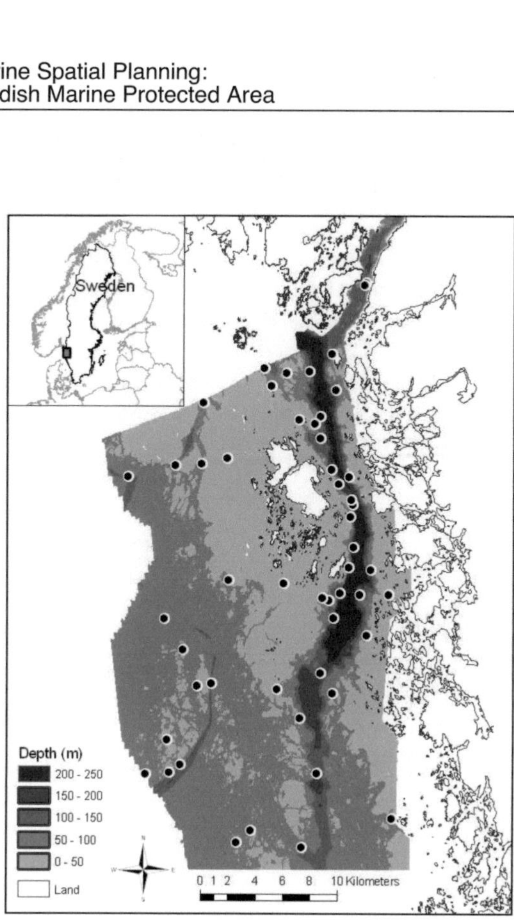

Fig. 1. Map of the study area showing the location of the ROV-surveyed sites (filled dots) and the bathymetry. The main trench running from north to south is known as the Koster Fjord. This fjord separates the mainland from the Koster Archipelago.

infundibuliformis were all part of the same complex and were thus counted as one variable). Where available, high-definition still photos were used to aid in the identification of taxa. In all, 192 taxa and taxon complexes were identified, out of which 178 were classified to species level. Bray-Curtis similarities were then calculated for all pairs of sites.

Lastly, sites were classified according to various thresholds for faunal similarity, rendering a total of nine classes (Table 1). The chosen classification was developed around a number of criteria, including class predictability (see Gonzalez-Mirelis et al. 2011), readiness to distinguish between classes solely by visual inspection (as it would be needed if the classification were to be used in future investigations) and closeness to classes in existing biotope classifications (such as that of Connor et al. 2004).

In the benthos, community composition and diversity have been found to be structured by water column processes (suspended particulate matter loading, food availability and hydrodynamic stress) as well as substrate properties, temperature and salinity. We developed a number of variables representing (rather local) geophysical processes that we assume to be proxies for one or more of the functionally relevant variables. Habitat data was derived from high-resolution multibeam data, including bathymetry and backscatter. The multibeam

Class label	Physical habitat description	Characterising taxa
AA	Offshore circalittoral rock	Axinellid sponges, *Terebratulina retusa*, Anomids, *Placostegus tridentatus*, Spirorbinids
AB	Lower circalittoral rock	Flustrids
AC	Offshore circalittoral biogenic reef	*Hathrometra sarsii*, *Lophelia pertusa*, *Mycale lingua*, *Ascidia obliqua*, *Filograna implexa*
BA	Lower circalittoral mixed sediment and rock	*Pandalus borealis*, *Liocarcinus* sp., *Sabella pavonina*, *Munida rugosa*
BB	Offshore circalittoral fine sediment	*Nephrops norvegicus*, Cerianthids
BC	Offshore circalittoral mixed sediment and rock	*Spirontocaris lilljeborgii*, *Lithodes maja*
BD	Lower circalittoral sediment	*Kophobelemnon stelliferum*, *Pachycerianthus multiplicatus*
C	Near-shore lower circalittoral sediment	*Pennatula phosphorea*
D	Near-shore lower circalittoral coarse sediment including shell hash	Gobids, *Pecten maximus*

Table 1. Description of biotopes

surveys were conducted in 2005 with a Simrad Multibeam EM 1002, at 95 kHz frequency. The set of predictor, abiotic variables included the following:

3.3.1.1 Depth

Depth has been found to be the primary habitat factor organising benthic communities, although its importance may be contingent upon spatial extent and subset of biota considered. In exposed habitats, depth can have a substantial effect on the amount of near-bed stress. Data on depth was obtained directly from 5 m resolution multibeam bathymetry data. It was resampled to the required resolution (15 m) using the mean of all (9) node values encompassed by each grid cell. Depth ranged from 30 to 262 m.

3.3.1.2 Substrate

The link between substrate type and biological composition of the benthos is robustly established, both regarding sediment type or granulometry (particularly important for infauna) and the availability of hard surfaces for organisms to settle on, like bedrock (important for epifauna). Along with depth, it represents a widely recognised driving factor of biological communities of the seabed. A classification of surficial geology (substrate type) was made available to us by the Swedish Geological Survey, who derived the classes from multibeam backscatter data. Categories are clay, gravel, rock and sand.

3.3.1.3 Surface area

Surface area refers to the total amount of available surface in the landscape and it is a function of the ruggedness of the terrain. It directly determines the amount of living space available, thus potentially influencing emergent macro-ecological properties, such as species richness, that do not retain explicit information about composition. It can however also be related to the presence of microscale features, such as overhangs and ledges, that are home to highly habitat-specific species, like in the case of the lamellibranch *Acesta excavata*. Here we measured total surface area of the grid cell in m^2 using the Surface Areas and Ratios from Elevation Grids v. 1.2 ArcView extension (Jenness, 2002). It was calculated using the 5 m resolution bathymetry raster and resampled to 15 m resolution using bilinear interpolation. It ranged from 225 to 900 m^2.

3.3.1.4 Aspect

The orientation of the slope may affect, in combination with other topographic attributes, current velocity and bed shear stress. Aspect was derived from the 15 m bathymetry, using a standard 3-by-3 running window. We used the standard eight compass directions, plus a class for horizontal areas.

3.3.1.5 Landform

The effect of geomorphology on biological composition is unclear (Howell, 2010), although for some species and over some spatial scales it may be significant. The causal link is thought to be related with current speed and habitat availability, as well as susceptibility to sediment accumulation (Ierodiaconou et al., 2011). We included geomorphological attributes in the form of landform categories. Landform can be calculated by classifying the landscape using Topographic Position Index (TPI) values at two different scales, where the TPI is the difference between a cell's elevation and the average elevation of a neighbourhood around that cell. It is a way of expressing jointly convexity and concavity. We used the Topography Toolbox for ArcGIS 9.2 developed by (Jenness, 2006), with a smaller rectangular neighbourhood of 5-by-5 cells, and a larger one of 11-by-11. Upland drainages were reclassified as shallow valleys and upper and open slope were lumped together. Categories were as follows: canyon, shallow valley, U-valley, plain, slope, local ridge, midslope ridge and high ridge.

GIS layers were obtained from the multibeam data sets and then their values were assessed, by spatial overlay, for the grab of 417 cells for which there was biological data available, thus constituting the sample base for the model. Note that every data point was made up of five values corresponding to the predictor variables and one value for the response, which was in turn taken from a 9-level response variable.

3.3.2 Fitting the model and evaluating its spatial predictions

Once a specific modelling method is selected, the most commonplace procedure is to split the sample base into two separate datasets, both comprising complete (predictor + response) data points. One is used for fitting the model and the other is reserved for evaluating model predictions. These are referred to as 'training set' and 'testing set'. The model, as induced by patterns in the training set, is used to make predictions for every single grid cell across the study area, typically involving a very large number of predictions and a potentially long computing time, contingent on the grid cell size:study area size ratio.

The particulars of this case study involved a restricted amount of data of unknown representativeness of the distribution of benthic diversity, potentially biased and noisy. Therefore, the model was chosen with great care that it would be capable of handling a challenging dataset and produce robust predictions. We fitted a Conditional Inference (CI) Tree-based forest which was found to outperform other decision tree-based models both in terms of classification accuracy and the ability to discriminate between classes. CI trees have been developed recently by Hothorn et al. (2006) (the reader should refer to this study for details of how the algorithm works). Here we will only point out that the framework uses a combination of machine learning principles and hypothesis testing that renders it robust and powerful, albeit computationally demanding. A split was implemented randomly within each level of the response variable to ensure the possibility of calculating a measure of accuracy for all classes.

A map of benthic biotope classes was produced by assigning the most likely class, as determined by the model, to every single grid cell across the modelling area (Figure 2). The area of the smallest polygon is, accordingly, $225m^2$ (one single cell) and is therefore not visible to the naked eye at the scale of the map. Tested against the known class of a reserved set of 104 sites, the model yielded an error rate of 28%. In other words, the model did significantly better than classifying every cell as the most prevalent class. A more accurate measure of error could have been attained using independently collected data, as well as understanding the spatial patterns of misclassified cases, but this was considered adequate for the purpose of this project.

Strictly, the output from our model should be regarded as a map of polygons *similar* in their environmental conditions, which are in turn defined on the basis of shared *expected* biological composition, where emphasis is added to draw attention to the fact that both can be measured in terms of (conditional) probabilities. Put differently, the model outputs are formally-defined biotopes. The good model fit and the achieved spatial resolution create an optimal scenario for addressing conservation planning questions.

3.3.3 Applications to marine spatial planning

The *Kosterhavet* area was designated a national park in 2009 with the conservation-related objective of long-term preservation of the marine ecosystems, habitats and species occurring naturally in the region, while ensuring the sustainable use of local biological resources, among other goals, less directly related to conservation.

Notably, the park is one where multiple uses are allowed. Various kinds of commercial and recreational fisheries have a stake in the area, alongside tourism, and to a lesser extent, shipping. There is a well established trawl fishery for shrimp (*Pandalus borealis*), and Norway lobster (*Nephrops norvegicus*), comprising around 30 boats, mostly under 12 m, as well as an additional small fleet of local fishermen targeting lobster (*Homarus gammarus*). The area is also important for recreation, with many hundreds of yachts and motor boats staying in the area over the summer. Tourism has increased 50% over the last decade with around 80,000 people visiting, mainly in July and August. In light of this, a pressing need has emerged to lay out, and ensure mechanisms of enforcement of, a management system within which competing demands inside the multiple-use park are adequately accommodated, while not compromising the conservation goals of the park.

This can only be achieved by means of adequate spatial planning, a tool now widely recognised to be suited for implementing an ecosystem-based approach to the management of ecosystems. Marine spatial planning involves the practice of zoning (to spatially and temporally designate areas for specific purposes), with the aim of reducing conflict both among different users competing for the same space, and more importantly, between users and the environment (Douvere, 2008), ensuring that the capacity of the ocean to provide goods and services remains undiminished. Far from being a straightforward task, the main challenge

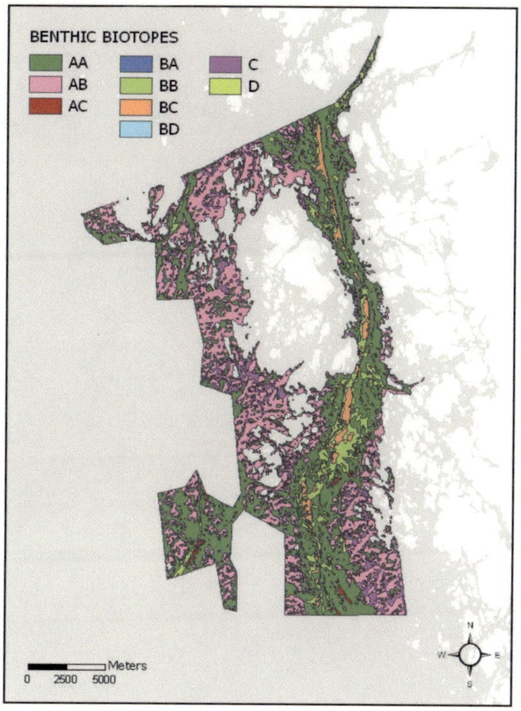

Fig. 2. Map of biotopes, as predicted for each 15 x 15 m cell and clipped to the national park boundary. AA: Offshore circalittoral hard substrate (Axinellid sponges, *T. retusa*, *P. tridentatus*), AB: Lower circalittoral hard substrate (flustrids turf), AC: Biogenic reef (*L. pertusa*, *H. sarsii*, *M. lingua*, *F. implexa*), BA: Circalittoral mixed rock and sediment (*P. borealis*, *S. pavonina*), BB: Circalittoral fine sand (*N. norvegicus*, cerianthids), BC: Circalittoral mixed rock and sediment (*B. tuediae*, *S. lilljeborgii*, *L. maja*), BD: Circalittoral fine sand (*K. stelliferum*, *F. quadrangularis*, *P. multiplicatus*), C: Lower circalittoral mud (*P. phosphorea*, *F. quadrangularis*), D: Lower circalittoral coarse sediments (*P. maximus*, gobid fishes).

that conservation spatial planning faces today is undoubtedly whether goals are perceived to be achieved in a way that minimizes, as far as possible, forgone opportunities for production (Margules & Pressey, 2000).

The zoning system currently in place involves three co-occurring management regimes (see Figure 3). Three sites are afforded the highest degree of protection by being designated as Seabed Protection Areas (SPAs) with a full trawl ban in place and where anchoring and use of other equipment that can damage the seafloor are prohibited. The remainder of the area of the national park is divided into two zones on the basis of depth, where areas above 60 m benefit from partial restrictions (partial protection zone, in Figure 3) with no commercial fisheries allowed, and areas deeper than 60 m are open for a specially-regulated fishery (see below).

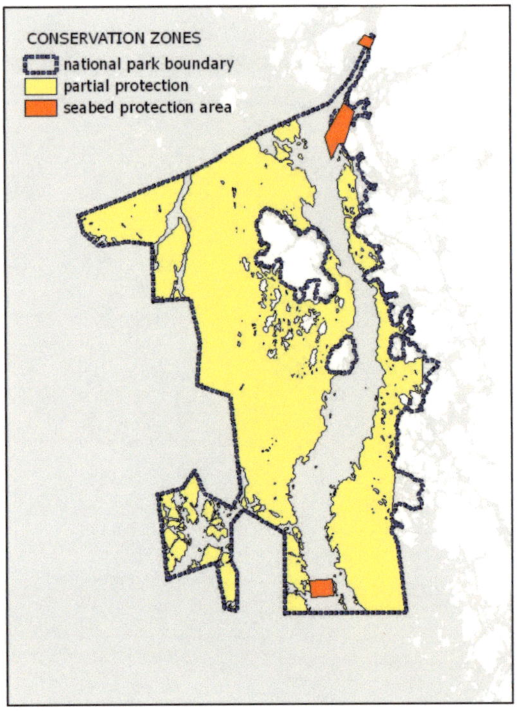

Fig. 3. Conservation zones, including: boundaries of the national park, partially protected zone and seabed protection areas.

The only commercial fishery that takes place inside the boundaries of the national park is a shrimp fishery, regulated by a special agreement between fishermen, researchers, fisheries organisations and authorities at different levels, in place since 2000, and is concentrated in the southern half of the fjord. Trawls must be of a specific size, lightweight and equipped with a sorting grid that allows escape of fish bycatch. Remarkably, education and sharing of knowledge are given special consideration in this agreement. Under its auspices, courses in marine ecology for fishermen, as well as courses in commercial fishing and fishery technology for researchers and officials, have been arranged in recent years. The fishery yields about 200 tonnes of shrimp per year.

Recreational lobster fishing by creels also takes places in the national park causing negligible impact. Immediately outside of the boundary to the west, trawling is allowed only for prawns and Norwegian lobster. Here too, species-specific sorting grids are required on the trawls. Further to the west, beyond the trawl limit, no trawling restrictions apply.

Given the goal set for the park, and the array of stakeholders involved, it is imperative to address the question: 'how much of each biotope is being afforded effective protection?', where a leap is taken from the idea of 'including all biotopes under some sort of protection level' to 'preserving ecosystems, habitats and species' (as stated in the management plan of the park) which we will, for the purpose of this chapter, assume to be possible.

3.3.3.1 Assessment of representation

The biotope map generated by the model enabled an assessment of the level of representativity of each managed zone within the park, also known as a gap analysis. As shown in Table 2, the partial protection zone includes mostly biotopes of type AB (98%) and C (89%), with no representation at all of any of the soft sediment biotopes (BA-BD); the most represented biotopes within SPAs are AC, *Lophelia* reefs (3%), which, it should be noted, are contributed by one single SPA, and AA (2%). Some of the biotopes not represented in the partially protected zone are indeed encompassed by the SPAs (BB, BC).

Class	Park (km^2)	Partial protection (%)	SPA (%)
AA	117.8	37.3	2.1
AB	90.5	98.0	0.6
AC	1.9	0.6	3.1
BA	0	0	0
BB	23.3	0	1.9
BC	7.2	0	0.5
BD	0.1	0	0
C	14.7	89.1	0
D	3.6	27.7	0

Table 2. Quantitative assessment of representation of biotopes within different conservation zones within the national park

It thus emerges that a zoning system based on depth alone is *guaranteed* to return a biased set of managed zones, because all soft sediment biotopes occur deeper than 60 m. This is a particularly acute issue in this area, where anthropogenic pressure is very unevenly distributed over biotopes, with BB and BC areas getting the bulk of it (compare fishing effort shown in Figure 4 and the biotope map, Figure 2). Therefore, not only are B type biotopes underrepresented in protected zones, but they are also the only ones at risk. Additionally, they are not afforded any conservation interest from the European Union, so the responsibility for the long-term preservation of the communities associated to this biotope rests solely with the national park. On the grounds of our analysis, increasing their representation in at least partially protected zones should be a priority.

The absence from the park as a whole of biotope BA is explained by the fact that a very small amount of it was predicted by the model overall (it amounted to less than 1 km^2). More importantly, the sites classified as BA were deemed to be heavily fished, and they featured only sparse epifauna. Careful examination of the classification and spatial predictions suggests that this class should be merged with either BB or BD.

One of the biggest gaps that has emerged through this analysis relates to class BD, an uncommon class in the region (<1 km^2 contained in the park, see Table 2), completely absent from any type of conservation zone. To make matters worse, the biotope is strongly associated with the seapen *Kophobelemnon stelliferum*, a species with conservation priority within Sweden.

3.3.3.2 An example implementation of systematic conservation planning

The above analysis has raised an important question: 'is seabed biodiversity adequately represented inside priority areas?' The answer of course depends on what is meant by 'adequate', which in turn raises more questions, namely: 'how much of each biodiversity surrogate (i.e. biotope) needs to be protected?', 'how should the protected sites be distributed so as to minimize conflict with users?' In general terms it is apparent that the management system in place, which was driven largely by depth and uniqueness (e.g. the coral reef SPA), has some drawbacks and can be improved.

Systematic conservation planning (SCP) involves finding cost-efficient sets of areas to protect biodiversity. SCP is a process that comprises, at a minimum, the following stages: (1) compile data on the biodiversity, or biodiversity surrogates, of the planning region; (2) identify conservation goals for the planning region, preferably in the form of quantitative, operational targets; and (3) select a set of conservation areas that collectively meet the representation targets assigned to the biodiversity features incurring the lowest possible cost. These are embedded in a larger process that includes the possibility of implementing conservation actions on the ground, as well as revisiting and adapting zoning plans based on monitoring data. The approach is highly effective because it is efficient in using limited resources to achieve conservation goals, it is defensible and flexible in the face of competing uses, and it is accountable in allowing decisions to be critically reviewed. Stage 3, probably the most critical of all, can be tackled by use of algorithms that can efficiently solve what has been called the "minimum-set problem" (Possingham et al., 2000): Minimize overall cost, subject to the constraint that all biodiversity targets are met (e.g. 20% of each biodiversity feature), where cost can be expressed as total size of the reserved area, revenue loss, etc. *Kosterhavet* provides for an optimal test case for a SCP approach, where biodiversity features can be readily formulated on the basis of the modelled biotopes.

Under this systematic framework, conservation planning becomes a data-driven process. Spatial data are required on all features that need to be considered, both those that contribute to achieving targets (which are, typically, biodiversity-related) and those requiring regulations, so that the true cost of allocating areas to a reserved zone can be accurately computed. The latter mainly refers to data on socioeconomic activities. When the same kind of information is available at the same level of detail throughout an area, it becomes possible to quantify the advantages and disadvantages of various zoning options and therefore legitimately compare these.

To demonstrate the possibilities of SCP we compiled fishing effort data from Vessel Monitoring System records for years 2007-2010. The fishing data were provided by the former Fisheries Board of Sweden, which is now the Swedish Agency for Marine and Water Management. Fishing positions were gridded to 1 ha cells and the maximum number of pings out of the whole period was used as an estimate of fishing effort for the location.

The software used to generate conservation networks was Marxan. Marxan (Ball et al., 2009), a decision support system designed to solve the minimum set problem, finds a number of near-optimal solutions using a heuristic algorithm called 'simulated annealing'. A planning unit layer (regular grid) comprising over 30,000 units (1 ha cells) was used. Planning units are the building blocks of reserve systems that are overlaid on maps of biodiversity features for conservation planning. Building from the amount of every feature in every planning unit, computed by spatial overlay, Marxan generated 100 different solutions to the problem of selecting the minimum number of planning units (with a spatial configuration constrained by a parameter that modifies overall boundary length) which achieves a total of at least 10% representation of all conservation features. The target amount used of 10% was arbitrary, though it is a commonly used figure. Note that the target does not have to be the same across all features, but importantly, in this approach it can be controlled by the user to prevent the inevitable imbalance that results from non-systematic approaches, as above. Cost was included as the fishing intensity layer described above so that the algorithm strived to avoid using cells located in important fishing grounds. Here we present only one possible solution (Figure 4).

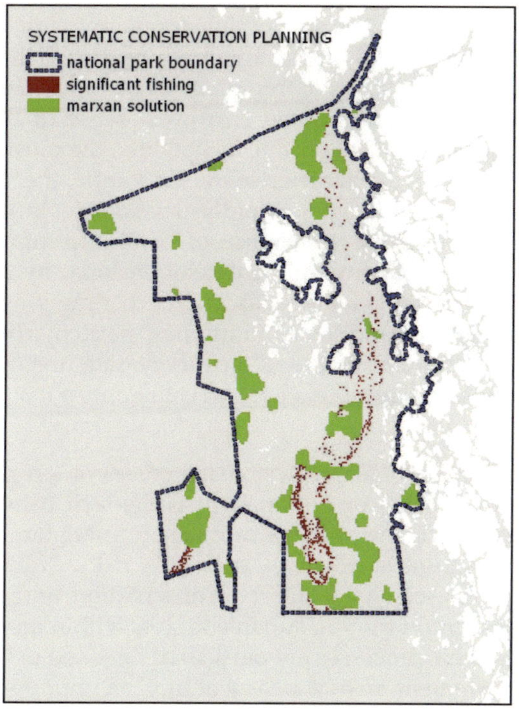

Fig. 4. Systematic conservation planning solution obtained with Marxan encompassing 10% of every biotope, as well as minimizing the use of areas heavily targeted by fishing.

Whether or not the implementation on the ground of these areas as a conservation network is feasible is the next question that should be addressed, and one which, importantly, will get a negotiation process started.

4. Conclusions and future research

If planning for conservation in the sense of designing networks of protected areas is usually a practice of hope, in the marine realm, where areas are prioritised at best on the basis of scant information and more or less questionable estimates of species occurrence to fill in the blanks, it is almost a practice of faith. A lack of robustness in the process of marine conservation and spatial planning not only jeopardises the efficacy of conservation areas, but also reduces vital societal support. While it will never be possible to make conservation planning decisions based on complete knowledge of the distribution of the full complement of biodiversity, from genes to ecosystems, a degree of robustness can be achieved, as we have shown, by means of

(1) selecting a level in the biodiversity hierarchy to operate in, (2) mapping the distribution of its units, and (3) applying spatial planning principles.

One question that we have not touched upon is to what extent the level of operation selected, in our case, biotopes, is a good surrogate for biodiversity at other levels, e.g. species, where management provisions and intergovernmental mandates apply, and this should be addressed in future research on the distribution of seabed biodiversity.

We have seen that two main approaches are possible for mapping the geographic distribution of ecological patterns on the seabed. The top-down approach is driven by geophysical attributes, relies heavily on hydroacoustic, remotely-sensed data and is highly effective for mapping the distribution of habitats, in the narrowest sense of the word. If the target, however, is to reproduce patterns of epibenthic megafauna, the most conspicuous organisms living attached to the seabed which characterise the landscape, the top-down approach may be suboptimal, and an alternative is being rapidly developed. In the bottom-up approach the process is driven by patterns of occurrence of benthic communities, which are then extrapolated using observed biota-environment relationships, by means of full-coverage, hydroacoustic (usually multibeam) data. This approach draws heavily from the field of distribution modelling of species and/or communities, particularly in the framework that involves a classification of communities based on survey data first, and the modelling of the obtained units later. The outcome from this process is not only a map, but a formal definition of biotopes.

Marine spatial planning and particularly systematic conservation planning make the most of thus generated maps of habitats or biotopes. It is the tacit convention that diversity at the species level (e.g. species richness) is the most appropriate target of conservation, even if the focus is placed on a subset of this (e.g. species at risk). Notwithstanding, ecosystem diversity *per se* has also been used as the target of conservation, and other types of ecological patterns or even processes, could also be suitable targets within an appropriate framework. Methods for mapping dynamic processes are particularly needed in the marine environment, including the pelagic environment, so that areas that fulfil an important role in the functioning of marine ecosystems (e.g. areas of upwelling, corridors for larval transport, etc., acting as sort of 'keystone spaces') can be incorporated in the process of marine spatial planning.

The issue at the core is unravelling the factors that explain the distribution of species, whether benthic or otherwise. We have focused on methodological issues, but the role that ecological theory plays within this process cannot be understated. A stronger footing on ecological theory will help develop better distribution models that produce more reliable spatial predictions, and it has been called for (Austin, 2007). Distribution modelling methods have usually been developed within the field of vegetation science and only later have they been adopted by benthic ecologists, causing a delay in the progress of benthic mapping science and in the development of tools to address questions of benthic biogeography, conservation planning, etc. A closer collaboration between benthic ecologists and vegetation scientists in the field of distribution modelling and biotope predictive mapping should help overcome the lag that has so far characterised the marine relative to the terrestrial science, and improve access to knowledge of distribution of biological diversity for all ecologists.

5. Acknowledgements

This research received funding from the Seventh Framework Programme of the EU (FP/2007-2013) under grant agreement no. 217246 made with the joint Baltic Sea research and development programme BONUS, from the Swedish Environmental Protection Agency from contract 08/391 PREHAB and FORMAS from contract 217-2006-357.

6. References

Allee, R., Dethier, M., Brown, D., Deegan, L., Ford, G., Hourigan, T., Maragos, J., Schoch, C., Sealy, K., Twilley, R., Weinstein, M. & Yoklavich, M. (2001). Marine and estuarine ecosystem and habitat classification, *Technical Report NMFS-F/SPO-43*, NOAA Technical Memorandum, Silver Spring, Maryland.

Austin, M. (2007). Species distribution models and ecological theory: A critical assessment and some possible new approaches, *Ecological Modelling* 200(1-2): 1–19.
URL:
http://www.sciencedirect.com/science/article/pii/S0304380006003140

Ball, I., Possingham, H. & Watts, M. (2009). *Spatial conservation prioritisation: Quantitative methods and computational tools*, Oxford University Press, Oxford, UK, chapter Marxan and relatives: Software for spatial conservation prioritisation, pp. 185–195.

Bergen, M., Weisberg, S., Smith, R. W., Cadien, D., Dalkey, A., Montagne, D., Stull, J., Velarde, R. & Ranasinghe, J. (2001). Relationship between depth, sediment, latitude, and the structure of benthic infauna assemblages on the mainland shelf of southern California, *Marine Biology* 138: 637–647.

Brown, C., Cooper, K., Meadows, W., Limpenny, D. & Rees, H. (2002). Small-scale mapping of sea-bed assemblages in the eastern English channel using sidescan sonar and remote sampling techniques, *Estuarine, Coastal and Shelf Science* 54: 263–278.

Brown, C., Hewer, A., Meadows, W., Limpenny, D., Cooper, K. & Rees, H. (2004). Mapping seabed biotopes at Hastings shingle bank, eastern English channel. part 1. assessment using sidescan sonar, *Journal of the Marine Biological Association of the UK* 84: 481–488.

Brown, C. J., Mitchell, A., Limpenny, D. S., Robertson, M. R., Service, M. & Golding, N. (2005). Mapping seabed habitats in the Firth of Lorn off the west coast of Scotland: evaluation and comparison of habitat maps produced using the acoustic ground-discrimination system, Roxann, and sidescan sonar, *ICES Journal of Marine Science: Journal du Conseil* 62(4): 790–802.
URL: *http://icesjms.oxfordjournals.org/content/62/4/790.abstract*

Buhl-Mortensen, P., Buhl-Mortensen, L., Dolan, M., Dannheim, J. & Kröger, K. (2009). Megafaunal diversity associated with marine landscapes of northern Norway: a preliminary assessment, *Norwegian Journal of Geology* 89: 163–171.

Buhl-Mortensen, P., Dolan, M. & Buhl-Mortensen, L. (2009). Prediction of benthic biotopes on a Norwegian offshore bank using a combination of multivariate analysis and GIS classification, *ICES Journal of Marine Science* 66: 2026–2032.

Connor, D. W., Allen, J., Golding, N., Howell, K., Lieberknecht, L., Northen, K. O. & Reker, J. (2004). The marine habitat classification for Britain and Ireland, version 04.05, *Technical Report ISBN 1 861 07561 8 (internet version)*, JNCC, Peterborough. http://www.jncc.gov.uk/MarineHabitatClassification/.

CORINE (1991). Biotopes manual. a method to identify and describe consistently sites of major importance for nature conservation data specifications, *Technical report*, European Communities Commission EUR 12587, 126 pp.

Douvere, F. (2008). The importance of marine spatial planning in advancing ecosystem-based sea use management, *Marine Policy* 32: 762–771.

Eastwood, P., Souissi, S., Rogers, S., Coggan, R. & Brown, C. (2006). Mapping seabed assemblages using comparative top-down and bottom-up classification approaches, *Canadian Journal of Fisheries and Aquatic Sciences* 63: 1536–1548.

Elith, J., H. Graham, C., P. Anderson, R., Dudík, M., Ferrier, S., Guisan, A., J. Hijmans, R., Huettmann, F., R. Leathwick, J., Lehmann, A., Li, J., G. Lohmann, L., A. Loiselle, B., Manion, G., Moritz, C., Nakamura, M., Nakazawa, Y., McC. M. Overton, J., Townsend Peterson, A., J. Phillips, S., Richardson, K., Scachetti-Pereira, R., E. Schapire, R., Soberón, J., Williams, S., S. Wisz, M. & E. Zimmermann, N. (2006). Novel methods improve prediction of species' distributions from occurrence data, *Ecography* 29(2): 129–151.
URL: *http://dx.doi.org/10.1111/j.2006.0906-7590.04596.x*

EUNIS (2005). European nature information system, *Technical report*.

Ferns, L. W. & Hough, D. (2002). Habitat mapping of the Bunurong coast (Victoria) - including the Bunurong Marine and Coastal Park, *Technical report*, Parks, Flora and Fauna Division, Department of Natural Resources and Environment, East Melbourne, Australia.

Ferrier, S., Drielsma, M., Manion, G. & Watson, G. (2002). Extended statistical approaches to modelling spatial pattern in biodiversity in north-east New South Wales. ii. community-level modelling, *Biodiversity and Conservation* 11: 2309–2338.

Ferrier, S. & Guisan, A. (2006). Spatial modelling of biodiversity at the community level, *Journal of Applied Ecology* 43: 393–404.

Field, J. G., Clarke, K. R. & Warwick, R. M. (1982). A practical strategy for analysing multispecies distribution patterns, *Marine Ecology Progress Series* 8: 37–52.
URL: *http://www.int-res.com/articles/meps/8/m008p037.pdf*

Foster-Smith, R., Davies, J. & Sotheran, I. (1999). Broad scale remote survey and mapping of sublittoral habitats and biota, *Technical report*, SeaMap Research Group, University of Newcastle-upon-Tyne, U.K.

Freitas, R., Silva, S., Quintino, V., Rodrigues, A. M., Rhynas, K. & Collins, W. T. (2003). Acoustic seabed classification of marine habitats: studies in the western coastal-shelf area of Portugal, *ICES Journal of Marine Science* 60(3): 599–608.
URL: *http://icesjms.oxfordjournals.org/content/60/3/599.abstract*

Gonzalez-Mirelis, G., Bergström, P. & Lindegarth, M. (2011). Interaction between classification detail and prediction of community types: implications for predictive modelling of benthic biotopes, *Marine Ecology Progress Series* 432: 31–44.

Gonzalez-Mirelis, G., Bergström, P., Lundälv, T., Jonsson, L. & Lindegarth, M. (2009). Mapping the benthos: Spatial patterns of seabed-dwelling megafauna in a Swedish fjord, as derived from opportunistic video data, *Journal of Marine Biodiversity* 39: 291–302.

Greene, H., Yoklavich, M. M., Starr, R. M., O'Connell, V. M., Wakefield, W., Sullivan, D. E., Jr., J. E. M. & Cailliet, G. M. (1999). A classification scheme for deep seafloor habitats, *Oceanologica Acta* 22(6): 663 – 678. <ce:title>Marine Benthic Habitats and their Living Resources: Monitoring, Management and Applications to Pacific Island Countries</ce:title>.
URL: *http://www.sciencedirect.com/science/article/pii/S0399178400889574*

HELCOM (1998). Red list of marine and coastal biotopes and biotope complexes of the Baltic Sea, *in* H. Nordheim & D. Boedeker (eds), *HELCOM - Baltic Sea Environment Proceedings*, Vol. 75, p. 115.

Hewitt, J. E., Thrush, S., Legendre, P., Funnell, G., Ellis, J. & Morrison, M. (2004). Mapping of marine soft-sediment communities: Integrated sampling for ecological interpretation, *Ecological Applications* 14: 1203–1216.

Holmes, K., Niel, K. V., Radford, B., Kendrick, G. & Grove, S. (2008). Modelling distribution of marine benthos from hydroacoustics and underwater video, *Continental Shelf Research* 28: 1800–1810.

Hothorn, T., Hornik, K. & Zeileis, A. (2006). Unbiased recursive partitioning: A conditional inference framework, *Journal of Computational and Graphical Statistics* 15: 651–674.

Howell, K. L. (2010). A benthic classification system to aid in the implementation of marine protected area networks in the deep/high seas of the ne Atlantic, *Biological Conservation* 143(5): 1041–1056.
URL: *http://www.sciencedirect.com/science/article/pii/S0006320710000285*

Ierodiaconou, D., Burq, S., Reston, M. & Laurenson, L. (2007). Marine benthic habitat mapping using multibeam data, georeferenced video and image classification techniques in Victoria, Australia, *Journal of Spatial Science* 52: 93–104.

Ierodiaconou, D., Monk, J., Rattray, A., Laurenson, L. & Versace, V. (2011). Comparison of automated classification techniques for predicting benthic biological communities using hydroacoustics and video observations, *Continental Shelf Research* 31(2), Supplement 1: S28–S38.
URL: *http://www.sciencedirect.com/science/article/pii/S027843431000021X*

Jenness, J. (2002). *Surface Areas and Ratios from Elevation Grid (surfgrids.avx) extension for ArcView 3.x, v. 1.2*, Jenness Enterprises, Available at:
http://www.jennessent.com/arcview/surface_areas.htm.

Jenness, J. (2006). *Topographic Position Index (tpi_jen.avx) extension for ArcView 3.x, v. 1.2*, Jenness Enterprises, Available at: http://www.jennessent.com/arcview/tpi.htm.

Jordan, A., Lawler, M., Halley, V. & Barrett, N. (2005). Seabed habitat mapping in the Kent group of islands and its role in marine protected area planning, *Aquatic Conservation: Marine and Freshwater Ecosystems* 15(1): 51–70.
URL: *http://dx.doi.org/10.1002/aqc.657*

Kendall, M. S., Jensen, O., Alexander, C., Field, D., McFall, G., Bohne, R. & Monaco, M. (2003). Benthic habitats of Gray's Reef National Marine Sanctuary, *Technical report*, NOAA/NOS/NCCOS/CCMA Biogeography.

Kloser, R. J., Bax, N. J., Ryan, T., Williams, A. & Barker, B. A. (2001). Remote sensing of seabed types in the Australian south east fishery; development and application of normal incident acoustic techniques and associated 'ground truthing', *Marine and Freshwater Research* 52: 475–489.

Kostylev, V., Todd, B., Fader, G., Courtney, R., Cameron, G. & Pickrill, R. (2001). Benthic habitat mapping on the Scotian Shelf based on multibeam bathymetry, surficial geology and seafloor photographs, *Marine Ecology Progress Series* 219: 121–137.

Lamarche, G., Lurton, X., Verdier, A.-L. & Augustin, J.-M. (2011). Quantitative characterisation of seafloor substrate and bedforms using advanced processing of multibeam backscatter: Application to cook strait, New Zealand, *Continental Shelf Research* 31(2, Supplement): S93–S109. <ce:title>Geological and Biological Mapping and Characterisation of Benthic Marine Environments</ce:title>.
URL: *http://www.sciencedirect.com/science/article/pii/S0278434310001925*

Margules, C. R. & Pressey, R. L. (2000). Systematic conservation planning, *Nature* 405(6783): 243–253.
URL: *http://dx.doi.org/10.1038/35012251*

McRea, J. E., Greene, H. G., O'Connell, V. M. & Wakefield, W. W. (1999). Mapping marine habitats with high resolution sidescan sonar, *Oceanologica Acta* 22(6): 679–686. <ce:title>Marine Benthic Habitats and Their Living Resources: Monitoring,

Management and Applications to Pacific Island Countries</ce:title>.
URL: *http://www.sciencedirect.com/science/article/pii/S0399178400889586*

Monk, J., Ierodiaconou, D., A.Bellgrove, E.Harvey & Laurenson, L. (2011). Remotely sensed hydroacoustics and observation data for predicting fish habitat suitability, *Continental Shelf Research* 31(2) Supplement 1: S17–S27.

Norris, J. G., Wyllie-Echeverria, S., Mumford, T., Bailey, A. & Turner, T. (1997). Estimating basal area coverage of subtidal seagrass beds using underwater videography, *Aquatic Botany* 58(3-4): 269–287. <ce:title>Geographic information systems and remote sensing in Aquatic Botany</ce:title>.
URL: *http://www.sciencedirect.com/science/article/pii/S0304377097000405*

Olenin, S. & Ducrotoy, J.-P. (2006). The concept of biotope in marine ecology and coastal management, *Marine Pollution Bulletin* 53: 20–29.

Parry, D., Kendall, M., Pilgrim, D. & Jones, M. (2003). Identification of patch structure within marine benthic landscapes using a remotely operated vehicle, *Journal of Experimental Marine Biology and Ecology* 285-286: 497–511.

Petersen, C. (1924). The necessity for quantitative methods in the investigation of the animal life on the sea-bottom, *Proceedings of the Zoological Society of London* 94: 687–694.

Possingham, H., Ball, I. & Andelman, S. (2000). *Quantitative methods for conservation biology*, Springer-Verlag, New York, chapter Mathematical methods for identifying representative reserve networks, pp. 291–305.

Rattray, A., Ierodiaconou, D., Laurenson, L., Burq, S. & Reston, M. (2009). Hydro-acoustic remote sensing of benthic biological communities on the shallow south east Australian continental shelf, *Estuarine, Coastal and Shelf Science* 84: 237–245.

Riegl, B., Korrubel, J. L. & Martin, C. (2001). Mapping and monitoring of coral communities and their spatial patterns using a surface-based video method from a vessel, *Bulletin of Marine Science* 69: 869–880.

Stevens, T. & Connolly, R. (2005). Local-scale mapping of benthic habitats to assess representation in a marine protected area, *Marine and Freshwater Research* 56: 111–123.

Stevens, T. & Connolly, R. M. (2004). Testing the utility of abiotic surrogates for marine habitat mapping at scales relevant to management, *Biological Conservation* 119(3): 351–362.
URL: *http://www.sciencedirect.com/science/article/pii/S0006320703004816*

Valentine, P., Todd, B. & Kostylev, V. (2005). Classification of marine sublittoral habitats, with application to the northeastern north America region, *American Fisheries Society Symposium* 41: 183–200.

Ward, T. J., Vanderklift, M., Nicholls, A. & Kenchington, R. (1999). Selecting marine reserves using habitats and species assemblages as surrogates for biological diversity, *Ecological Applications* 9: 691–698.

Williams, A., Bax, N. J., Kloser, R. J., Althaus, F., Barker, B. & Keith, G. (2009). Australia's deep-water reserve network: implications of false homogeneity for classifying abiotic surrogates of biodiversity, *ICES Journal of Marine Science: Journal du Conseil* 66(1): 214–224.
URL: *http://icesjms.oxfordjournals.org/content/66/1/214.abstract*

9

Management Strategies to Limit the Impact of Bottom Trawling on VMEs in the High Seas of the SW Atlantic

Julio Portela et al.*
Spanish Institute of Oceanography, TRAGSATEC, General Secretariat of the Sea, Spain

1. Introduction

For the past nine years, the issue of protecting biodiversity in the deep sea in areas beyond national jurisdiction - the high seas (HS) - has been widely debated by the United Nations General Assembly (UNGA) and in other international fora. The UNGA adopted a series of resolutions, beginning with Resolution 59/25 in 2004, which called on high seas fishing nations and regional fisheries management organisations (RFMOs) to take urgent action to protect vulnerable marine ecosystems (VMEs) from destructive fishing practices in areas beyond national jurisdiction (Rogers & Gianni, 2010; UNGA, 2004). In December 2006 the UNGA adopted resolution 61/105 on Sustainable Fisheries, calling on flag states, RFMOs and arrangements to immediately act for the sustainable management of fish stocks and to protect VMEs from destructive fishing practices (Portela et al., 2010; UNGA, 2007). In 2009, the UNGA adopted Resolution 64/72 reaffirming the 2006 resolution and made it clear that the measures called for in Resolution 61/105 should be implemented, consistent with the 2009 FAO *International Guidelines for the Management of Deep-Sea Fisheries in the High Seas* (FAO Deepwater Guidelines), by flag states and RFMOs. Resolution 64/72 placed particular emphasis on conducting impact assessments of bottom fisheries on the high seas (UNGA, 2009: Paras 119–120).

Recent relevant studies have concluded that bottom fishing may damage or destroy long-lived epifaunal animals such as corals and sponges, reducing the three-dimensional complexity of the seabed and leading to decreased species diversity and faunal biomass (Althaus et al., 2009; Clark & Rowden, 2009; Koslow et al., 2001; Orensanz et al., 2008; Reed et al., 2005; Rogers & Gianni, 2010; Stone, 2006). Due to its characteristics, bottom trawling is likely to have the most serious adverse impacts on vulnerable deep-sea benthic species (Rogers & Gianni, 2010; Weaver et al., 2011). Several authors (Bensch et al., 2008; Coggan et

*Juan Acosta, Javier Cristobo, Araceli Muñoz, Santiago Parra, Teodoro Ibarrola, José Luis Del Río, Raúl Vilela, Pilar Ríos, Raimundo Blanco, Bruno Almón, Elena Tel, Victoria Besada, Lucía Viñas, Virginia Polonio, María Barba and Pedro Marín
Spanish Institute of Oceanography, TRAGSATEC, General Secretariat of the Sea, Spain

al., 2007; Collie et al., 2000a, 2000b; Durán Muñoz et al., 2008, 2009; Kaiser, 1998; Kaiser et al., 1998, 2006; Murillo et al., 2010; Portela et al., 2010; Rogers & Gianni, 2010) have studied the potential disturbance of the seabed by bottom otter trawls and the possible negative effects on the structure of benthic communities.

Following UNGA recommendations and the FAO Deepwater Guidelines (FAO, 2009), the Spanish Institute of Oceanography (Instituto Español de Oceanografía [IEO]) conducted between October 2007 and April 2010 a series of 13 multidisciplinary research surveys on the HS of the SW Atlantic onboard the multipurpose R/V "Miguel Oliver". The main objectives of these campaigns were: i) the quantitative and qualitative description of the biotopes, ecosystems or communities identified as VMEs; (ii) the identification of the vulnerable organisms eventually found in the study area; and (iii) to assess the potential negative impact of bottom trawl fishing on them.

The study resulted in: i) a detailed cartographic and bathymetric mapping of the area; ii) a description of the geological substratum and of the benthic features; iii) the identification and description of the VMEs; iv) the delineating of candidate sites for protected areas supported by geological, geomorphological and biological criteria; v) an analysis of the overlap between fishing activities and VMEs; vi) the analysis of the abundance and distribution of the main commercial species; vii) comparisons between fished and unfished areas using seafloor observations with towed cameras, observations of fishing impacts with a remotely operated vehicle (ROV), sampling of seabed communities in impacted versus non-impacted areas, and documenting by-catch of benthic invertebrates in fishing gear; and viii) the analysis of hydrographic conditions and pollutants.

The project swath-mapped for the first time large areas of the Argentine Continental Margin (ACM) off the Argentinean Economic Exclusive Zone (EEZ) from 41°30´S to 48°S, obtaining full data coverage of the seafloor in this region between the outermost continental shelf and the middle slope down to 1600 m water depth contour (Figure 1A). This large area of the ACM included two main specific regions: the southernmost region (45°S to 48°S), corresponding to a segment of the outer continental shelf and to the upper, middle and low continental slope segments (Figure 1B); and the northern region, (41°30´S to 45°S) (Figure 2A, B and C), covering only part of the upper and middle continental slope, since the continental shelf is located within the Argentinean EEZ and therefore out of the scope of the study.

Sedimentation and morphology of the outer continental shelf and slope of the ACM are strongly influenced by the Falkland/Malvinas Current (FMC) at depths under than 2500 m. The FMC is generated as a branch of the Antarctic Circumpolar Current (ACC) and flows towards NNE along the ACM (Legeckis & Gordon, 1982). The ACC divides into 2 branches when reaching the South Falkland/Malvinas slope, the oriental branch being the most intense. Over the ACM the current is about 100 km wide and its eastern limit flows parallel to the 200 m isobath (Anon., 2008). The FMC greatly influences the outer continental shelf and slope of the ACM, with an estimated flux between 1 and 2 Sv† towards NNE (Piola & Rivas, 1997; Piola, 2008) and velocities between 5 and 10 cm s^{-1} over middle and outer shelf (Peterson 1992). This important marine dynamic has been corroborated by the sand waves

† The sverdrup is a unit of measure of volume transport (0.001 km^3/s)

(found in the outer continental shelf at 150 m depth oriented SSW-NNE), together with the megaripples (found in the medium slope between 1000 and 2000 m oriented E-W), and the presence of erosive sea bottoms and striking scarps (in the southern zone of our study area between 1000 and 1500 m) (Acosta et al., submitted).

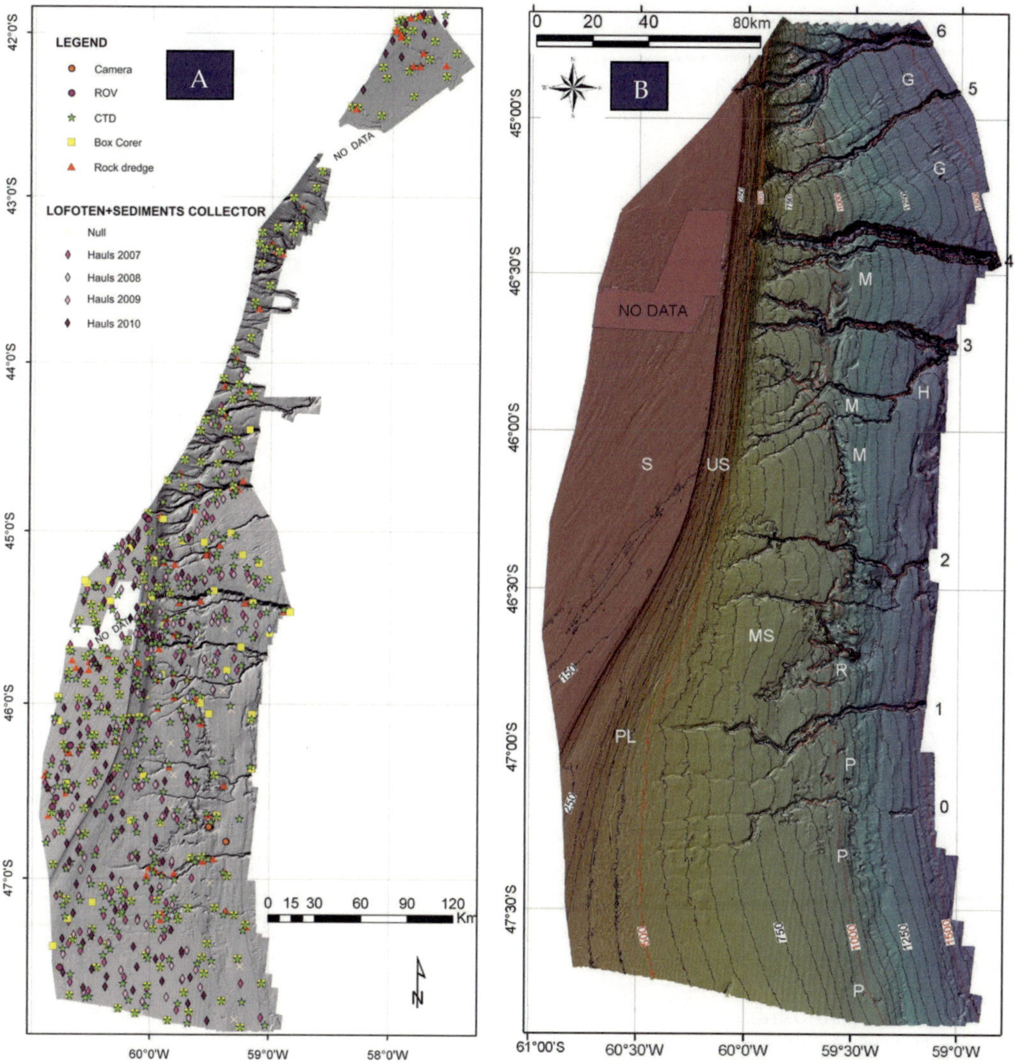

Fig. 1. **A**) Study area with the swath-mapped region and location of the different stations. **B**) Outer continental shelf and slope from 45ºS to 48ºS (southern region). G: gullies; M: moats; MS: middle continental slope; P: pockmarks; PL: iceberg plough marks; R: ridge; S: continental shelf; US: upper continental slope

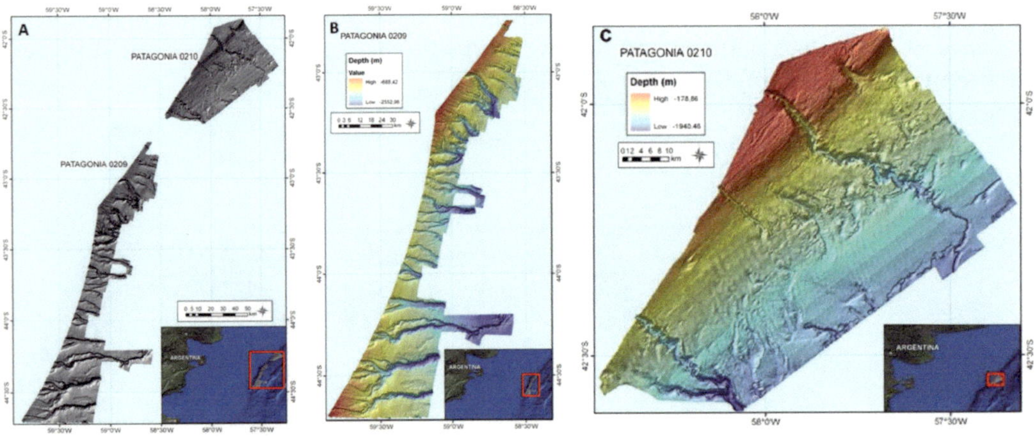

Fig. 2. **A**) Upper and middle continental slope between 41°30´S and 45°S (northern region) studied in 2009 (**B**) and 2010 (**C**)

Vulnerability is related to the likelihood that a population, community, or habitat will experience substantial alteration from short-term or chronic disturbance, and the likelihood that it would recover and in what time frame. These are, in turn, related to the characteristics of the ecosystems themselves, especially biological and structural aspects. VME features may be physically or functionally fragile. The most vulnerable ecosystems are those that are both easily disturbed and very slow to recover, or may never recover (FAO, 2009). Because of their abundance and sensitivity to damage from fishing activities, cold water coral reefs, coral gardens, sponge beds, and rock outcrops are among the most vulnerable ecosystems.

Bottom trawls can damage the physical structures of the seafloor, particularly impacting rare or fragile corals, sponges and other "sessile" seabed organisms. Several studies (Clark & Koslow, 2007; Rogers, 1999; Stone, 2006; Weaver et al., 2011) have shown that fishing practices impact on the abundance and diversity of the fauna associated with sedentary living communities. Bottom fishing gears destroy long-lived epifaunal organisms such as corals, sponges and other large protruding epifaunal species on the seabed, reducing the three dimensional complexity of the bottom which provides shelter for number of marine species (Etnoyer & Morgan, 2003, 2005; Freese et al., 1999; Stone, 2006). The bottom´s increased complexity provides feeding, spawning and refugia sites for invertebrates and fish (Mortensen et al., 2005; Reed, 2002).

Concerning pollutants, it is important to remember that sediments accumulate contaminants, depending on their morphologic and geochemical characteristics. In general, the concentrations of pollutants increase as particle size decreases (Rowlatt, 1995). Contaminants do not remain permanently linked to the sediments, as they can be released to the water column and become available for the marine biota as a result of physical, chemical and biological processes (Chapman & Wang, 2001). Consequently, as contaminants can have a special impact on VMEs, their study should be included in every descriptive programme for the monitoring and control of marine pollution and environmental impact.

Focusing on marine ecosystems, this chapter mainly addresses geomorphological, benthic and fishing impact matters. However, due to their eventual secondary effects on VMEs, a brief description of the hydrographical and contamination aspects is also presented here.

2. Material and methods

The interpretation of the requirements of Resolutions 61/105 and 64/72 has been largely based on the FAO Deepwater Guidelines, as well as on the criteria defined by the North Atlantic Fisheries Organization (NAFO) and habitat codes from the European Nature Information System (EUNIS) database. The FAO Deepwater Guidelines were developed as practical guidance on what was required to enable fisheries managers to develop sustainable ecosystem-based deep-sea fisheries on the HS, in accordance with international law and agreements.

2.1 Geomorphology

Swath bathymetry data were acquired using a multibeam echosounder Simrad EM-302, transmitting frequencies between 26 and 34 kHz, swath opening of up to 150°, 288 beams and 432 soundings per swath. Data were logged and processed with SIS, Neptune and C-Floor software packages, obtaining a grid resolution of 50 m with full seafloor coverage (meeting the International Hydrographic Organization [IHO] standards for marine hydrographical surveys). Navigation was provided by a differential GPS Simrad GN33 integrated within the ship central navigation system (MDM400). Analyses and representation of bathymetric data were performed with ArcGIS software. Fledermaus advanced visualization software has been used to provide 3D views.

Very high resolution seismic reflection profiles were obtained using a Simrad TOPAS PS 18 functioning on the parametric principle, along swath bathymetry lines, at intervals of 9 and 10 km. This source has two primary frequencies, 15 and 18 kHz and the possibility of obtaining a secondary frequency of 0.5 and 5 kHz. Recorded profiles had a very high vertical resolution (less than 1 m) and penetration ranged from 50 to 200 m in unconsolidated sediments. Seafloor photographs were taken at selected sites using a Nikon D700 digital camera mounted on a framework equipped with a submarine navigation system consisting of a Sea Bird Electronics SBE911plus CTD with an altimeter Benthos PS916D.

2.2 Benthos and sediments

Different gears were used to sample benthic fauna in the study area: a Lofoten bottom trawl fishing gear was used over soft bottoms and bathyal plains during fishing operations, whereas rock and megabox corer dredges served to specifically prospect particular bottom environments whose structures and composition had previously been geomorphologically characterized and identified.

Sedimentological samples were collected mainly using net collectors attached to the Lofoten fishing gear during fishing cruises. An USNEL-type megabox corer was also used (maximum breakthrough of 60 cm; effective sampling area of 0.25 m^2 [50 cm × 50 cm]) for the benthos surveys. In addition, a few samples were taken using a Bouma-type box corer (effective sampling area of 0.0175 m^2 [10 cm × 17.5 cm]). Both box corer types are designed to take undisturbed samples from the top of the seabed, and are suitable to sample almost every type of sediment. Temperature and redox profiles (Eh) of sediment were immediately performed for the box corer sample after each station. In the laboratory, the granulometrical analysis of the sediment was carried out by dry sieving the coarse fraction (>62 µm) and by sedimentating of the fine fraction (<62 µm; Buchanan, 1984). The organic matter content of the sediment was estimated by the loss in weight of dried samples (100°C, 24 h) after combustion (500°C, 24 h).

2.3 Interactions between bottom trawling and VMEs

The 13 research cruises sampled a total of 433 stations (375 bottom trawls and 58 rock and box corer dredges) in the study area, and the presence of VMEs and/or fragile organisms was recorded in 176 of them.

GIS techniques served to aggregate data by cells of 2000 m length, covering the entire area of study. Each cell contains information related to environmental conditions: bottom floor characteristics, geomorphology, mean sea bottom temperature, slope and depth, and biological data (presence of sensitive and/or VME indicator species).

Data were integrated in R statistical software and the BIOMOD package was run to generate an optimized distribution model. To do so, surface range envelope (SRE) method (Busby, 1991) was used for selecting 500 random pseudo-absences in dissimilar environmental conditions, and random forest (RF) method (Breiman, 2001) was subsequently chosen to perform the presence/absence distribution modelling. Predictive performance of the model was checked using the receiver operating characteristic (ROC) curve through multiple cross-validation procedures, splitting original data three times in two random subsets for calibration (80% data) and evaluation (20% data) using as model predictive performance index the mean ROC value obtained from the three repetitions. To detect and quantify overlapping areas, model results were integrated with the fishery footprint, which included more than 14,000 vessel monitoring system (VMS) positions of Spanish vessels fishing in these waters registered between 2001 and 2008.

Moreover, during the whole study, three of the cruises aimed estimation of stock abundance and biomass indices of the main commercial species. A stratified random design with stratum boundaries defined by latitude and depth ranges was applied. Scheduled fishing stations (hauls of 30 min) were performed using a LOFOTEN-type gear of 35 mm codend mesh size. Abundance and biomass indices were calculated using a swept area model. The study area was divided into 13 depth strata further subdivided into 2571 grids of around 5 nm^2. The position of the hauls was randomly chosen prior to surveying each stratum and hauls were allocated according to several criteria previously defined. As above said, the fishing gear was also used as a benthic sampler, complementary to the more specific megabox corer and rock dredges used in the other surveys with no fishing objectives. Density maps of the main commercial species were calculated from catch values (kg), using the *ArcGis Density* tool and applying the Kernel quadratic function (Silverman, 1986) in a way that density values are expressed in kg/0.5h/square surface units, in the present case, 0.0125 x 0.0125.

2.4 Hydrography and contamination

A Seabird-25 CTD probe was used to characterize the hydrographical conditions of the study area. After processing procedures, a total of 406 valid CTD profiles were analysed. The CTD was systematically deployed at fishing stations below 500 m, but not always at greater depths. At each cast, the CTD was deployed to 5 m depth and stabilised for approximately 3 min. It worked in auto-contained mode at frequency of 8 scans·s^{-1}. Once stable, the CTD was brought back to the surface and started profiling at a constant speed of 1 m·s^{-1}. Due to the schedule of the surveys, the CTDs were mainly associated to the spring-summer period.

SeaBird software and standard calibration values were used for pre-processing data and converting to physical units. Subsequently, MatLab served for quality control and post-processing tasks. Temperature and salinity fields at different depths were calculated by using the variational inverse method (Brasseur et al., 1996), initially designed to solve problems with high resolution vertical profiles but irregular horizontal coverage, and implemented in the DIVA package (Troupin et al., 2008) and ODV-4 software (Schlitzer, 2011). DIVA and ODV-4 allowed analysing and interpolating data taking into account coastlines and bathymetry features to structure and subdivide the domain on which the estimation is performed. Calculations were optimized and performed on a finite element mesh adapted to the specific gridding domains.

For the study of contaminants, surface sediment samples (layer between 0 and 5 cm) were collected with a megabox corer grab. The total fraction of sediment (fraction below 2 mm) was analysed for both metals and polycyclic aromatic hydrocarbons (PAHs). Trace metals analysis was performed using the total digestion of the samples with a hydrofluoric acid and aqua-regia mixture in a microwave oven, followed by neutralization with boric acid. Nitric acid digestion was applied for mercury analysis. Quantification was carried out by atomic absorption spectrometry (flame, graphite furnace and cold vapour) (Beiras et al., 2011). Thirteen individual PAHs were measured following the method described by Viñas et al. (2002). The samples were Soxhlet extracted using a mixture of organic solvents. The recovered extracts were treated with activated copper for elemental sulphur removal. A clean-up step was performed by column chromatography using deactivated alumina. The concentration and composition of the PAHs were determined by high-performance liquid chromatography–fluorescence detection (HPLC-FLD) with wavelength programming.

Quality control was assured by the frequent use of certified reference materials as well as by the regular participation in international intercalibration exercises, where our IEO laboratory always obtained satisfactory results.

3. Results

3.1 Geomorphology

About 59,105 km^2 of the external shelf (off the Argentinean EEZ) and upper and middle slope of the ACM down to 1600 m depth contour were mapped (Figure 1A). The study differentiated two main regions (southern and northern), as described in section 1.1.

3.1.1 Southern region (45°S - 48°S)

CONTINENTAL SHELF: The outer continental shelf investigated is dominated by sediment ridges oriented in NNE-SSW direction, oblique to the shelf break and changing in its northern part to a N-S orientation. The sand ridge crests were 15 m height and separated by 3 to 4 km. They were considered as relictual of the post-glacial transgressions. Figure 3 shows the presence of low rocky outcrops that seem to be prograded by the sand ridges aforementioned. These rocky outcrops are concordant with flat rocks and rock fragments of sandstones abundantly colonized by different species, as revealed by analysed samples and ROV video recordings.

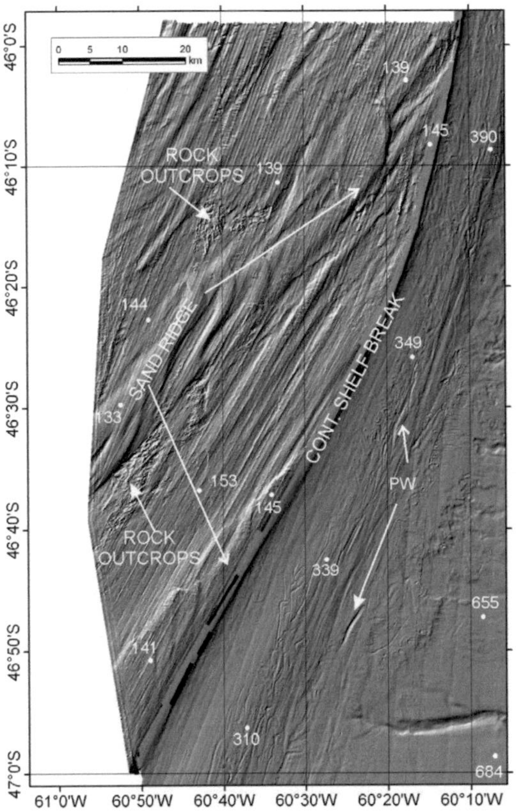

Fig. 3. Rock outcrops and sand ridges in the Patagonian Continental Shelf between 46°10′ and 46°40′S. PW: plough marks; White dots: water depth in m

CONTINENTAL SLOPE: The upper slope is 6 to 20 km wide, has a declivity of 4° to 1°20′ and descends from the continental shelf break (128-200 m depth) to depths between 250 and 750 m. Seven submarine canyons (0-6 in Figure 1B) entrench the upper slope south of 45° 40′S, belonging to the Patagonian Submarine Canyon System. Within this area, these canyons and their multiple branches dissect eastwards the upper and middle continental slopes across terraces and steps. At approximately 3.5 km depth, these canyons are collected by a morphologically very diverse slope-parallel SSW–NNE oriented channel, known as the Almirante Brown transverse canyon (Lastras et al., 2011).

South of 45°20′S the base of the upper slope is scarred by iceberg plough marks. López-Martinez et al. (2011), inferred that the plough marks were probably eroded by icebergs carried northwards by the FMC during the last glaciation (Figure 4).

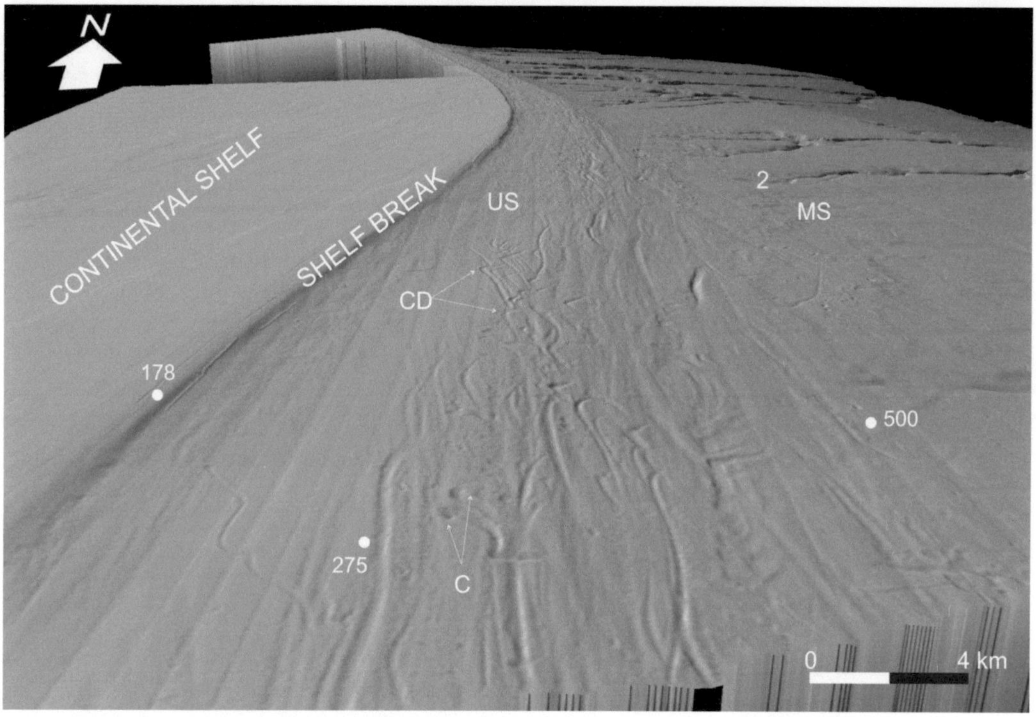

Fig. 4. The base upper slope shows erosion by iceberg plough marks during the last glaciation. US: upper slope; MS: middle slope; C: iceberg craters or pit marks; CD: abrupt change of direction of plough marks; White dots: water depth in meters.

The middle continental slope generally had a gradient of <1° and extended from depths between 250 and 750 m to depths comprised between 1600 and 2000 m. This middle slope is the creation of bottom current erosion during the opening of the Drake Passage in the Eocene-Oligocene and the northwards extension of the FMC (Hernández-Molina et al., 2009, 2010; Acosta et al., submitted).

Secondary features carved on the middle slope include plough marks cut by icebergs during the last glaciation on its inner edge at depths of 250 to 550 m south of 46°20′S (López-Martinez et al., 2011), seven submarine canyons, three sets of gullies, pockmarks created by the expulsion of gas, thin-skinned tectonics south of 47°20′S, sediment drifts and a belt of bottom current erosion along its seaward edge. Modifying the morphology of the middle slope we also found carbonate mounds capped by cold-water corals, siliceous sponges and bryozoan bioherms (Muñoz et al., submitted).

3.1.2 Northern region (41°30'S - 45°S)

The southernmost part of the northern region exhibits numerous canyons and gullies in the middle part of its courses, located in the upper and middle slope (Figure 2B) from 179 to

2253 m depth. Figure 5 shows a 3D scheme showing the uneven development of the canyons, as well as their longitudinal extension only covering their middle sections.

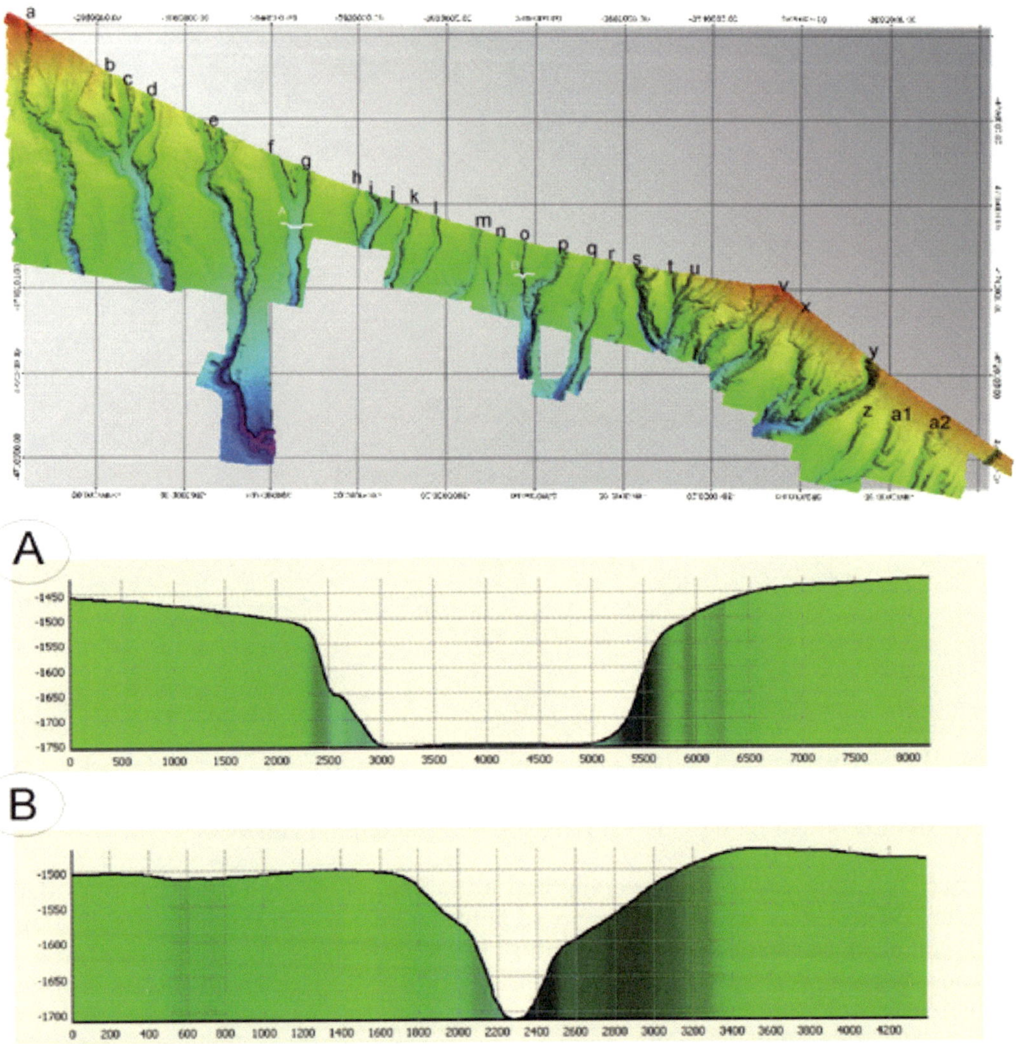

Fig. 5. Color coded DTM of Patagonia 02/09 area (landscape representation). a-y = Canyons. Lower panel A-B = Bathymetric cross section of f/g and o canyons. Note the different cross section in U and V

In the northernmost part, the study depths ranged from 178 m in the outer continental shelf to 1940 m in the middle slope. Figure 2C shows the outer continental shelf with both the head of two main canyons and the presence of a striking field of megaripples and iceberg plough marks. Megaripples with amplitude of more than 25 meters, base widths of 1000 m and crest lengths of more than 25 km are a singular and unique feature covering approximately 780 km^2 of the slope surface, and orientated NW-SE. The existence of this huge field of tractive sediment forms may be interpreted as the result of the action of the FMC front of more than 25 km generating this mega field of sediment waves on the seafloor.

3.2 Benthos

The benthic megafauna caught during the cruises, including both invertebrates as well as phyla Chordata and Hemichordata, showed dominance of the phyla Cnidaria, Porifera and Echinodermata. These phyla are largely considered as vulnerable ecosystem indicators, according to the latest international standards established by the United Nations (UN) and the Oslo-Paris convention (OSPAR). In the Southwest Atlantic studied area, we found a significant diversity of species, habitats and ecosystems, whose special characteristics meet the vulnerability criteria established by international organizations (FAO, UN and OSPAR).

3.2.1 Cold-water coral reefs

Cold-water coral reefs are self-sustained and spatially well defined coral framework sediment systems measuring from tens of metres to several kilometres in lateral extension and up to 40 m in thickness, thus influencing local current regimes (Dorschel et al., 2007). Most cold-water coral reefs occur in deep waters, between 200 and 1500 m, where light is reduced or absent (Freiwald, 2011). Cold-water corals belong to the phylum Cnidaria and include, among others: anthozoan stony corals (O. Scleractinia), black corals (O. Antipatharia), soft corals (O. Alcyonacea), sea pens (O. Pennatulacea), sea fans (O. Gorgonacea) and hydrocorals or lace corals (Class Hydrozoa; O. Anthoathecate) (Cairns, 2007). *Oculina, Lophelia, Solenosmilia, Madrepora* and *Enallopsamia* are known stony colonial coral genera that constitute the majority of deep sea coral reef structures around the world. In our surveys, the most frequent species was *Bathelia candida,* exclusively distributed on southern south America, from Rio Grande (south Brazil) to south Chile (Cairns, 1982; Kithara et al., 2009). This species, less known than *Lophelia pertusa* (EUNIS codes A5.631 and A6.611), but ecologically also very important as bioconstructor. *Solenosmilia variabilis* was also found in minor quantities in the samples.

In the study area, the largest biomass of cold-water corals was located at depths between 400 and 1000 m, sometimes in low slope areas of sandy bottoms, forming both small aggregates and reefs of few metres. In the study area, *Bathelia candida* provided habitat for a great diversity of invertebrates and fishes (ICES, 2003). Dead specimens accounted for an important percentage of the community. These stony corals were colonized by many other species. Associated fauna was dominated by filter-feeders, cnidarians, sponges, molluscs and brachiopods, but also echinoderms and crustaceans (Figure 6).

The most representative species found among Cnidarians were: Order Scleractinia, with *Caryophillia* spp, and *Desmophyllum dianthus,* growing up on the colonial scleractinia, while *Flabellum* spp. *Javania* spp. were found on sediment or small peebles and mollusca debris; Order Alcyonacea, with *Alcyonium* sp., *Anthomastus* sp. among other taxa; Order Gorgonacea, mainly represented by Primnoidae family species as *Plumarella* sp., *Covexella* spp., *Primnoella* spp., *Thouarella* spp., *Dasystenella* sp., and *Fannyella* spp., among other taxa. *Paragorgia* sp. was another abundant sea fan growing on scleractinians, always of small or medium size depending on the substratum surface size and stability.

Order Anthoatecatha, represented by many species of the family Stylasteridae, such as *Adelopora pseudothyron, Errina antarctica, Errina inferolabiata, Errinopsis* spp., *Errinopora cestoporina, Cheiloporidion pulvinatum, Crypthelia* spp., *Sporadopora dichotoma, Sporadopora* sp., *Lepidopora* spp., *Conopora pauciseptata* and *Stylaster densicaulis.*

Fig. 6. **A**) View of the outer living zone and inner dead coral framework zone of *Bathelia candida* in the southeast end of the ridge, south of Canyon 2. The polyps (marked in red) evidence the high proportion of living coral on the reef. **B**, **C**) Underwater photographs showing a high density of associated species

Phylum Echinodermata: *Gorgonocephalus chilensis*, *Astrotoma agassizi*, *Odontaster* sp., *Henricia* sp., *Hippasterias* spp. and *Notasterias* sp. Porifera (*Rosella* spp., *Tedania* spp., *Mycale* spp., *Myxilla spp.*, *Esperiopsis* sp.).

Sub-phylum Crustacea: *Munida spinosa*, *Thymops birsteini*.

Phylum Mollusca: *Bathydomus longisetosus*, *Miomelon* sp., indeterminate Nudibranchia, *Marseniopisis* spp., *Hiatella* sp., *Limopsis* spp., among other.

Brachiopods, Picnogonida and Ascidians (*Polycarpa* sp., *Pyura bouvetensis*) were also found in the surveys' samples.

3.2.2 Coral gardens

The study photographed cold-water corals both on a hummocky area of the upper slope on the ridge south of Canyon 2 and on individual glacial erratics on the eastern edge of the Perito Moreno Terrace, but also sampled them at stations scattered throughout the middle slope. The bottom photographs showed that the hummocky terrain on the upper slope between 300 and 500 m was due to the presence of soft corals. These organisms were so abundant that the biologists named these sites "coral gardens" (Figure 7). Coral gardens meet the criteria for classification as VMEs (habitats which can occur within deep seabed EUNIS types A6.1 to A6.9). The biological diversity of the coral garden community was very high and contained many species of corals belonging to different taxonomic groups, such as leather corals (order Alcyonacea); gorgonians (order Gorgonacea); black corals (order Antipatharia) -with an exclusive species in the study area, *Dendrobathypathes grandis*-; hard corals (order Scleractinia); and hydrocorals (family Stylasteridae, order Anthoathecata).

Fig. 7. Coral Gardens bottoms in the study region. These areas could be similar in structural complexity to tropical coral reefs with which they shared several important characteristics, including complex vertical relief and high taxonomic diversity. Views can give an idea of the high biodiversity, mainly Primnoidae with the very abundant species *Thouarella viridis* (green sea fan)

Primnoidae was the dominant gorgonian family in the coral gardens of the study area, represented by several genera and species (Cairns & Bayer, 2005, 2009). The habitat included relatively large numbers of sponge species (with orders Hadromerida and Poecilosclerida especially well represented), although they were not a dominant component of the community. Other commonly associated fauna included echinoderms such as basket stars (*Gorgonocephalus chilensis*), brittle stars (*Astrotoma agassizi*, *Ophiacantha* spp., *Ophiactis* spp.), crinoids, molluscs, crustaceans and deep-water fishes, such as grenadier (*Macrourus carinatus*).

3.2.3 Sponge beds

Sponge beds or sponge aggregates in deep water (EUNIS Code: A6.62) consisted mainly of two Porifera classes: Hexactinellida and Demospongiae. Their presence was important between 250 and 1300 m (Bett & Rice, 1992), in areas where the water temperature varies between 4° and 10°C and where currents flow at approximate 0.5 knots. Since sponges have a preference for deep habitats similar to those of cold-water corals, it is common to find both ecosystems coexisting in the same locations. In the study area the presence of deep water hexactinellid sponges belonging to the genus *Rossella*, provided a three-dimensional structure to the seabed on which other species live, hunt or find refuge against predators and existing currents.

Carnivorous sponges (Figure 8) generally colonize hydrothermal vents and abyssal zones, but in the study area they lived at depths not exceeding 1500 m. The different expeditions provided samples in which we identified several species, some of them new to science, belonging to the genera *Asbestopluma*, *Chondrocladia*, *Eucheliplima* and *Cercicladia*, a new genus (Ríos et al., 2011).

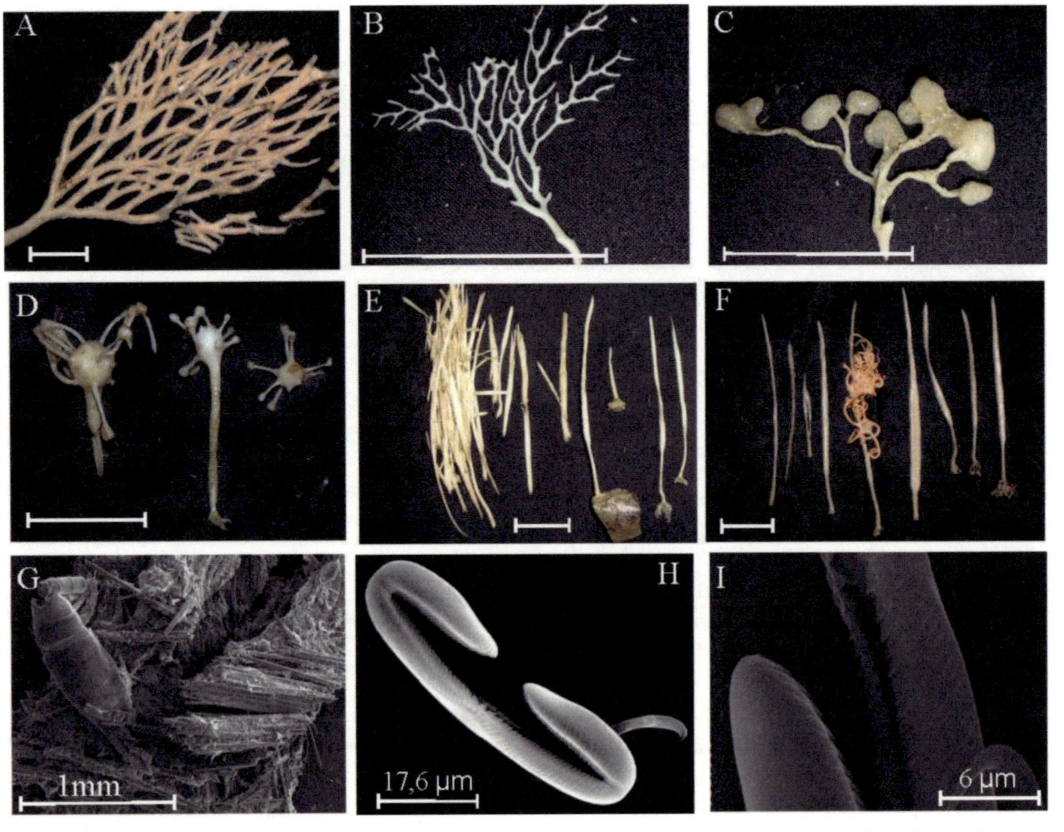

Fig. 8. Study carnivorous sponges. **A**. *Asbestopluma* sp. **B**. *Asbestopluma* sp. **C**. *Chondrocladia* sp. **D**. *Chondrocladia* sp. **E**. *Eucheliplima* sp. **F**. *Eucheliplima* sp. and brittle stars linked to one of the specimens. (Scale **A-F** 5 cm). **G**. Crustacean attached to the sponge spicules of *Eucheliplima* sp. **H**. *Eucheliplima* sp. Spicules: placochela and sigma. **I**. Placochela central detail

3.2.4 Deep marine rocky environments

The deep marine rocky environments can be considered as VMEs because they host rare or endemic species; they are important for the survival, functionality or recovery of fish populations; or due to their fragility and high biodiversity. Our study detected such ecosystem characteristics in different areas. Even if the species' identification of the rock dredge samples is currently ongoing, the preliminary results evidence a high biodiversity on these rocky areas (supported by a number of underwater images taken by the ROV and the submarine digital cameras) (Figure 9). In fact, they host a high number of species belonging to different zoological phyla, including those traditionally considered as vulnerable or protected, such as Porifera, Cnidaria and other invertebrate taxa (ophiuroids, crinoids, asteroids, bryozoans, tunicates, etc).

Fig. 9. High biodiverse rocky areas of the continental shelf (135 m). On mud-sand sediments, rocks are colonized by a wide variety of organisms from different zoological groups, competing for the available space. The image shows a large brittle star (*Gorgonocephalus* sp. accompanied of different species of the gorgonian family Primnoidae, asteroids of the genus *Porania*, crinoids, ophiuroids (*Ophiacantha vivipara*), and other taxa (i.e. Porifera and Bryozoans)

3.3 Sediments

Sediment data of the 156 sampled stations showed a predominance of fine sands throughout the study area, with low contents of organic matter, and poorly to moderately well sorted. In general, the thinnest sedimentary types with a higher organic content were found in the southernmost study area, below parallel 46°30'S (Figure 10). Below are detailed the different sedimentary types and their spatial distribution along the whole study area, according to the bathymetric range.

Depths less than 200 m. In the Patagonian continental shelf (59 sampled stations) the sediments were characterized by the presence of all the studied sandy types, even if 30% of the stations were dominated by medium sand and 16% by very fine sand. With a low average of organic content (1.02% ± 0.54%), the sorting varied between good and poor for a moderate average value of the sorting coefficient (S0 = 1.36 ± 0.16). Spatially, larger particle size sediments were found in the central zone of the continental shelf (latitudes between 45°70'S and 46°40'S), separated in the north by fine sands and in the south by also fine sand sediments but with the highest organic content and mud percentage (Figure 10).

Depths from 201 to 300 m. One of the worst studied bathymetric strata, with few sampled stations. The sediment was characterized by the presence of fine sands (Q50 = 0.19 mm ± 0.05 mm) with the lowest organic content of the whole study area, averaging 0.92% ± 0.60%. The sediment was moderately sorted (average S0 = 1.34 ± 0.08). In contrast to the continental shelf, the coarse sediments were almost absent in this stratum, with both the particle diameter moderately diminishing and the organic content moderately increasing from north to south (Figure 10).

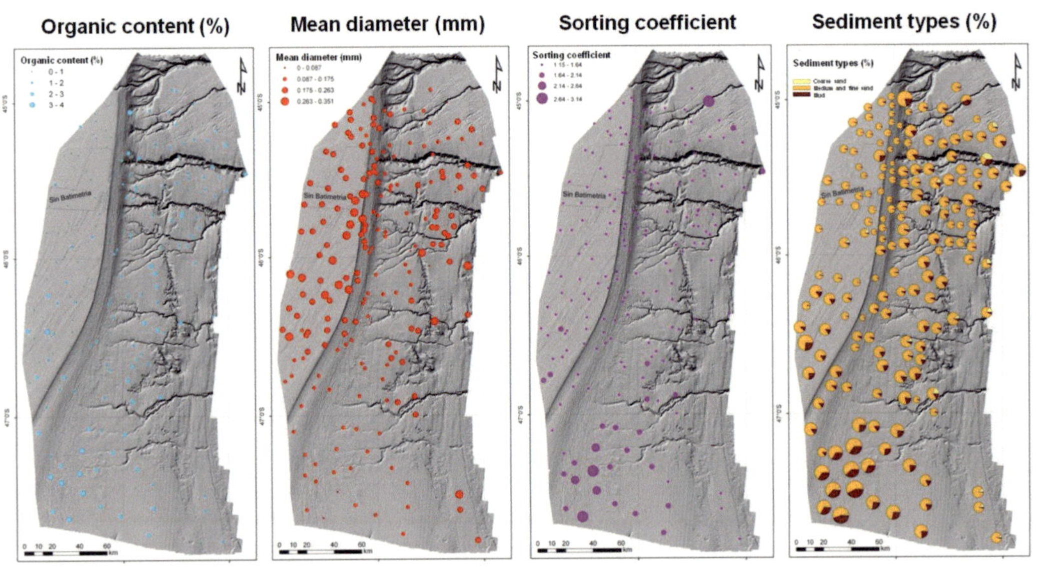

Fig. 10. Spatial distribution of the sediment organic content (%), mean diameter (Q50 in mm), sorting coefficient (S0) and sediment types (%).

Depths from 301 to 400 m. Samples of this stratum come from 14 stations. Again, fine sand is the characteristic type of sediment of this bathymetric range (Q50 = 0.160 mm ± 0.04 mm) and the average organic content is the second lowest value of the whole study (1.01% ± 0.64%). The average value of the sorting coefficient (S0 = 1.31 ± 0.13) evidenced a moderate sorting and was the lowest value of the whole study area. Spatially, the thinnest layers of sediments were associated with high organic matter contents, increasing from north to south (Figure 10).

Depths from 401 to 500 m. This stratum was characterized by the thinnest and muddiest sediments of all the bathymetric strata sampled. This bathymetric range was dominated by the sediments mainly composed of very fine sands (Q50 = 0.120 mm ± 0.04 mm). The average organic content was moderated (2.09% ± 1.09%) and the sediments were the second less sorted (S0 = 1.76 ± 0.58). As observed in the previous bathymetric stratum, the southern stations had the thinnest sediment layers with the highest percentage of organic matter in sediment surface (Figure 10).

Depths from 501 to 700 m. In this stratum samples of a total from 21 stations have been obtained and as in the previous bathymetric range, the sediment is dominated by the presence of very fine sands, with an average mean diameter of 0.130 ± 0.03 mm. The organic content of the sediment was moderated, and is the highest for the whole study (2.19 ± 0.84 %). The sorting of the sediment, in average, was moderated (S0 = 1.59 ± 0.35), with a great range of variation. Spatially, the thinnest sediments in this stratum are found with the highest organic content as the latitude increases (Figure 10).

Depths from 701 to 1000 m. This stratum gathered the highest sampling intensity, due to its largest bathymetric extent. Sandy sediments predominated, and most of them were composed of fine sand (Q50 = 0.160 mm ± 0.02 mm). The average of the organic content was low (1.75% ± 0.47%) and sediments were moderately to moderately well sorted (average S0 = 1.49 ± 0.17). From the spatial point of view, and contrarily to previous strata, the reduction of the granulometry and the increase of the organic content was not so clear as we moved southwards. In addition, this bathymetric stratum had some stations with high percentages of coarse sands in their granulometric composition (Figure 10).

Depths from 1001 to 1500 m. A total of 24 samples were gathered, most of them north to parallel 46°S (Figure 10). This deep stratum was dominated by sandy sedimentary types, mainly composed of fine sands (Q50 = 0.190 mm ± 0.05 mm). The average of the sediment organic content was low (1.44% ± 0.37%) and sediments were moderately to moderately well sorted (average S0 = 1.39 ± 0.11). No clear spatial pattern of distribution was observed in this stratum, where the thinnest sediments (<62 μm) of the whole study were found. As in the previous stratum, we observed high percentages (over 11%) of coarse sands (>500 μm) in the granulometric composition of some stations (Figure 10).

Depths major than 1501 m. A total of 6 samples were gathered in the deepest zone, mainly inside canyons and submarine gullies. The deepest stratum was characterized by the presence of sandy sedimentary types, dominated by fine sands (Q50 = 0.200 mm ± 0.08 mm). The average organic content was low (1.68% ± 0.43%), and the sediments were the worst sorted of the study (average S0 = 3.07 ± 2.99). A study of the sediments' spatial distribution did not make sense, because the stations were located inside the big canyons and submarine gullies transversely crossing the whole study area. There was a high heterogeneity of the sediments between samples, with some stations dominated by thick sands (>500 μm) in their granulometric composition (from 17% to 39.5%), whereas three stations had more than 23% of mires (<62 μm) (Figure 10).

3.4 Interactions between bottom trawling and VMEs

The ROC method described by Zweig & Campbell (1993) was used to measure performance of the chosen model. The area under the curve (AUC) score estimated a model accuracy of 0.913 (1 means perfect prediction and values ≤0.5 mean that prediction does not deviate from random assignment). The most significant variables influencing the distribution of sensitive habitats and/or organisms were, in order of importance: sea bottom temperature, depth, sediment mean grain size and slope.

Output was a VMEs distribution probability map ranging from 0 to 1. ROC threshold chose the cut-off limit that maximized the percentage of presence/absence cells correctly predicted for the evaluation data. Thus, only areas with probabilities over the cut-off limit, set in 0.585, were interpreted as "true presence" areas. The locations of these areas were represented by superimposing onto the fishery footprint, showing the overlapping areas in Figure 11B, under the legend "fisheries impact".

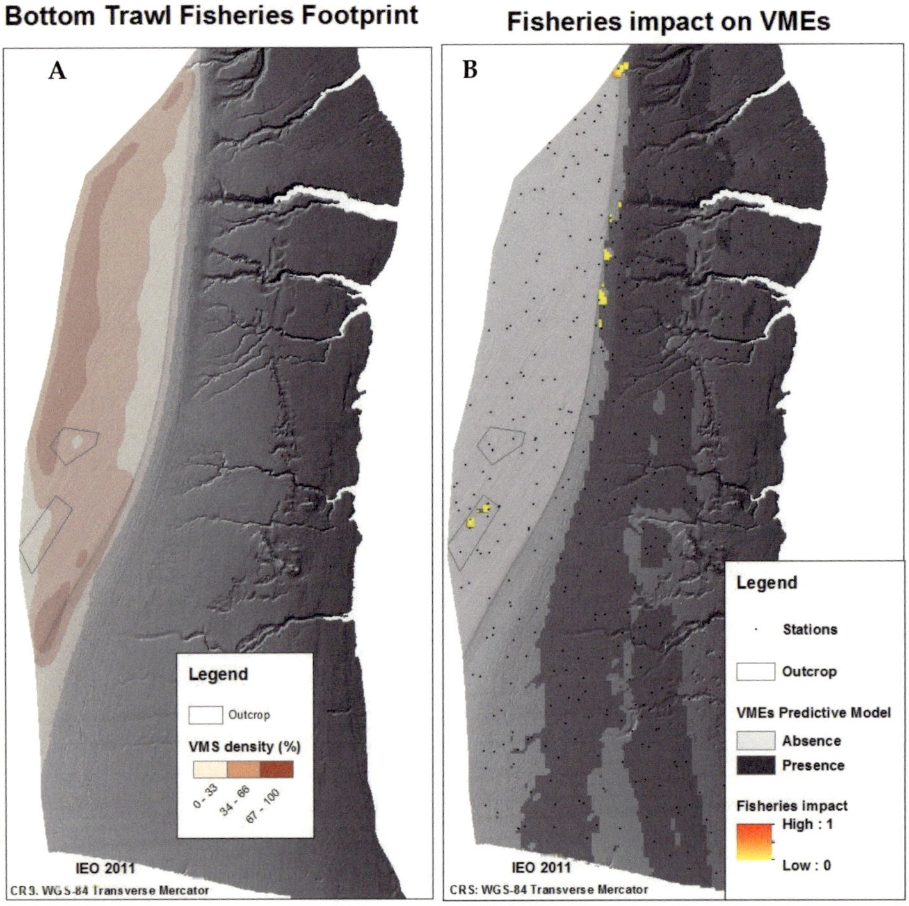

Fig. 11. **A**) Footprint of the fishery in the area between 45°S-48°S. **B**) Impact of bottom trawling on VMEs

Regarding biomass assessment and distribution of the main commercial fished species, grenadier (*Macrourus carinatus Günther, 1878*) was identified as the most caught species, with the highest estimated biomass. Grenadier was fished between 200 and 1500 m, but the highest catches were recorded between 500 and 1000 m. The bathymetric distribution range of this species is located away from the usual fishing grounds for the Spanish bottom trawl fishing fleet (shallower) (Figure 11A) so, in spite of being the most abundant species, it is not a target species for the Spanish fleet, which mainly targets shortfin squid (*Illex argentinus Castellanos, 1960*) and Argentine hake (*Merluccius hubbsi Marini, 1933*) in the study area. The maximum catches of shortfin squid, distributed throughout the continental shelf and slope, mainly occurred in hauls between 300 and 500 m depth. Catch distribution of hake clearly indicated its shallower bathymetric range (<200 m), corresponding to the trawls over the continental shelf. Figure 12 shows the catch distribution (kg·haul^{-1}) and the density maps for these three species.

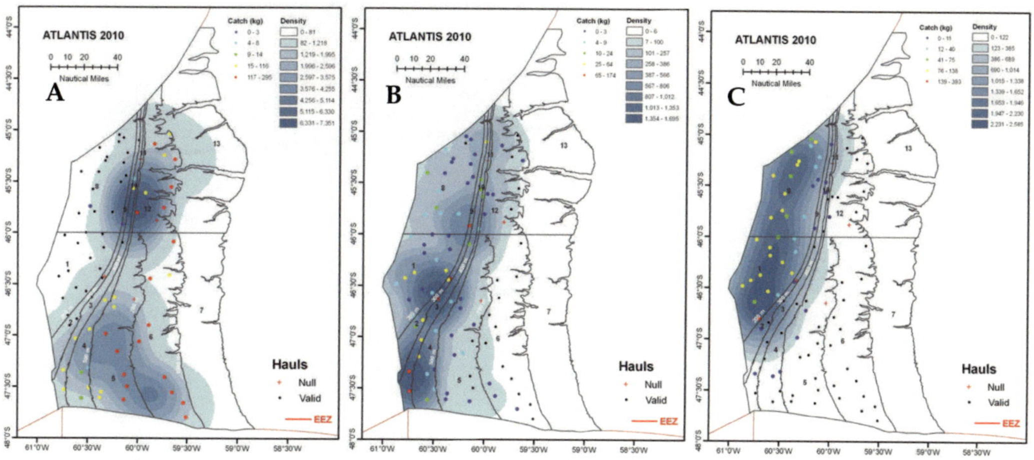

Fig. 12. Catch distribution (kg/haul) and density maps of grenadier (**A**), shortfin squid (**B**) and Argentine hake (**C**) obtained in the 2010 survey

3.5 Hydrography and contamination

Surface temperature and salinity intervals ranged from 8°C to 15°C and from 33.4 to 34.2, respectively. Warmer and less salty waters were located over the continental shelf, while colder and saltier waters were distributed over the slope, in the southeast part of the study area. Surface temperatures showed seasonal differences. Below the thermocline, waters became more homogeneous (temperature from 2.5°C to 5°C, salinity from 34.0 to 34.5), with subtle differences compared to those of warmer waters close to the slope. CTD data evidenced that the water characteristics resulted from the mixing of the sub-Antarctic waters with the water from the coastal continental discharges. As shallower waters are continuously modified by the atmospheric exchanges, the structure of the thermohaline field at the studied area showed seasonal differences in their stratification, mainly associated with the timing of the surveys, but also with bathymetry (platform/slope/ocean) and latitude (Brandini et al., 2000; Anon., 2008).

DIVA analysis showed small differences and patterns, that had probably been masked by a traditional objective analysis, such as the slight temperature and salinity differences found in the southern part of the latitudinal section (Figure 13) associated to sub-Antarctic waters. Horizontal field at surface (Figure 14) clearly showed the frontal area linked to the continental slope: warmer and less salty waters near the coast and colder ones eastwards, out of the continental slope. Similar patterns were observed at depth, with the FMC flowing along the continental slope and transporting sub-Antarctic waters northwards between 55°S and 39°S-36°S (Bianchi et al., 1982; Forbes & Garrafo, 1988; Lusquiños & Schrott, 1983;). The principal core of the FMC is located between 150 and 600 m depth, showing typical temperatures between 4°C and 5°C and salinities between 34.1 and 34.2 (Piola & Gordon, 1989). This water mass interacts, almost permanently, with shelf bottom waters, which acquire sub-Antarctic characteristics. Our results are in agreement with the hydrographical descriptions found in the bibliography for this area.

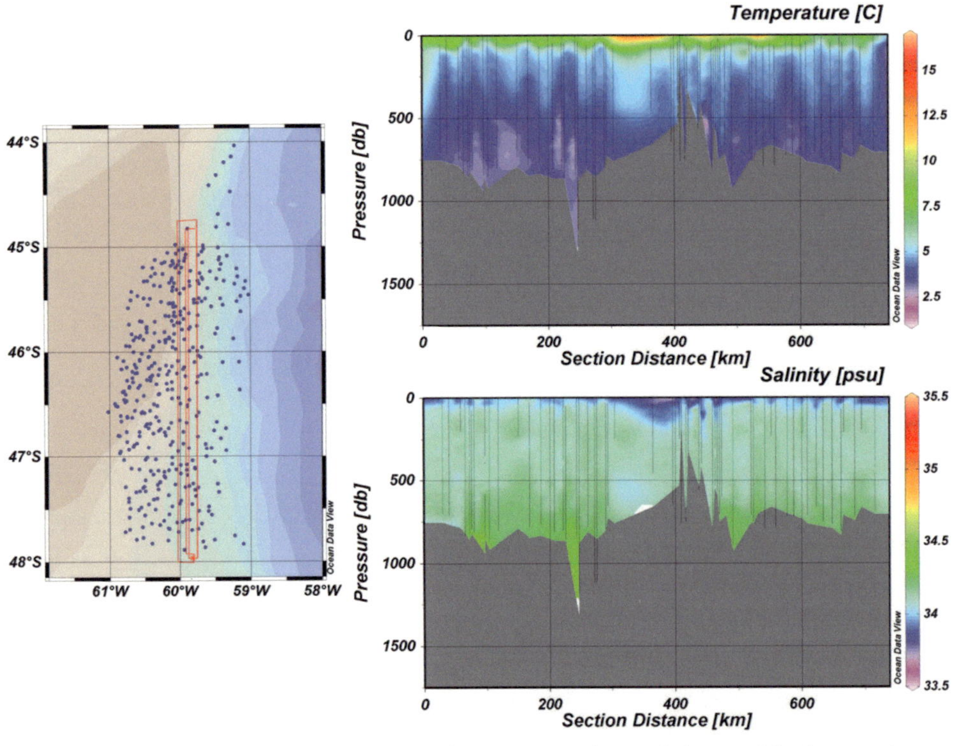

Fig. 13. Latitudinal section (South to North) showing zonal variability and bathymetric stratification. Shallower waters are fresher and warmer, probably due to coastal discharges. Southern parts of the section show colder and saltier waters coming from the Antarctic Ocean

In lower latitudes, and also over the slope from north to south, the observed water characteristics were in agreement with those of the Brazil current that transports subtropical waters (S>35.0 and T>16°C) up to its convergence with the FMC defining the Subtropical Confluence (36°S -38°S) (Gordon & Greengrove, 1986).

Metals and PAHs occur naturally in oceans, rivers and soils. However, the continuous increase of their concentration in estuaries and coastal zones, and therefore in marine organisms, is mainly due to anthropogenic inputs (Harvey, 1997; Kumar et al., 2008). Some metals, like Fe, Cu, Zn and Mn, are essential for organisms, but they become toxic when exceeding a certain concentration. Other elements, such as Hg, Pb, Cd, As and Cr, are not necessary for any metabolic function and become toxic even at very low concentrations. One of the most serious consequences of their continued presence is their accumulation through the food chain. On the other hand, hydrocarbons are extremely toxic to marine life and, those having a high molecular weight produce highly carcinogenic metabolites when they are incorporated into the biota (Long et al., 1998).

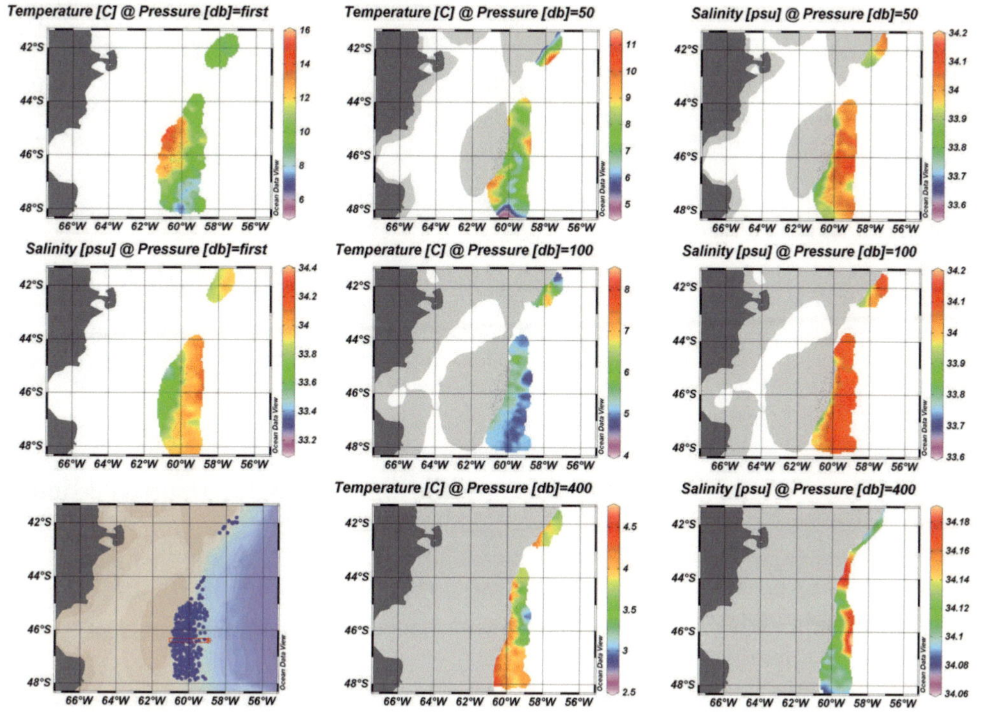

Fig. 14. Horizontal mapping of temperature and salinity at different depths

Low concentrations of trace metals were found in most sediments of the sampled stations (Table 1). Sampling sites were many kilometres from anthropogenic inputs, therefore it was not expected that sediments would be contaminated at most of the stations. In general, higher levels of metals have been obtained in the muddy samples with the highest percentage of organic matter, corresponding to the area to the South of study. In Figure 15, spatial distribution of Cu, Zn, Mn and Al are presented. Al, Fe or Li were determined to recognize natural variations in the trace metal concentration due to natural variation in sediment texture and mineralogy. Trace metal concentrations varied considerably, as it would be expected with a varied mineralogy. The enrichments in different areas are due to complex natural transport and biogeochemical processes, not to anthropogenic sources.

PAH concentrations (Table 2) were very low, even under the Limit of Quantification (LQ) for some of the analysed PAHs in most of the samples. However, we must specify that for most of the samples only the lighter (low molecular weight) and more hydrosoluble PAHs could be quantified, while the heavier and more hydrophobic ones (generally originated in combustion processes) were hardly detected. This situation is typical of areas situated far from the impacts of anthropogenic inputs. Some PAH isomer pair ratios (Phen/Ant and Fl/Pyr) were used to identify possible sedimentary sources and the different surveys were represented in different colours (Figure 16). As expected, due to the distance of the study area from the coast, no samples evidenced PAH due to grass, wood or coal combustion, but three different areas can be identified in Figure 16: samples from the area sampled in campaign 12/07, with petrogenic PAHs origin; samples from the area sampled in campaign 01/08, with pyrolitic-petrogenic PAHs; and samples from surveys 10/08 and 11/08, with PAHs with a more petroleum combustion origin.

	Al	As	Cd	Cu	Cr	Fe	Li	Mn	Ni	Pb	Zn	Hg
Mean	4.97	6.99	0.082	7.40	53.04	2.16	25.09	377	9.46	7.64	39.08	0.011
Median	4.81	5.71	0.069	7.56	39.70	1.86	25.00	373	9.41	7.68	38.60	0.010
Range	1.57-8.10	2.65-18.90	0.022-0.291	<2.00-14.50	8.03-230	1.18-6.74	7.74-46.80	204-734	1.62-18.80	2.37-30.90	17.70-72.40	<0.001-0.113

Table. 1. Summary of the concentrations of trace metals (mg kg $^{-1}$ dry weight, except Al and Fe, in percent) in surface sediments

	Phen	Ant	Fl	Pyr	BaA	Chrys	BeP	BbF	BkF	BaP	BghiP	dBahA	IP
Me-dian	2.13	0.49	1.45	1.97	0.49	0.79	0.98	0.65	0.49	0.39	0.90	0.75	1.24
Range	<LQ-13.2	<LQ -2.71	<LQ -3.41	0.29-14.9	<LQ -1.45	<LQ -1.95	<LQ -3.57	<LQ -3.68	<LQ -1.13	<LQ -1.03	<LQ – 7.13	<LQ – 0.75	<LQ -1.24
LQ	0.33	0.12	0.69	0.21	0.27	0.39	0.78	0.48	0.36	0.27	0.57	0.51	1.08

Table. 2. Summary of the concentrations of PAHs (µg kg^{-1} dry weight) in surface sediments. The 13 analysed PAHs were: phenanthrene (Phen), anthracene (Ant), fluoranthene (Fl), pyrene (Pyr), benz[a]anthracene (BaA), chrysene (Chrys), benzo[e]pyrene (BeP), benzo[b]fluoranthene (BbF), benzo[k]fluoranthene (BkF), benzo[a]pyrene (BaP), benzo[ghi]perylene (BghiP), dibenz[a,h]anthracene (dBahA), and indeno[1,2,3-cd]pyrene (IP). LQ: Limit of Quantification

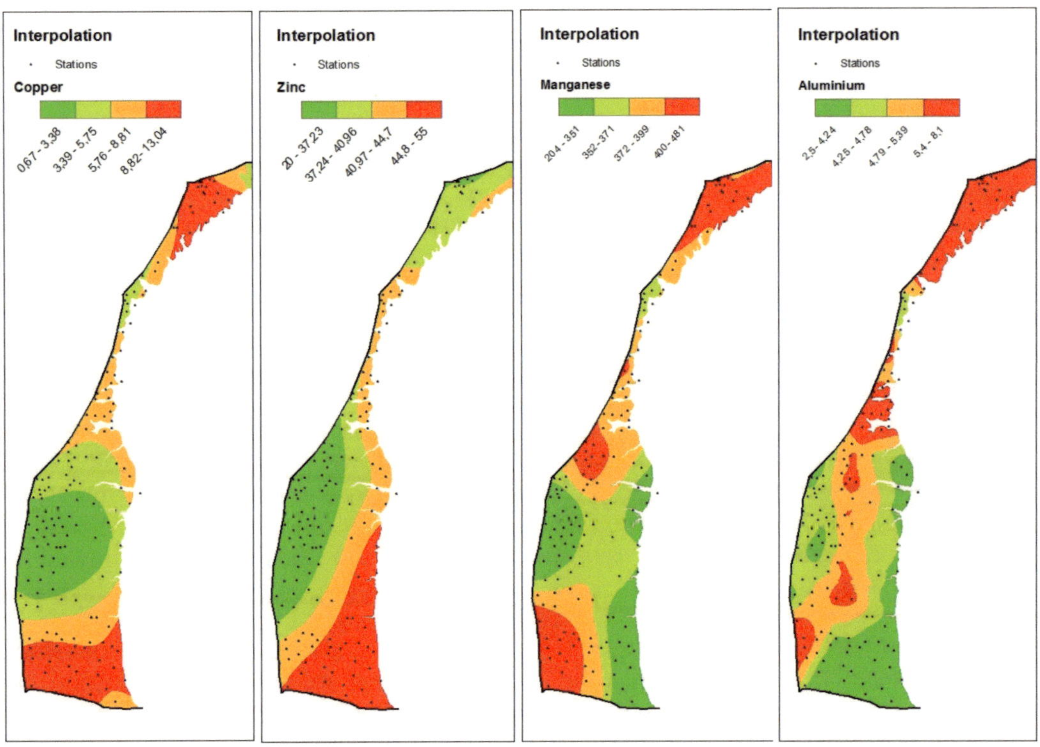

Fig. 15. Spatial distribution of copper, zinc, manganese and aluminium. Values are given in mg kg $^{-1}$ dry weight, except for Al (in percent)

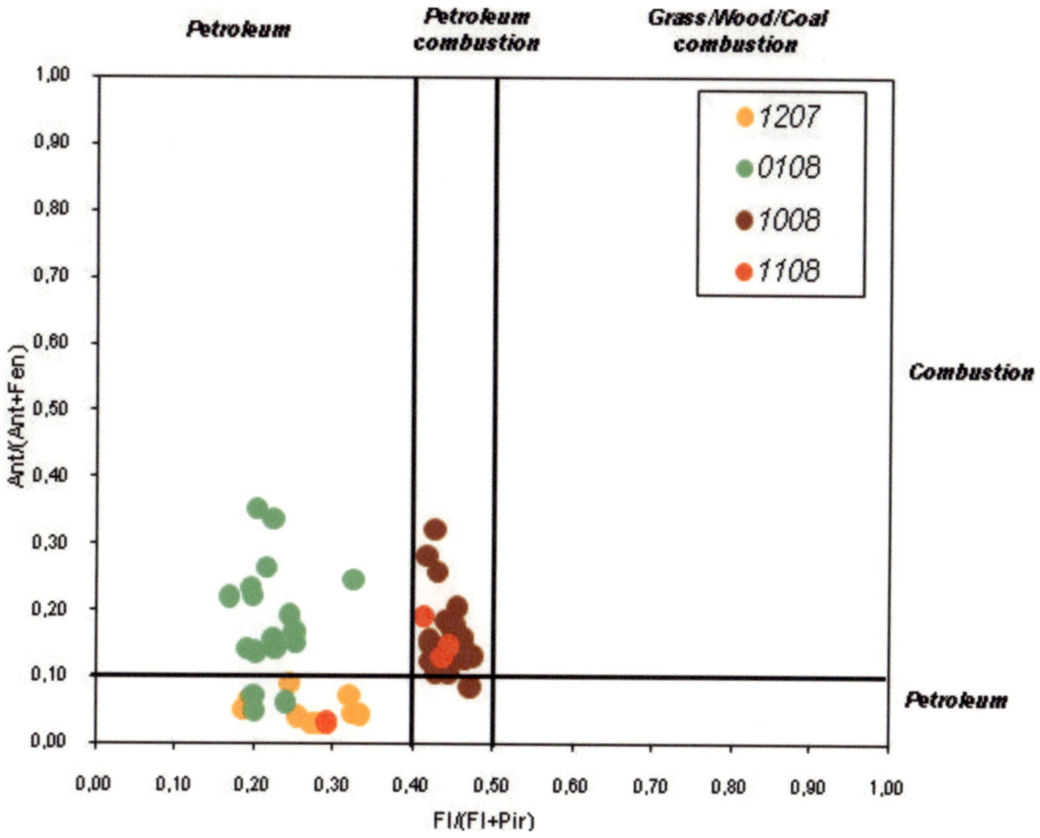

Fig. 16. Isomer ratios to identify PAH sources

4. Conclusion

The series of 13 multidisciplinary research surveys on the HS of the SW Atlantic have allowed the identification of the following main geologic and geomorphologic types that could be considered as mega-habitats (FAO, 2008) or ecotopes (UNGA, 2009):

i. Rock outcrops on the continental shelf made of rock slabs (sandstone) and a deep outcrop located in the middle slope, between canyons 1 and 2 (EUNIS codes A6.1, A6.7);
ii. Mounds (EUNIS code A6.75) located on the top edge of a bend in canyon 2, morphologically corresponding to carbonate mounds or recifal structures;
iii. Rough and hummocky sea floor associated with seepage of fluids (EUNIS code A6.9) located on the upper slope and in the northern third of the study area;
iv. Cold seeps and mud volcanoes in the middle slope of the southern third of the study area, where there were numerous pockmarks (mega-pockmarks due to their size);

v. Numerous canyons and gullies that have been mapped throughout the whole study area. The complex geological and hydrological structure associated with these canyons makes them considered as mega-regions in which habitats are likely to contain vulnerable species.

Biological diversity associated with reefs of *Bathelia candida* is around three times those of the surrounding seabed, indicating that these reefs create biodiversity hotspots and increase densities of associated species (ICES, 2003). In general, deep-sea bottoms vulnerable habitats (as those described above) or special geological structures (as seamounts or canyons) are structurally complex habitats known to sustain high densities and diversity of megafaunal organisms. Our results seem to support the hypothesis that the deep-sea environment is a small organism habitat. In the study area, these habitats host a much higher biodiversity compared to adjacent areas. The complex three-dimensional habitat structures (coral gardens, sponge beds and *Bathelia candida* reefs in particular) provide a multitude of micro-niches favouring the sustainability of an associated animal community by providing enhanced feeding possibilities, hiding places and nursery areas.

In the study area, vulnerable species are mainly distributed in areas located between 400 and 1000 m depth, where the sea bottom temperature ranges from 2.7ºC to 4.5ºC and the seabed sediment grain size is up to 2.3 (Q50). Alternatively, vulnerable species can also be found in outcrop areas of the continental shelf and in some restricted zones of the slope and at the head of canyons. Density maps (Figure 12) of the main commercial species display their geographical distribution and the areas where fishing operations targeting them take place. The Spanish fishing fleet mainly targets hake, distributed along the continental shelf, and shortfin squid, found in the upper slope at depths <500 m. Grenadier, the most abundant species in the study area, is not a target species for this fleet, because it mainly lives deeper than 500 m, where the presence of VMEs was found to be more abundant.

The analysis of the fishery footprint and the defined distribution of VMEs evidenced that Spanish bottom trawling fleet on the HS of the SW Atlantic has a negligible impact on identified vulnerable species or VMEs in the study area. The lack of commercial fishing data of other bottom trawling fleets also operating in this region (i.e. Poland, former USSR and GDR, etc.), makes it impossible to ascertain if VMEs are known or likely to occur in HS areas where bottom fishing takes place, or whether fishing practices are significantly damaging them. Moreover, we do not know if the observed absence of VMEs in the Spanish bottom trawling fishing grounds is a consequence of the impact of previous fishing activities or is due to other reasons.

The research undertaken and its main findings led to the delineating of several areas to be protected, according to biological and geological criteria adopted for the quantitative, qualitative and geographic description of the areas with the presence of organisms classified as vulnerable. Indeed, a specific scientific advice was made to the Spanish Government to protect areas meeting both biological and geological criteria.

Nine areas or regions (Figure 17) along the Patagonian Shelf and slope were identified as VMEs and were designated as candidate areas for closure (a total of ~41,300 km^2). According to this scientific advice, the Spanish Government implemented a fishing ban for the Spanish bottom trawling fleets in the high seas of the SW Atlantic on 1st July, 2011.

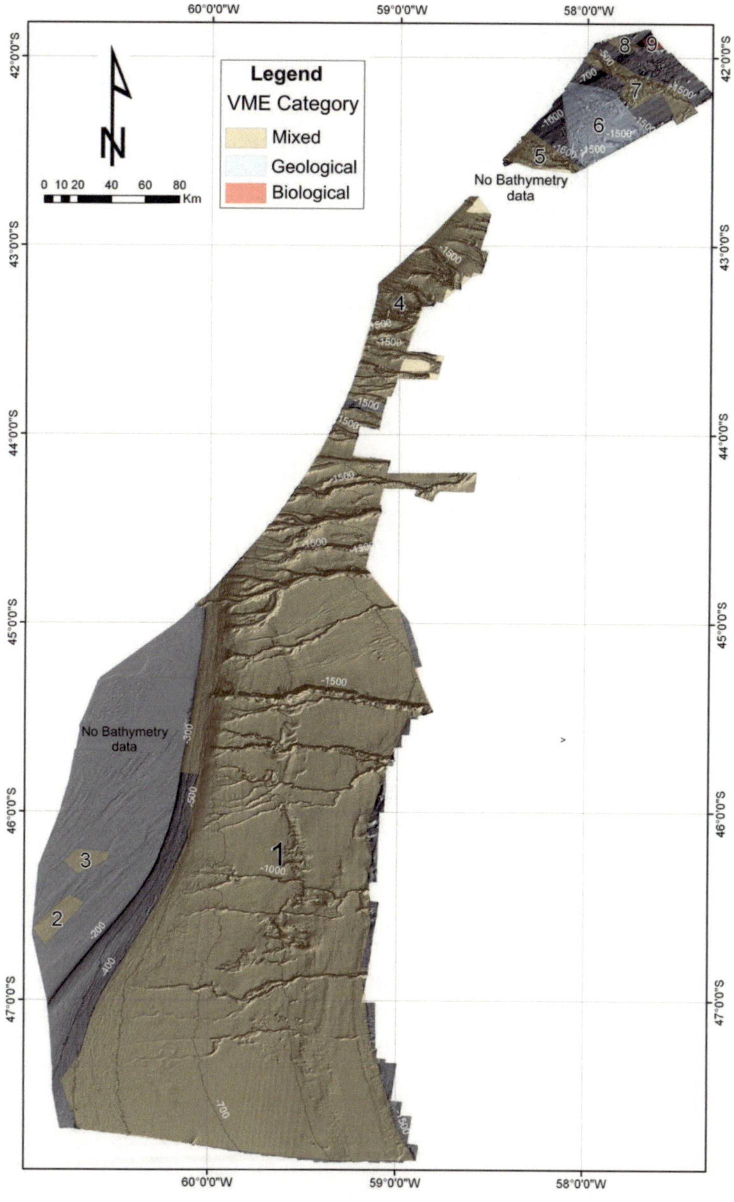

Fig. 17. Candidate sites for protected areas in the HS of SW Atlantic

5. Acknowledgment

We wish to thank the crew of the R/V "Miguel Oliver" (owned by the Spanish Secretariat of the Sea [SGM]) and her captain, for the professionalism and the courtesy extended towards us during the research cruises. We also wish to express our gratitude to all the people involved in the thirteen research surveys, namely the scientific and technical personnel who made this work possible.

6. References

Acosta, J., Muñoz, A., Cristobo, J.M., Druet, M., Uchupi, E. & Atlantis Group (2011, in press). Geomorphology and shallow structure of a segment of the Argentine Patagonian margin. *Marine Geology* (submitted)

Althaus, F., Williams, A., Schlacher, T.A., Kloser, R.J., Green, M.A., Barker, B.A., Bax, N.J., Brodie, P. & Schlacher-Hoenlinger, M.A. (2009). Impacts of bottom trawling on deep-coral ecosystems of seamounts are long-lasting. *Marine Ecology Progress Series* 397: pp. 279-294

Anon. (2008). Foro para la Conservación del Mar Patagónico y Áreas de Influencia (2008). *Estado de conservación del Mar Patagónico y áreas de influencia – versión electrónica.* [En línea]. Puerto Madryn, Argentina, Edición del Foro. Available from: www.marpatagonico.org

Beiras, R., Duran, I., Parra, S., Urrutia, M.B., Besada, V., Bellas, J., Viñas, L., Sánchez-Marín, P., González-Quijano, A., Franco, M.A., Nieto, O. & González, J.J. (2011). Linking chemical contamination to biological effects in coastal pollution monitoring. *Ecotoxicology* DOI 10.1007/s10646-011-0757-3

Bensch, A., Gianni, M., Gréboval, D., Sanders, J.S. & Hjort, A. (2008). Worldwide review of bottom fisheries in the high seas. *FAO Fisheries and Aquaculture Technical Paper.* No. 522. Rome, FAO. 145p

Bett, B.J. & Rice, A.L. (1992). The influence of hexactinellid sponge (*Pheronema carpenteri*) spicules on the patchy distribution of macrobenthos in the Porcupine Seabight (bathyal NE Atlantic). *Ophelia* 36 (3): 217-226

Bianchi, A., Masonneau, M. & Oliviera, R. (1982). Análisis estadístico de las características TS del sector austral de la plataforma continental argentina. *Acta Oceanográfica Argentina.* 3 (1). 93-118

Brandini, F.P., Boltovskoy, D., Piola, A., Kocmur, S., Rottgers, R., Abreu, P.C. & Lopes, R.M. (2000). Multiannual trends in fronts and distribution of nutrients and chlorophyll in the southwestern Atlantic (30-62 ° S). *Deep-Sea Research Part I*, 47, 1015-1033

Brasseur, P., Beckers, J.-M., Brankart, J.-M. & Schoenauen, R. (1996). Seasonal temperature and salinity fields in the Mediterranean Sea: Climatological analyses of an historical data set. *Deep-Sea Research*, 43, 159-192

Breiman, L. (2001). Random forests. Machine Learning, 45(1): 5–32 Buchanan, J.B. (1984). Sediment analysis. In: IBP Handbook No. 16. *Methods for the study of marine benthos.* Blackwell Scientific Publications, Oxford. pp. 41-65

Buchanan, J.B. (1984). Sediment analysis. In: IBP Handbook No. 16. *Methods for the study of marine benthos.* Blackwell Scientific Publications, Oxford. pp. 41-65

Busby, J.R. (1991). BIOCLIM - A Bioclimatic Analysis and Prediction System. In: Margules, C.R.& M.P. Austin (eds.) *Nature Conservation: Cost Effective Biological Surveys and Data Analysis.* pp. 64-68. Canberra: CSIRO

Cairns, S.D. (1982). Antarctic and Subantarctic scleractinia. Biology of the Antarctic Seas XI. *Antarctic Research Series*, vol 34; Paper I, pp 1-74

Cairns, S.D. (2007). Deep Water Corals: An overview with special reference to Diversity and Distribution of deep water scleractinian corals. *Bulletin of marine Science*, 81 (3); 311-322

Cairns, S.D. & Bayer, F.M. (2005). A review of the genus *Primnoa* (Octocorallia: Gorgonacea: Primnoidae), with the description of two new species. *Bulletin of Marine Science*, 72 (2); 225-256

Cairns, Stephen D. & Frederick M. Bayer (2009). A Generic Revision and Phylogenetic Analysis of the Primnoidae (Cnidaria: Octocorallia). *Smithsonian Contributions to Zoology*, number 629, iv + 79 pages, 19 figures, 4 tables

Chapman, P.M. & Wang, F. (2001). Assessing sediment contamination in estuaries. *Environmental Toxicology and Chemistry* 20: pp. 3-22

Clark, M.R. & Koslow, J.A. (2007). Impacts of fisheries on seamounts. Pp. 413–441 in *Seamounts: Ecology, Fisheries, and Conservation*. T.J. Pitcher, T. Morato, P.J.B. Hart, M.R. Clark, N. Haggan, and R.S. Santos (eds), Blackwell Fisheries and Aquatic Resources Series 12, Blackwell Publishing, Oxford, UK.

Clark, M.R. & Rowden, A.A. (2009). Effect of deepwater trawling on the macro-invertebrate assemblages of seamounts on the Chatham Rise, New Zealand. *Deep-Sea Research* I 56: pp. 1540–1554

Coggan, R., Populus, J., White, J., Sheehan, K., Fitzpatrick, F. & Piel, S. (eds) 2007. Review of Standards and Protocols for Seabed Habitat Mapping. *MESH*

Collie, J.S., Escanero, G.A. & Valentine, P.C. (2000a). Photographic evaluation of the impacts of bottom fishing on benthic epifauna. *ICES Journal of Marine Science*, 57: pp. 987–1001

Collie, J. S., Hall, S.J., Kaiser, M.J. & Poiner, I.R. (2000b). A quantitative analysis of fishing impacts on shelf-sea benthos. *Journal of animal ecology* (2000) 69, pp. 785-798

Dorschel, B., Hebbeln, D., Foubert, A., White, M. & Wheeler, A. (2007). Hydrodynamics and cold-water coral facies distribution related to recent sedimentary processes at Galway Mound west of Ireland. *Marine Geology*, 244, 184–195

Durán Muñoz, P., Murillo, F.J., Serrano, A., Sayago-Gil, M., Parra, S., Díaz Del Río, V., Sacau, M., Patrocinio, T. & Cristobo, J. (2008). A Case Study of available methodology for the identification of Vulnerable Ecosystems/Habitats in bottom deep-sea fisheries: Possibilities to apply this method in the NAFO Regulatory Area in order to select Marine Protected Areas. *NAFO SCR Doc*. 08/06 Serial No. N5491. 20 pp

Durán Muñoz, P., Sayago-Gil, M., Cristobo, J., Parra, S., Serrano, A., Díaz Del Río, V., Patrocinio, T., Murillo, F.J.; Palomino, D. & Fernández-Salas, L.M. (2009). Seabed mapping for selecting cold-water coral protection areas on Hatton Bank, Northeast Atlantic. – *ICES Journal of Marine Science*, 66: pp. 2013–2025

Etnoyer, P. & Morgan, L.E. (2003). Occurrences of habitat forming deep-sea corals in the Northeast Pacific Ocean. NOAA Office Habitat Protection, Silver Spring, MD

Etnoyer, P. & Morgan, L.E. (2005). Habitat-forming deep-sea corals in the Northeast Pacific Ocean. Freiwald A, Roberts JM (eds), 2005, *Cold-water Corals and Ecosystems*. Springer-Verlag Berlin Heidelberg, pp 331-343

FAO (2008). Technical consultation on international guidelines for the management of deep-sea fisheries in the high seas. Rome (Italy), 4-8 February 2008. TC: DSF/2008/Inf. 3. 33 pp

FAO (2009). International Guidelines for the Management of Deep-Sea Fisheries in the High Seas. Food and Agricultural Organisation of the United Nations, Rome, Italy, 73pp. Available from: http://www.fao.org/docrep/011/i0816t/i0816t00.htm

Forbes, M.C. & Garrafo, Z. (1988). A note on the mean seasonal transport on the Argentinian shelf. *Journal of Geophisical Research*, 93:2311–2319

Freese, L., Auster, P.J., Heifetz, J. & Wing, B.L. (1999) Effects of trawling on seafloor habitat and associated invertebrate taxa in the Gulf of Alaska. Marine Ecology Progress Series 182: 119–126

Freiwald, A. (2011) Cold-Water Coral Reefs. In: Hopley, D. (Ed), Encyclopedia of Modern Coral Reefs: Structure, form and process. *Springer-Verlag Earth Science Series*, p. 225-229. DOI 10.1007/978-90-481-2639-2

Gordon, A.L. & Greengrove, C.L. (1986). Geostrophic circulation of the Brazil-Falkland Confluence. *Deep-Sea Research*, 33, 573-585.

Harvey, R.G. (1997). Polycyclic Aromatic Hydrocarbons. Wiley-VCH, Inc, New York, USA

Hernández-Molina, F.J., Paterlini, M., Violante, R., Marshall, P., Isasi, M., de Somoza, L. & Rebesco, M. (2009). Contourite depositional system on the Argentine Slope: an exceptional record of the influence of the Antarctic water masses. *Geology* 37: 507-510

Hernández-Molina, F.J., Paterlini, M., Somoza, L., Violante, R., Arecco, M.A., de Isai, M., Rebesco, M., Uenzelmann-Neben, G., Neben, S. & Marshall, P. (2010). Giant mounded drifts in the Argentine continental margin: origins, and global implications for the history of thermohaline circulation. *Marine and Petroleum Geology* 27: 1508-1530

ICES (2003). Environmental status of the European Seas. A quality status report prepared by the International Council for the Exploration of the Sea, Copenhagen. Available from: http://www.ices.dk/reports/germanqsr/23222_ICES_Report_samme.pdf

Kaiser, M.J. (1998). Significance of Bottom-Fishing Disturbance. *Conservation Biology*, pp. 1230–1235. Volume 12, No. 6, (December 1998)

Kaiser, M.J., Edwards, D.B., Armstrong, P.J., Radford, K., Lough, N.E.L., Flatt, R.P. & Jones H.L. (1998). Changes in megafaunal benthic communities in different habitats after trawling disturbance. *ICES Journal of Marine Science* 55: pp. 353–362

Kaiser, M.J., Clarke, K.R., Hinz, H., Austen, M.C.V., Somerfield, I.P. & Karakassis, J. (2006). Global analysis of response and recovery of benthic biota to fishing. *Marine Ecology Progress Series* 311: 1–14, 2006

Kithara, M.V., Capítoli, R.R. & Horn Filho, Norberto, O. (2009). Distribuição das espécies de corais azooxantelados na plataforma e talude continental superior do Sul do Brazil. *Iheringia, Sér. Zool*, Porto Alegre, 99 (3): 223-236

Koslow, J.A., Gowlett-Holmes, K., Lowry, J.K., O'Hara, T.O., Poore, G.C.B. & Williams, A. (2001). Seamount benthic macrofauna off southern Tasmania: community structure and impacts of trawling. *Marine Ecology Progress Series*, 213: 111–125

Kumar, K.S., Sajwan, K.S., Richardson, J.P. & Kannan, K. (2008). Contamination profiles of heavy metals, organochlorine pesticides, polycyclic aromatic hydrocarbons and alkylphenols in sediment and oyster collected from marsh/estuarine Savannah GA, USA. *Marine Pollution Bulletin* 56: pp.136-149

Lastras, G., Acosta, J., Muñoz, A. & Canals, M., (2011). Submarine canyon formation and evolution in the Argentine Continental Margin between 44°30'S and 48°S. *Geomorphology*, vol 128, pp 116-136

Legeckis, R. & Gordon, A.L. (1982). Satellite observations of the Brazil and Falkland currents, 1975 to 1976 and 1978. *Deep-Sea Research* 29, 375-401

Long, E.R., Field, L.J. & MacDonald, D.D. (1998). Predicting toxicity in marine sediments with numerical sediment quality guidelines. *Environmental Science and Technology* 17 (4):714-727

López-Martinez, J., Muñoz, A., Dowdeswell, J.A., Linés, C. & Acosta, J. (2011). Relict sea-floor ploughmarks records deep-keeled Antarctic icebergs to 45°S on the Argentine margin. *Marine Geology*, 288, 43-48

Lusquiños, A.J. & Schrott, A.G. (1983). Corrientes en el Mar Epicontinental argentino en invierno. Technical Report 74. Subsecretaría de Estado de Ciencia y Técnica. Programa Nacional de Recursos Naturales Renovables, Buenos Aires, 74 pp

Mortensen, P.B., Buhl-Mortensen, .L, Gordon, D.C., Fader, G.B.J., McKeown, D.L. & Fenton, D.G. (2005). Effects of fisheries on deep-water gorgonian corals in the Northeast Channel, Nova Scotia (Canada). *Am. Fish. Soc. Symp.* 41:369–382

Muñoz, A., Cristobo, J., Ríos P., Druet, M., Polonio, V., Uchupi, E., Acosta, J. & Atlantis Group (2011), in press. Sediment Drifts and Cold-Water Coral Reefs in the Patagonian Upper and Middle Continental Slope. *Marine and Petroleum Geology* (submitted)

Murillo, F. J., Durán Muñoz, P., Altuna, A. & Serrano, A. (2010). Distribution of deep-water corals of the Flemish Cap, Flemish Pass, and the Grand Banks of Newfoundland (Northwest Atlantic Ocean): interaction with fishing activities. *ICES Journal of Marine Science*, 68 (2): pp. 319-332

Orensanz, J.M., Bogazzi, E. & Parma, A.M. (2008). Impacto de la pesca sobre el subsistema bentónico. In: Foro para la Conservación del Mar Patagónico y Áreas de Influencia (2008). *Estado de conservación del Mar Patagónico y áreas de influencia - versión electrónica.* [En línea]. Puerto Madryn, Argentina, Edición del Foro. Available from: www.marpatagonico.org

Peterson, R.G. (1992). The Boundary Currents in the western Argentine Basin. *Deep-Sea Research* 39, 623-644

Piola, A.R. & Gordon, A.R. (1989). Intermediate waters in southest south atlantic. *Deep-Sea Research*, 36:1–16

Piola, A.R. & Rivas, A.L. (1997). Corrientes en la plataforma continental. In: *El Mar Argentino y sus Recursos Pesqueros*. Tomo 1. Antecedentes Históricos de las Exploraciones en el mar y las Características Ambientales. Boschi, E.E. (Ed.), 119-132, Instituto Nacional de Investigación y Desarrollo Pesquero (INIDEP), Mar de Plata

Piola, A.R., (2008). Oceanografía física. In: Foro para la Conservación del Mar Patagónico y Áreas de Influencia (2008). *Estado de conservación del Mar Patagónico y áreas de influencia - versión electrónica*. [En línea]. Puerto Madryn, Argentina, Edición del Foro. Available from: www.marpatagonico.org

Portela J.M., Pierce G.J., del Río J.L., Sacau M., Patrocinio T. & Vilela R. (2010). Preliminary description of the overlap between squid fisheries and VMEs on the high seas of the Patagonian Shelf. *Fisheries Research*, Vol.106, No.2 (November 2010), pp. 229–238

Reed, J.K. (2002). Deep-water *Oculina* coral reefs of Florida: biology, impacts, and management. In: *Biology of Cold Water Corals*, Watling, L. & Risk, M. (eds), *Hydrobiologia* 471: 43–55, Kluwer Academic Publishers

Reed, J.K., Shepard, A.N., Koenig, C.C., Scanlon, K. M. & Gilmore, R. G. Jr. (2005). Mapping, habitat characterization, and fish surveys of the deep-water *Oculina* coral reef Marine Protected Area: a review of historical and current research. In: *Cold-Water Corals and Ecosystems*, Freiwald, A., Roberts, J.M. (eds), 443–465, Springer-Verlag, Berlin Heidelberg

Ríos, P., Kelly, M. & Vacelet, J. (2011) *Cercicladia*, a new genus of carnivorous sponge without chelae, from Macquarie Ridge, southern New Zealand, and the Patagonian margin, south Atlantic (Porifera, Cladorhizidae). *Zootaxa* 3131: 52–62

Rogers, A.D. (1999). The biology of *Lophelia pertusa* (Linnaeus 1758) and other deep-water reef forming corals and impacts from human activities. *Int. Rev. Hydrobiol.* 84:315- 406

Rogers, A.D. & Gianni, M. (2010). The Implementation of UNGA Resolutions 61/105 and 64/72 in the Management of Deep-Sea Fisheries on the High Seas. Report prepared for the Deep-Sea Conservation Coalition. International Programme on the State of the Ocean, London, United Kingdom, 97pp. Available from : http://www.stateoftheocean.org/pdfs/61105-Implemention-finalreport.pdf

Rowlatt, S.M. (1995). Results of the 1990/1991 baseline study of contaminants in the North Sea sediments. *ICES Cooperative Research Report*, 208

Schlitzer, R. (2011). Ocean Data View (ODV). Available from : http://odv.awi.de/

Silverman, B.W. (1986). Density Estimation for Statistics and Data Analysis. New York : Chapman and Hall. London.

Stone, R.P. (2006). Coral habitat in the Aleutian Islands of Alaska: depth distribution, fine-scale species associations, and fisheries interactions. *Coral Reefs* 25: 229–238

Troupin, C., Beckers, J.M., Ouberdous, M. & Sirjacobs, J. (2008). Diva User's Guide. Available from http://modb.oce.ulg.ac.be

UNGA (2004). Resolution 59/25 Sustainable fisheries, including through the 1995 Agreement for the Implementation of the Provisions of the United Nations Convention on the Law of the Sea of 10 December 1982 relating to the Conservation and Management of Straddling Fish Stocks and Highly Migratory Fish Stocks, and related instruments. UNGA A/RES/59/25. Available from: http://www.un.org/Depts/los/general_assembly/general_assembly_reports. htm, 16pp

UNGA (2007). Resolution 61/105 Sustainable fisheries, including through the 1995 Agreement for the Implementation of the Provisions of the United Nations Convention on the Law of the Sea of 10 December 1982 relating to the Conservation and Management of Straddling Fish Stocks and Highly Migratory Fish Stocks, and related instruments. UNGA A/RES/61/105. Available from: http://www.un.org/Depts/los/general_assembly/general_assembly_reports. htm, 21pp

UNGA (2009). Resolution 64/72. Sustainable fisheries, including through the 1995 Agreement for the Implementation of the Provisions of the United Nations Convention on the Law of the Sea of 10 December 1982 relating to the Conservation and Management of Straddling Fish Stocks and Highly Migratory Fish Stocks, and related instruments. UNGA A/RES/64/72. Available as General Assembly document A/64/L.29. Available from: http://daccess-dds-ny.un.org/doc/UNDOC/GEN/N09/466/15/PDF/ N0946615.pdf?OpenElement, 26pp

Viñas, L., Franco M.Á. & González, J.J. (2002). Distribution of PAHs in surficial sediments of the Vigo estuary, Spain.Central axis and adjacent shelf. *Polycyclic Aromatic Compounds* 22 (2):161-173

Weaver, P.P.E., Benn, A., Arana, P.M., Ardron, J.A., Bailey, D.M., Baker, K., Billett, D.S.M., Clark, M.R., Davies, A.J., Durán Muñoz, P., Fuller, S.D., Gianni, M., Grehan, A.J., Guinotte, J., Kenny, A., Koslow, J.A., Morato, T., Penney, A.J., Perez, J.A.A., Priede, I.G., Rogers, A.D., Santos, R.S. & Watling, L. (2011). The impact of deep-sea fisheries and implementation of the UNGA Resolutions 61/105 and 64/72. Report of an international scientific workshop, National Oceanography Centre, Southampton, 45 pp. Available from: http://hdl.handle.net/10013/epic.37995

Zweig, M.H. & Campbell, G. (1993). Receiver-operating characteristic (ROC) plots: a fundamental evaluation tool in clinical medicine. Clin. Chem. 39, 561–577

10

Hydrocarbon Contamination and the Swimming Behavior of the Estuarine Copepod *Eurytemora affinis*

Laurent Seuront

[1]*School of Biological Sciences, Flinders University,*
[2]*South Australian Research and Development Institute, Aquatic Sciences,*
[3]*National Center of Scientific Research, UMR LOG 8187,*
[4]*Center for Polymer Studies, Department of Physics, Boston University,*
[1,2]*Australia*
[3]*France*
[4]*USA*

1. Introduction

Rivers and estuaries have allowed and sustained human settlement and related activities for thousands of years (e.g. Morhange et al., 2005). These water bodies have been used for activities such as irrigation, industry, transportation, fisheries, tourism and the related development of industrial and recreational facilities, and have subsequently been polluted by waste discharges, intentional and accidental spills, urban, industrial and agricultural runoffs (McLusky and Elliott, 2004; Shannon et al., 2011). These various sources of pollution all carry a wide range of inorganic and organic pollutants such as polycyclic aromatic and monoaromatic hydrocarbons (Vane et al., 2011), heavy metals (Jose et al., 2011), radioactive compounds (Villa et al., 2011) and a range of pharmaceuticals and pesticides (Arias et al., 2011; Munaron et al., 2011) that accumulate in water, sediment and living organisms (Whaltham et al., 2011; Schnitzer et al., 2011; Barua et al., 2011). Consequences have been observed from the molecular to the ecosystem level through e.g. immunological, cellular, reproductive and developmental impairments, and teratogenic effects that have widely been reported for a wide range of aquatic species (Noaksson et al., 2005; Thompson et al., 2007; Galante-Oliveira et al., 2009; Yang et al., 2010).

Changes in motion behavior, as a response to exposure to organic or inorganic pollutants, have been observed in a range of aquatic invertebrates such as *Artemia salina* (Venkateswara Rao et al., 2007), *Balanus amphitrite* (Faimali et al., 2002, 2006; Amsler et al., 2006), *Brachionus calyciflorus* (Janssen et al., 1994; Charoy et al., 1995; Charoy & Janssen, 1999), *Chironomus* sp. (Gerhadt & Janssens de Bisthoven, 1995; Janssens de Bisthoven et al., 2004), *Choroterpes picteti* (Macedo-Sousa et al., 2008), *Corophium volutator* (Kirkpatrick et al., 2006a; Kienle & Gerhardt, 2008), *Crangonyx pseudogracilis* (Kirkpatrick et al., 2006b), *Daphnia magna* (Bailleul & Blust, 1999; Shimizu et al., 2002; Understeiner et al., 2003; Goto & Hiromi, 2003; Gerhadt et al., 2005; Ren et al., 2007, 2008; Ren & Wang, 2010; Duquesne & Küster, 2010),

Echinogammarus meridionalis (Macedo-Sousa et al., 2007, 2008), *Gammarus fossarum* (Xuereb et al., 2009a,b), *Gammarus pulex* (Gerhadt, 1995; Gerhadt et al., 2007), *Hippolyte inermis* (Untersteiner et al., 2005), *Hydropsyche pellucidula* (Macedo-Sousa et al., 2008), but still rarely on copepods (Sullivan et al., 1983; Seuront & Leterme, 2007; Seuront, 2010a,b, 2011a,b; Cailleaud et al., 2011). Despite the central role played by copepods in aquatic ecosystems (Schmitz, 2008; Matthews et al., 2011), the amount of work devoted to copepod chemoreception (e.g. Doall et al., 1998; Bagøien & Kiørboe, 2005; Goetze & Kiørboe, 2008; Yen et al., 2011), and recent evidence for copepods to modify their swimming behavior in response to exposure to hydrocarbon compounds (Seuront & Leterme, 2007; Seuront, 2010a,b, 2011a,b) and 4-nonylphenol and nonylphenol-ethoxy-acetic-acid (Cailleaud et al., 2011), little is still known on the potential for chemical contaminants to affect copepod swimming behavior.

The quantitative assessment of changes in copepod swimming behavior is, however, critical as it is the main driver of encounter rate probability (Visser, 2007; Kiørboe, 2008; Seuront, 2011a), which in turn controls key processes such as mating and feeding rates, and predator avoidance, hence individual fitness and population dynamics. As a consequence, behavioral changes may be used as important indicators for ecosystem health. While they are driven by biochemical processes, they also reflect the fitness of individual organisms as well as potential consequences at the population level, such as altered abundance of a species in a ecosystem. Behavioral responses to water contamination, or more generally changes in water properties, have been shown to be a sensitive non-invasive sub-lethal end-point with short-response times for toxicity bioassays (e.g. Cailleaud et al., 2011; Seuront, 2011a,b), compared to community-related measures which require changes in species composition before an impact is detected. In ecotoxicology, behavioural approaches allow for repeated measures and time-dependent data analysis, and have the advantage of being of similar sensitivity and efficiency than biochemical and physiological responses (Gerhadt, 2007, 2011), and more sensitive than mortality responses (Garaventa et al., 2010).

In the drastic instance of oil spills that particularly threaten coastal waters (Varela et al., 2006; Lee et al., 2009; Rumney et al., 2011), previous studies have shown acute and deleterious effects in both meroplanktonic (Fisher & Foss, 1993; Epstein et al., 2000; Shafir et al., 2003) and holoplanktonic organisms (Samain et al., 1980; Jernelov et al., 1981; Cowles & Remillard, 1983a,b; Guzmán del Próo et al., 1986; Tawfiq & Olsen, 1993; Pavillon et al., 2002; Chen & Denison, 2011), which ultimately lead to biomass decrease and structure change at the community level. The identification and assessment of water contamination is, however, far more complex in situations where zooplankton communities are exposed to sub-lethal concentrations of the water-soluble fraction of pollutants that affect zooplankton physiology, feeding and fecundity as most marine bioassays still rely on exposure times of 24 to 48-h to determine the concentration of a test chemical at which 50% of neonates die or are immobilized; or the number of individuals that died (e.g. Barata et al. 2002; Calbet et al. 2007).

In this context, the objectives of this chapter were (i) to assess the ability of *E. affinis* adult males and females to detect and avoid patches of contaminated water, (ii) to evaluate whether their three-dimensional swimming behavior is affected by hydrocarbon contamination of estuarine waters and (iii) to illustrate the ability of fractal analysis (i.e. the fractal properties of three-dimensional swimming paths and the cumulative probability distribution function of move lengths) to detect the stress potentially induced by hydrocarbon contamination. Note that a specific attention has been given to low concentrations of the water-soluble fraction of diesel oil (0.01%, 0.1% and 1%), that are shown to be well below the lethal concentration for *E. affinis*, to assess the impact of a chronic exposure to low concentrations of petroleum hydrocarbons.

2. Methods

2.1 Study species, sampling and acclimatization

Eurytemora affinis is one of the most abundant zooplankton species in the brackish part of Northern Hemisphere estuaries, usually localized around the Maximum Turbidity Zone (Soetaert & Van Rijswijk, 1993), and plays a significant role in estuarine food webs as an important food supply for many fishes, shrimps and mysids (Fockedey & Mees, 1999). *E. affinis* individuals were collected from the Seine estuary using a WP2 net (200-µm mesh size) at a temperature of 15°C in the low salinity zone (S=4 PSU) at low tide near the 'Pont de Normandie' (49°28'26N, 0°27'47W). Specimens were gently diluted in 30-litre isotherm tanks using *in situ* estuarine water and transported to the laboratory where adult males and both non-ovigerous and ovigerous females were immediately sorted by pipette under a dissecting microscope, and kept separately in 20-liter aquaria filled with filtered (Whatman GF/C glass-fibre filters, porosity 0.45µm) *in situ* estuarine water for 24-h until the behavioral experiments took place.

2.2 Hydrocarbon contaminant: The water-soluble fraction of diesel oil

The product considered as a potential contaminant of coastal waters was commercial diesel fuel oil. The water-soluble fraction of commercial diesel oil (WSF) was prepared stirring 1.8 l of filtered *in situ* seawater (Whatman GF/C filters) with 0.2 l of commercial diesel fuel oil for 2 h at 100g. The mixed solution was allowed to stand for 24 h without stirring to separate the oil layer from the oil-saturated water. WSF stock solutions were siphoned into autoclaved, acid-rinsed glass containers and diluted with uncontaminated seawater at 'high' (1%), 'medium' (0.1%) and 'low' (0.01%) concentrations. The water-soluble fraction of oil and their derivatives products contain a mixture of polycyclic aromatic hydrocarbons (PAH), monoaromatic hydrocarbons often referred to as BTEX (benzene, toluene, ethylbenzene and xylenes), phenols and heterocyclic compounds, containing nitrogen and sulphur (Saeed & Al-Mutairi, 1999; Elordui-Zapatarietxe et al., 2008; Rodrigues et al., 2010).

Technical limitations hampered the assessment of the precise chemical nature of the WSF stock solutions. The range of WSF concentrations used in the present work has, however, specifically been chosen to investigate the sub-lethal effects related to natural background concentrations of pollutants (Ohwada et al., 2003; Hashim, 2010). More specifically, among those compounds, BTEX are the main class of hydrocarbons found in WSF (Carls & Rice, 1990; Saeed & Al-Mutairi, 1999), and naphthalene is one of the most abundant polycyclic aromatic hydrocarbons dissolved in oil contaminated waters (Corner et al., 1976) and has been widely used in toxicological assays (Corner et al., 1976; Berdugo et al., 1977; Harris et al., 1977; Calbet et al., 2007). BTEX and naphthalene concentrations are respectively in the range 450-35000 $\mu g\ l^{-1}$ and 30-26000 $\mu g\ l^{-1}$ in 100% water-soluble fraction (Saeed & Al-Mutairi, 1999; Rodrigues et al., 2010). The 'high' (1%), 'medium' (0.1%) and 'low' (0.01%)

concentrations used in the present work hence correspond to concentrations in the range 4.5-350 μg l^{-1}, 0.45-35 μg l^{-1} and 0.045-3.5 μg l^{-1} for BTEX, and 0.3-260 μg l^{-1}, 0.03-26 μg l^{-1} and 0.003-2.6 μg l^{-1} for naphthalene.

2.3 Acute responses (mortality and narcosis) to WSF contamination

The water-soluble fraction concentrations considered here are well below the lethal concentrations observed for a range of copepod species (Barata et al., 2002; Calbet et al., 2007; Seuront & Leterme, 2007; Seuront, 2011a). As to my knowledge, no information is available on the effects of WSF on *Eurytemora affinis*, toxicity assays were conducted to assess the acute responses (mortality and narcosis) produced by the water-soluble fraction of commercial diesel oil. Acute responses (mortality and narcosis) to WSF of the copepod *E. affinis* were investigated by 24-h and 48-h incubations at 15°C (12/12 light/dark cycle) in contaminated 0.2 μm filtered estuarine water. Groups of ten adult *E. affinis* males and non-ovigerous females were placed separately in 1-liter Pyrex glass bottles (Schott) sealed with a Teflon screw lid and filled with the appropriate test solution. The effect of the water-soluble fraction of diesel oil on *E. affinis* mortality and narcosis was inferred for a range of concentrations (50, 25, 10, 5, 1, 0.1 and 0.01%). Each treatment was triplicated, and triplicate uncontaminated control bottles were used to assess baseline mortality. At the end of the incubations, copepods were sieved through a 200 μm nylon mesh, washed with filtered estuarine water, and transferred into Petri dishes, where their activity was monitored using a stereomicroscope. To discriminate mortality from narcotization, copepods were examined after a period of 4-h in uncontaminated filtered estuarine water to assess the degree of recovery (Berdugo et al., 1977; Calbet et al., 2007). The lethal concentration $LC50$ (concentration at which 50% of the specimens died) was subsequently estimated from the nonlinear allosteric decay of the survival rate S following:

$$S = S_{max} LC50^{\alpha} / (C_{WSF}^{\alpha} + LC50^{\alpha}) \quad (1)$$

where S_{max} is the maximum survival rate (%), C_{WSF} the experimental WSF concentration (50%, 25%, 10%, 5%, 1%, 0.1% and 0.01%) and α a fitting parameter (Barata et al., 2002).

The survival responses of *E. affinis* males and non-ovigerous females were highly significantly fitted by Eq. (1) (Fig. 1). Specifically, after a 24-h exposure, the survival rates of both males and non-ovigerous females (Fig. 1a,b) were very high, with $LC50_{male} = 18.9\%$ and $LC50_{female} = 20.0\%$. In contrast, after a 48-h exposure, the survival rates of both males and females decay much faster with increasing WSF concentration (Fig. 1c,d), with $LC50_{male} = 6.5\%$ and $LC50_{female} = 7.5\%$. No significant differences were found in $LC50$ between males and females ($P > 0.05$). In addition, no narcotic effects were observed; the mortality and narcosis were very similar, i.e. $LC50_{mort} = 6.5\%$ and $LC50_{narc} = 6.3\%$ for males, and $LC50_{mort} = 7.5\%$ and $LC50_{narc} = 7.7\%$ for females. Note that the lethal concentration $LC50$ estimated here are substantially higher than the $LC50$ obtained for *Temora longicornis* adult females after 24-h ($LC50_{24} = 1.5\%$) and 48-h ($LC50_{48} = 1.3\%$) toxicity assays conducted over the same range of WSF concentrations; see Seuront (2011a), his Figure 1. While further work is needed to infer the origin of the observed differences in the lethal concentration of *E. affinis* and *T. longicornis*, it is likely to be related to the highly polluted nature of the estuarine environment where *E. affinis* proliferate. The behavioral properties of ovigerous females are not considered in the present work. It is nevertheless stressed that in contrast to adult males and non-ovigerous females, a clear narcotic effect was observed, with $LC50_{mort} = 25.2\%$ and $LC50_{narc} = 12.0\%$ after a 24-h exposure and $LC50_{mort} = 8.1\%$ and $LC50_{narc} = 4.0\%$ after a 48-h exposure. The behavioral experiments described below were conducted with the same WSF stock solutions than the above-mentioned toxicity assays for mortality and narcosis.

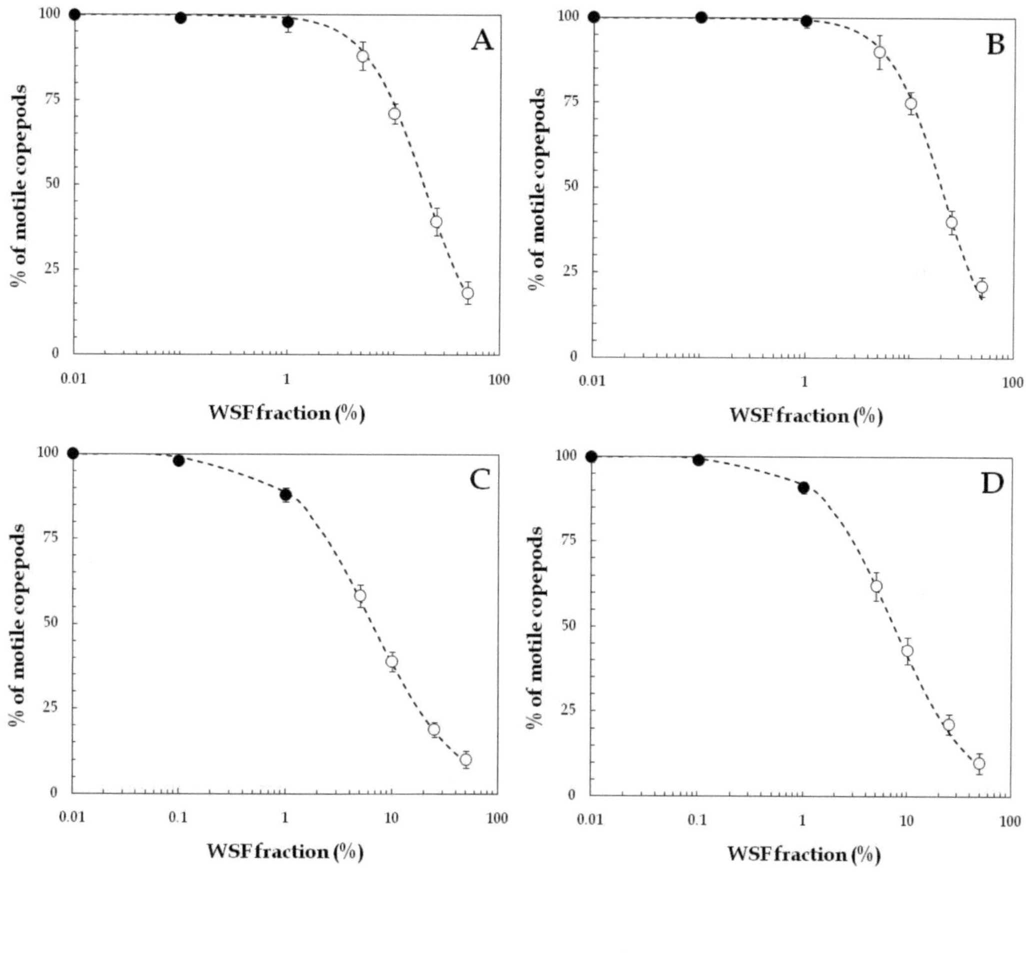

Fig. 1. Survival response of *Eurytemora affinis* adult males (A,C) and non-ovigerous females (B,D) to a range of WSF concentrations (50%, 25%, 10%, 5%, 1%, 0.1% and 0.01%) over 24-h (A, B) and 48-h (C,D) toxicity assays. The dashed lines are the best fits of the allosteric decay model, see Eq. (1). Error bars are standard deviations. The black dots correspond to the WSF concentrations used in the behavioral experiments.

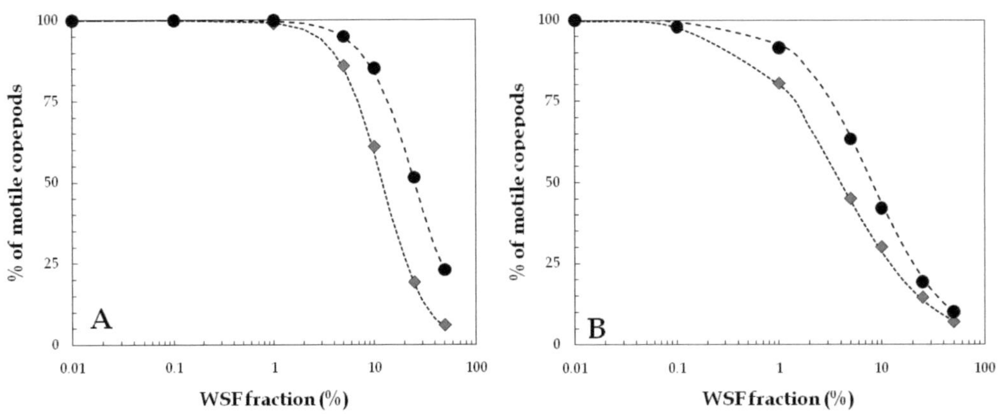

Fig. 2. Survival (black dots) and narcotization (grey diamonds) responses of *Eurytemora affinis* ovigerous females to a range of WSF concentrations (50%, 25%, 10%, 5%, 1%, 0.1% and 0.01%) over 24-h (A) and 48-h (B) toxicity assays. The dashed lines are the best fits of the allosteric decay model, see Eq. (1).

2.4 Behavioral experiments

Two behavioral assays were developed to assess the impact of the water soluble fraction of diesel oil on the swimming behavior of *Eurytemora affinis* adult males and non-ovigerous females. The first one was based on the ability of *E. affinis* to detect and avoid local patches of contaminated water. The second behavioral assay evaluates whether the three-dimensional swimming behavior of *E. affinis* is affected by contaminated water and demonstrates that fractal analyses based on either the geometric properties of their swimming paths or the cumulative probability distribution of their move lengths can detect the stress potentially induced by hydrocarbon contamination. Prosome lengths of males and females used during the behavioral experiments respectively ranged between 0.83 and 0.86 mm, and 0.87 and 0.92 mm.

2.4.1 Patch avoidance experiments

The first behavioral assay developed in the present work is based on the ability of *E. affinis* adult males and non-ovigerous females to avoid a patch of WSF contaminated estuarine water. Behavioral experiments were conducted in a 3.375-liter (15 × 15 × 15 cm) glass chamber. Patches of contaminated water were created in the centre of the chamber as a contaminant point source by dispensing contaminated fluid from a fine pipette (Eppendorf) to flow down into a 5-mm diameter permeable sphere (20-µm pore size) that was suspended in the experimental container. Patches were created through the slow injection (<0.5 ml min^{-1}) of contaminated water in the porous sphere. Prior to the behavioral experiments, the amount of contaminated water to be injected to create patches of different diameters (i.e. 1, 2, 4, 5.5 and 7 cm) was determined through the injection of fluorescein-stained GF/C filtered and autoclaved estuarine water into the experimental container. Note that the patches created were spherical and characterized by an isotropic distribution of contaminant across and a non-significant increase in size due to molecular diffusion for behavioral experiments lasting up to a hour; see Seuront (2010a) for more details.

For each WSF treatment, behavioral experiments were conducted in triplicate for each patch size (1, 2, 4, 5.5 and 7 cm), and an equivalent amount of uncontaminated estuarine water water was injected into the experimental chambers for control observations (Seuront, 2010a). Preliminary experiments conducted with and without the injecting porous sphere did not exhibit any significant difference in the occurrence of *E. affinis* males and non-ovigerous females in the middle of the experimental chamber (χ^2-test, $P > 0.05$); this is consistent with previous observations conducted on *E. affinis* non-ovigerous adult females and *Temora longicornis* adult females (Seuront, 2010a). Before each behavioral experiment, 30 individuals of either *E. affinis* non-ovigerous adult females or adult males were transferred to the experimental filming set-up, filled with uncontaminated estuarine water and the corresponding contaminated patch, and allowed to acclimatize for 15 min (Seuront, 2006, 2010a). A new group of 30 individuals was used for each treatment.

2.4.2 Behavioral response to hydrocarbon contaminated estuarine water

These experiments were specifically designed to study the detailed properties of the swimming behavior of *E. affinis* adult males and non-ovigerous females under conditions of increasing WSF contaminations, and in the absence of the local gradients related to point source contamination. These experiments were also conducted in a 3.375-liter (15 × 15 × 15 cm) glass chamber at 5 PSU under control conditions (uncontaminated estuarine water) and under conditions of contamination by the water-soluble fraction of diesel oil (WSF). For each treatment, the water was contaminated with 'high' (1%), 'medium' (0.1%) and 'low' (0.01%) WSF concentrations. Before each behavioral experiment, 30 individuals of either *E. affinis* non-ovigerous adult females or adult males were transferred into the experimental vessel filled with control or WSF contaminated estuarine water, and allowed to acclimatize for 15 min (Seuront, 2006). A new group of 30 individuals was used for each treatment.

2.5 Behavioral observations and image analysis

All behavioral experiments were replicated three times. WSF stock and working solutions were prepared 24 h before the behavioral experiments took place. The control experiments and the treatments were randomized, as well as the replicates of each experimental condition. The experimental chamber was rinsed with acetone, GF/C filtered estuarine water and distilled water, and allowed to dry between trials to remove any chemical compound. Prior to each experiment, the experimental individuals were transferred in a filming set-up filled up with uncontaminated seawater and the corresponding treatments, and were allowed to acclimatize for 15 min (Seuront, 2006). All experimental individuals were used only once, and no narcosis or mortality was ever observed on any of the tested individuals. The three-dimensional trajectories of *E. affinis* adult males and non-ovigerous females were recorded at a rate of 25 frames s^{-1} using two orthogonally oriented and synchronized infrared digital cameras (DV Sony DCR-PC120E) facing the experimental chamber. Six arrays of 72 infrared light-emitting diodes (LEDs), each mounted on a printed circuit board about the size of a business card (i.e. 9.3 cm long × 4.9 cm wide) connected to a 12-V DC power supply, provided the only light source from the bottom of the chamber. The cameras overlooked the experimental chamber from the side, and the various components of the set-up were adjusted so that the copepods were adequately resolved and in focus. The two cameras represented the $x-z$ and the $y-z$ planes of the experimental chamber; 3-D swimming paths were obtained by combining information from the 2-D views. All the experiments were conducted in a temperature-controlled room at 18°C in the dark and at night to avoid any potential behavioral artifact related to the diel cycle of the copepods (Seuront, 2011b). Selected video clips were captured (DVgate Plus) as MPEG movies and converted into QuickTime TM movies (QuickTime Pro), after which the x, y and z coordinates of swimming pathways were automatically extracted and subsequently combined into a 3D picture using LabTrack software (DiMedia, Kvistgård, Denmark). The time step was always 0.04 s, and output sequences of (x,y,z) coordinates were subsequently used to characterize the motion behavior.

2.6 Behavioral analysis

2.6.1 Swimming speed

The swimming speed v (mm s^{-1}) over consecutive tracking intervals was estimated as $v = f \times d_t$, where f is the sampling rate of the camera (f = 25 frame s^{-1}), and d_t the distance (mm) between two points in a three-dimensional space. The distance d_t (mm) was computed from the (x, y, z) coordinates as $d_t = [(x_{t+1} - x_t)^2 + (y_{t+1} - y_t)^2 + (z_{t+1} - z_t)^2]^{1/2}$, where (x_t, y_t, z_t) and $(x_{t+1}, y_{t+1}, z_{t+1})$ are the positions of a copepod at time t and $t + 1$, respectively. Average swimming speed and their standard deviations were measured over the duration of each individual track.

2.6.2 Patch avoidance and patch escape behavior

The avoidance response of *E. affinis* adult males and non-ovigerous females to patches of contaminated water was expressed by both their patch avoidance and escape behaviors, i.e. the percentage of individuals that respectively avoided contaminated patches and escaped patches after entering them (Seuront, 2010a). Avoidance was identified as a sharp change in the direction of travel or the combination of a sharp change in swimming direction and an increase in swimming speed. Patch avoidance behavior was quantified by the distance d_a at which individuals avoided contaminated patches, the avoidance turning angle $\alpha_{i,j}$ and avoidance velocity $u_{i,j}$ for a contaminated patch of diameter i ($i = 1$, 2, 4, 5.5 and 7 cm) and concentration j ($j = 0.01\%$, 0.1% or 1%). Patch escape behavior was quantified by the escape turning angle $\beta_{i,j}$ and escape velocity $v_{i,j}$ for a contaminated patch of diameter i and concentration j. The significance of turning angles $\alpha_{i,j}$ and $\beta_{i,j}$, and velocities $u_{i,j}$ and $v_{i,j}$ was assessed inferring if $\bar{\alpha}_{i,c} = \alpha_{i,j}$ and $\bar{\beta}_{i,c} = \beta_{i,j}$, and if $\bar{u}_{i,c} = u_{i,j}$ and $\bar{v}_{i,c} = v_{i,j}$, where $\bar{\alpha}_{i,c}$ and $\bar{\beta}_{i,c}$ and $\bar{u}_{i,c}$ and $\bar{v}_{i,c}$ are the mean turning angles and swimming speed estimated outside and inside uncontaminated control patches of diameter i. Note that no significant differences were observed between $\bar{\alpha}_{i,c}$ and $\bar{\beta}_{i,c}$ (Wilcoxon-Mann-Whitney U-test, P > 0.05) and between $\bar{u}_{i,c}$ and $\bar{v}_{i,c}$ (Wilcoxon-Mann-Whitney U-test, P > 0.05), i.e. $\bar{\alpha}_{i,c} = \bar{\beta}_{i,c}$ and $\bar{u}_{i,c} = \bar{v}_{i,c}$. The turning angle θ_t between two successive moves was defined as $\theta_t = 180 - 180\theta / \pi$, where $\theta = arccos(\vec{A}\vec{B} / \|A\|\|B\|)$, $\vec{A}$ and $\vec{B}$ the vectors between the locations $X_1(x_t, y_t, z_t)$ and $X_2(x_{t+1}, y_{t+1}, z_{t+1})$ and between locations $X_2(x_{t+1}, y_{t+1}, z_{t+1})$ and $X_3(x_{t+2}, y_{t+2}, z_{t+2})$, and $\|A\|$ and $\|B\|$ the lengths of the vectors $\vec{A}$ and $\vec{B}$.

2.6.3 Patch entrance rate and proportional residence time

The avoidance of contaminated patches was further quantified by the patch entrance rate and the proportional residence time. The patch entrance rate is the ratio $F_{i,j} = 100N_{i,j} / N_{i,c}$, where $N_{i,j}$ is the number of individuals that encountered a contaminated patch of diameter i and concentration j during the 30-min behavioural experiment, and $N_{i,c}$ is the number of individuals that entered an uncontaminated patch of diameter i during the 30-min control experiments. Finally, the patch residence time was estimated as the time individual copepods spent in uncontaminated and contaminated patches. The proportional residence time ($PRT_{i,j}$) is the ratio of the time spent in a contaminated patch of diameter i and concentration j to the total time in spent in an uncontaminated patch of diameter i.

2.6.4 Complexity of swimming paths

The complexity of swimming paths was assessed using fractal analysis. In contrast to standard behavioral metrics such as turning angle and net-to-gross displacement ratio (NGDR), fractal analysis and the related fractal dimension D have the desirable properties to be independent of measurement scale and to be very sensitive to subtle behavioral changes that may be undetectable to other behavioral variables (Seuront & Leterme, 2007; Seuront et al., 2004a,b; Seuront &Vincent, 2008; Seuront, 2010b, 2011b). Fractal analysis has been applied to describe the complexity of zooplankton and ichtyoplankton swimming paths (Coughlin et al., 1992; Bundy et al., 1993; Dowling et al., 2000; Seuront et al., 2004a,b,c; Uttieri et al., 2005, 2007, 2008; Seuront, 2006, 2010b, 2011a,b; Seuront &Vincent, 2008, Ziarek et al., 2011). The fractal dimensions of *E. affinis* swimming paths were estimated using two different, but conceptually similar, methods to ensure the reliability of fractal dimension estimates; see e.g. Fielding (1992) and Hastings and Sugihara (1993), and Seuront (2010b) for a review.

The box dimension method relies on the "l cover" of the object, i.e. the number of boxes of length l required to cover the object. A more practical alternative is to superimpose a regular grid of boxes of length l on the object and count the number of boxes occupied by a subset of the object. This procedure is repeated using different values of l. The volume occupied by a swimming path is then estimated using a series of counting boxes spanning a range of volumes down to some small fraction of the entire volume. The number of occupied boxes increases with decreasing box size, leading to the following power-law relationship:

$$N(l) \propto l^{-D_b} \tag{2}$$

where l is the box size, $N(l)$ is the number of boxes occupied by the swimming path, and D_b is the box fractal dimension. The fractal dimension D_b is estimated from the slope of the linear trend of the log-log plot of $N(l)$ versus l.

The mass dimension method counts the number of pixels occupied by an object in cubes ($\delta \times \delta$) sampling windows as $N_O(\delta)$. The mass $m(\delta)$ of occupied pixels is then defined as:

$$m(\delta) = \frac{N_O(\delta)}{N_T(\delta)} \tag{3}$$

where $N_O(\delta)$ and $N_T(\delta)$ are the number of occupied pixels and the total number of pixels within an observation window of size δ. These computations are repeated for various values of δ, and the mass dimension D_m is defined as :

$$m(\delta) \propto \delta^{D_m} \tag{4}$$

The fractal dimension D_m is estimated from the slope of the linear trend of the log-log plot of $m(\delta)$ versus δ. Practically, the mass $m(\delta)$ can be estimated using cubes of increasing size δ starting from the centre of the experimental domain (Seuront, 2010b). Note that increasing in the box size l (Eq. 2) and δ (Eq. 4) may result in exclusion of a greater proportion of pixels along the periphery of the domain. Under an assumption of three-dimensional isotropy, this issue can be circumvented applying a toroidal edge correction (Seuront, 2010c). However, to avoid potential biases related to both the anisotropy of the swimming paths and the initial position of the overlying three-dimensional grid of orthogonal boxes, for each box size l and δ the grid was rotated in 5° increments from $\alpha = 0$ to $\alpha = 45°$ in the $x-y$ plane and from $\beta = 0$ to $\beta = 45°$ ° in the $x-z$ plane. The resulting distributions of fractal dimensions D_b and D_m were averaged, and the resulting dimension $\bar{D}_b$ and $\bar{D}_m$ used to characterize the complexity of a swimming path.

The appropriate range of scales l (Eq. 2) and δ (Eq. 4) to include in the regression anlyses was chosen following the R^2-SSR criterion (Seuront et al., 2004a). Briefly, I consider a regression window of varying width ranging from a minimum of 5 data points to the entire data set. The windows are slid along the entire data set at the smallest available increments, with the whole procedure iterated $n-4$ times, where n is the total number of available data points. Within each window and for each width, we estimated the coefficient of determination (r^2) and the sum of the squared residuals for the regression. I subsequently used the values of l (Eq. 2) and δ (Eq. 4), which maximized the coefficient of determination and minimized the total sum of the squared residuals (Seuront et al., 2004a), to define the scaling range and to estimate the related dimensions D_b and D_m. Note that D_b and D_m are bounded between 1 for a linear swimming path, and 2 for a path so complex that it fills the whole space available.

2.6.5 Complexity of instantaneous successive displacements

By analogy with a self-organized critical system that builds up stress and then releases the stress in intermittent pulses (Seuront & Spilmont, 2002), the level of stress arising from each experimental condition was described by a power law, which states that the cumulative probability distribution function of move length L greater than a determined length l follows (Seuront & Leterme, 2007; Seuront, 2011b):

$$N(l \leq L) \propto l^{-\phi} \tag{5}$$

where the move lengths L correspond to the distances travelled by *E. affinis* individuals every 0.04 s and ϕ a scaling exponent (referred to as a 'stress exponent' hereafter) describing the distribution. The exponent ϕ is estimated as the slope of $N(l \leq L)$ vs. l in log-log plots, and has been shown to decrease under stressful conditions for both vertebrates (Alados et al., 1996; Seuront & Cribb, 2011) and invertebrates (Seuront & Leterme, 2007; Seuront, 2011b). It is stressed, however, that Eq. (5) differs from the power law previously used to identify Lévy flights in a range of marine organisms including microzooplankton (Bartumeus et al., 2003), gastropods (Seuront et al., 2007) and fish (Sims et al., 2008; Humphries et al., 2010):

$$P(l_d = l) \propto l^{-\mu} \tag{6}$$

where l_d is the displacement length, l a threshold value, and μ ($1 < \mu \leq 3$) characterizes the power law behavior of the tail of the distribution. The move lengths L (Eq. (5)) differ from the flight path lengths l_d (Eq. (6)) which are defined as sequences of straight-line movements between the points at which significant changes in direction occurred; a significant change in direction is considered when the direction of the current flight segment (joining two successive recorded positions) and the direction of the previous flight segment is more than 90° (Bartumeus et al., 2005; Reynolds et al., 2007).

2.7 Statistical analyses

The distribution of the behavioral parameters was significantly non-normal (Kolmogorov-Smirnov test, $P < 0.01$), even after $\log_{10}$ or square-root transformations. Non-parametric statistics were then used throughout this work. Comparisons between behavioral parameters inside and outside patches were conducted using the Wilcoxon-Mann-Whitney *U*-test (WMW test; Zar, 2010). The effects of the size of uncontaminated control patches were compared using the Kruskal-Wallis test (KW test; Zar, 2010). The effects of contaminated patch size and concentration were investigated using the Scheirer-Ray-Hare extension of the Kruskal-Wallis test (SRH test; Sokal & Rohlf, 1995). Appropriate multiple comparison procedures were subsequently used to test for differences between patch diameter and patch concentration. Multiple comparisons between WSF treatments were conducted using the Kruskal-Wallis test, and a subsequent multiple comparison procedure based on the Tukey test was used to identify distinct groups of measurements (Zar, 2010). Correlation between variables was investigated using Kendall's coefficient of rank correlation, τ (Kendall & Stuart, 1966).

3. Behavioral response to point sources of hydrocarbon contamination

3.1 Patch avoidance and patch escape

The swimming paths of *Eurytemora affinis* adult males and non-ovigerous females were not affected by uncontaminated control patches (Fig. 3a,b). However, both males and females consistently avoided contaminated patches (Fig. 3c,d; Fig. 4), irrespective of patch size. The distance d_a at which individuals exhibited patch avoidance was not affected by patch size ($P > 0.05$) or patch concentration ($P > 0.05$), but were significantly smaller ($P < 0.05$) for *E. affinis* adult males (1.7 ± 0.1 mm, $\bar{x} \pm \text{SD}$) than for non-ovigerous females (2.2 ± 0.2 mm).

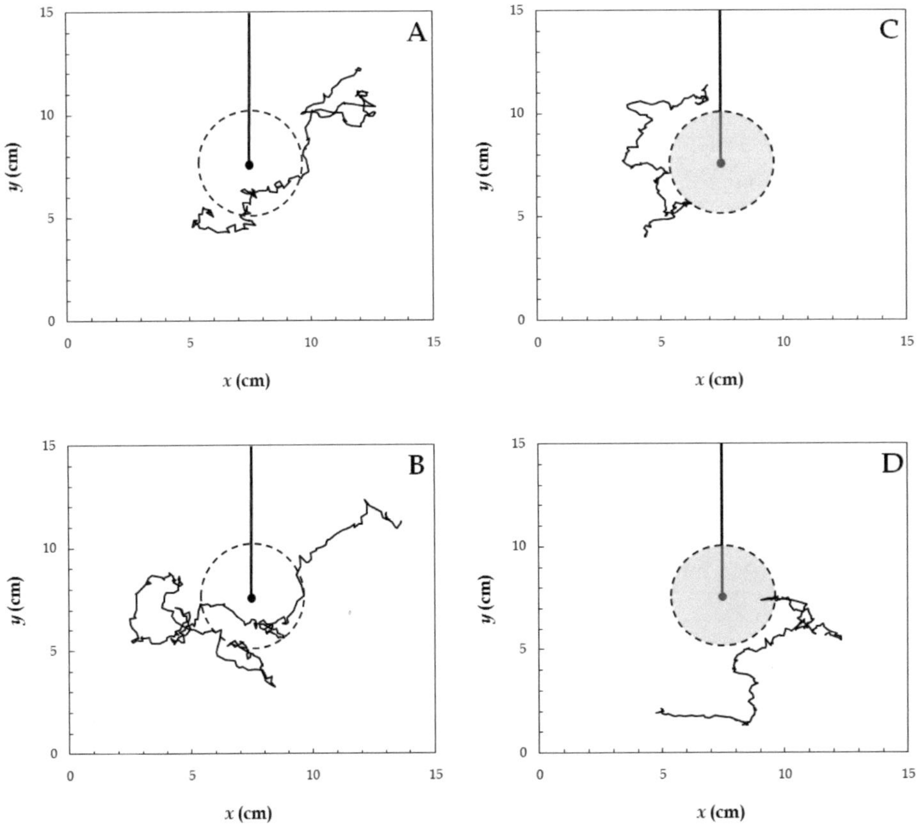

Fig. 3. Two-dimensional projections of three-dimensional swimming paths of *Eurytemora affinis* adult males (a,c) and non-ovigerous females (b,d) in response to uncontaminated (a,b) and contaminated patches (c,d) 4 cm in diameter. Contaminated patches correspond to 0.01%, 0.1% and 1% dilutions of the water-soluble fraction of diesel oil in filtered estuarine water, while uncontaminated patches were created using uncontaminated GF/C filtered and autoclaved estuarine water.

No avoidance and no escape behaviors were observed for the different sizes of uncontaminated patches (Fig. 3a,b; Fig. 4). The percentages of individuals that exhibited a patch avoidance behaviour (Fig. 3c,d) did not significantly differ with patch sizes and concentrations ($P > 0.05$) for both males (97.1 ± 2.3 %) and females (95.2 ± 2.1 %). The avoidance turning angle $\alpha_{i,j}$ and avoidance velocity $u_{i,j}$ did not differ with the size or the concentration of the contaminated patches ($P > 0.05$) for males ($\alpha_{i,j} = 68.1 \pm 3.2\,°$ and $u_{i,j} = 4.9 \pm 0.9$ mm s^{-1}) and females ($\alpha_{i,j} = 65.2 \pm 2.3\,°$ and $u_{i,j} = 8.5 \pm 0.6$ mm s^{-1}). No significant differences were found in avoidance turning angle $\alpha_{i,j}$ between males and females ($P > 0.05$). In contrast, avoidance velocity $u_{i,j}$ was significantly higher for females than for males ($P < 0.05$). The avoidance turning angle and velocity were significantly higher ($P < 0.01$) than the mean turning angle $\bar{\alpha}_{i,c}$ and swimming speed $\bar{u}_{i,c}$ estimated for uncontaminated patches, i.e. $\bar{\alpha}_{i,c} = 36.7 \pm 1.2\,°$ and $\bar{u}_{i,c} = 2.2 \pm 0.5$ mm s^{-1} for males and $\bar{\alpha}_{i,c} = 37.3 \pm 1.2\,°$ and $\bar{u}_{i,c} = 1.7 \pm 0.3$ mm s^{-1} for females.

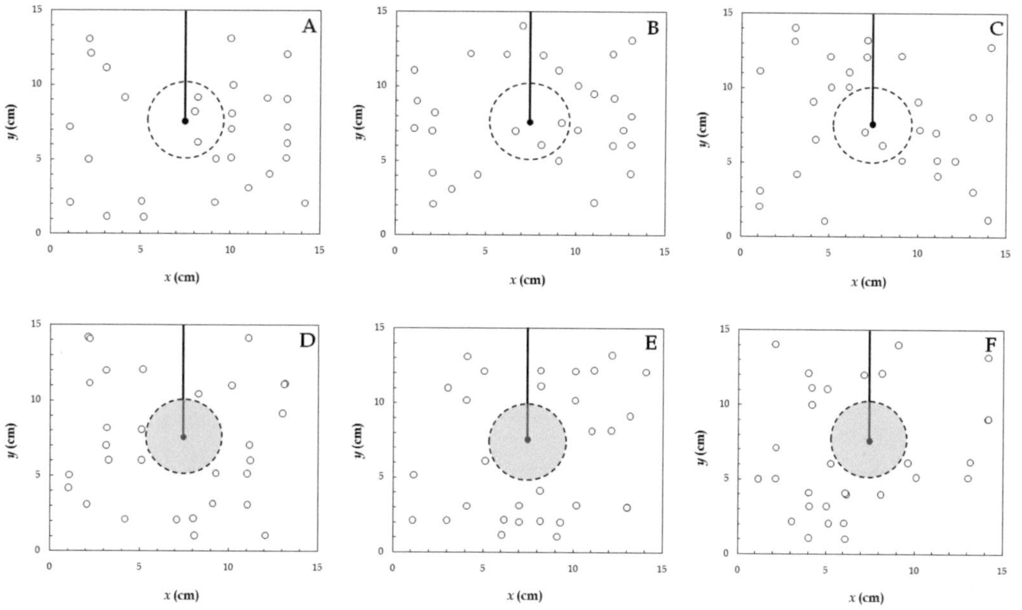

Fig. 4. Two-dimensional snapshots of the position of 30 *Eurytemora affinis* adult males (open dots) in relation to the position of an uncontaminated patch (a-c) and a patch contaminated with the water-soluble fraction of diesel oil at a concentration of 0.1% (d-f). Both patches were 4 cm in diameter.

The escape turning angle $\beta_{i,j}$ did not differ with the size or the concentration of the contaminated patches ($P > 0.05$) for males ($\beta_{i,j} = 70.1 \pm 3.3$ °) and females ($\beta_{i,j} = 68.4 \pm 2.1$ °). In contrast, the escape velocity $v_{i,j}$ significantly increased ($P < 0.05$) with WSF concentration for both males and females, with $v_{i,j} = 4.7 \pm 1.1$ mm s^{-1}, $v_{i,j} = 5.4 \pm 1.2$ mm s^{-1} and $v_{i,j} = 6.5 \pm 0.9$ mm s^{-1} for males and $v_{i,j} = 8.2 \pm 1.0$ mm s^{-1}, $v_{i,j} = 8.8 \pm 1.1$ mm s^{-1} and $v_{i,j} = 9.8 \pm 1.1$ mm s^{-1} for females for estuarine water contaminated at 'low' (0.01%), 'medium' (0.1%) and 'high' (1%) concentrations of the soluble-fraction of diesel oil. The escape turning angle $\beta_{i,j}$ and escape velocity $v_{i,j}$ were significantly higher ($P < 0.01$) than the mean turning angle $\overline{\beta}_{i,c}$ and swimming speed $\overline{v}_{i,c}$ estimated inside uncontaminated patches, i.e. $\overline{\beta}_{i,c} = 37.4 \pm 1.4$ ° and $\overline{v}_{i,c} = 2.1 \pm 0.4$ mm s^{-1} for males and $\overline{\beta}_{i,c} = 37.6 \pm 1.3$ ° and $\overline{v}_{i,c} = 1.8 \pm 0.4$ mm s^{-1} for females. No significant differences were found in avoidance turning angle $\beta_{i,j}$ between males and females ($P > 0.05$). The avoidance velocity $v_{i,j}$ was, however, significantly higher for females than for males ($P < 0.05$).

Despite the dependence of both the avoidance and escape velocities $u_{i,j}$ and $v_{i,j}$ to the concentration of the soluble-fraction of diesel oil, the smaller avoidance distance, avoidance velocity and escape velocity observed for males suggest that male sensory abilities may be less acute than female ones. The behavioral responses of both males and females nevertheless converge towards an adaptation to avoid and escape WSF contaminated patches, hence minimize the exposure time to a source of contamination.

3.2 Entrance rate in uncontaminated vs. hydrocarbon-contaminated patches

The percentage of adult males entering contaminated patches (Fig. 5a) was highly significantly affected by patch concentration ($P < 0.01$), but not patch size ($P > 0.05$), leading to $F_{i,1} = 12.3 \pm 0.6$ %, $F_{i,0.1} = 21.7 \pm 1.2$ % and $F_{i,0.01} = 28.7 \pm 1.5$ %. In contrast, the ratio $F_{i,j}$ was not affected by patch size ($P > 0.05$) and patch concentration ($P > 0.05$) for non-ovigerous females (Fig. 5b), with $F_{i,1} = 10.3 \pm 0.6$ %, $F_{i,0.1} = 10.2 \pm 1.3$ % and

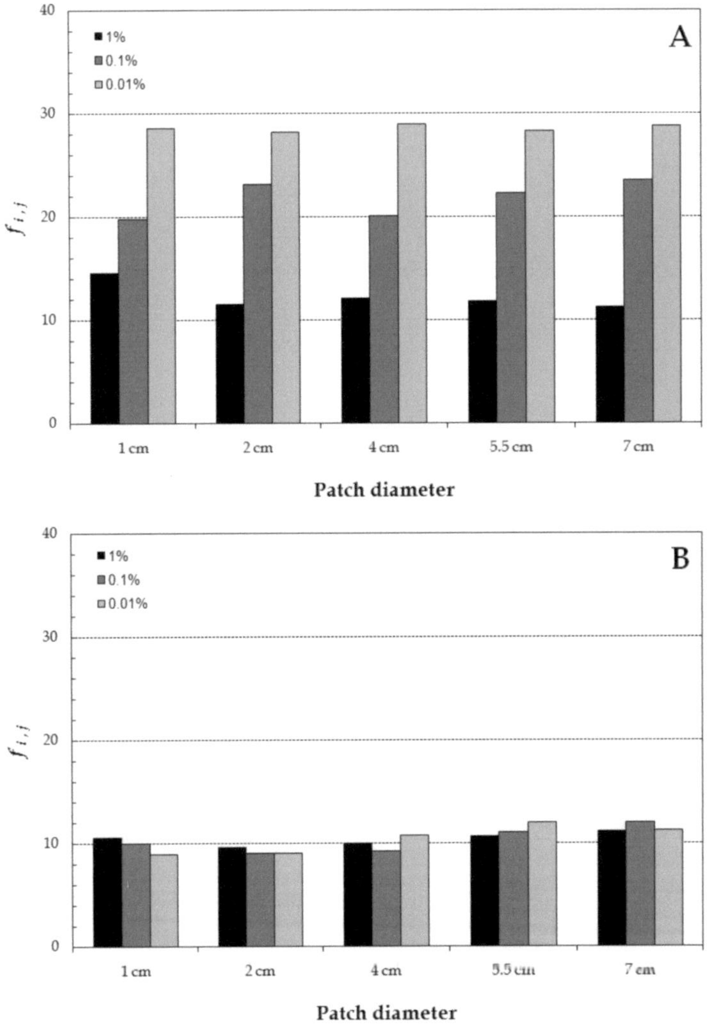

Fig. 5. The patch entrance rate $F_{i,j}$ between the number of *Eurytemora affinis* individuals that entered a contaminated patch of diameter *i* and concentration *j* and the number of individuals that entered an uncontaminated patch of diameter *i*, shown for adult males (a) and non-ovigerous females (b). Contaminated patches correspond to 0.01%, 0.1% and 1% dilutions of the water-soluble fraction of diesel oil in filtered estuarine water, while uncontaminated patches were created using uncontaminated GF/C filtered and autoclaved estuarine water.

$F_{i,0.01} = 10.4 \pm 1.4$ %. Note that for each patch size, the percentage of adult males entering contaminated patches significantly decreases with patch concentration ($P < 0.01$). This is consistent with the significant exponential increase observed in $F_{i,j}$ with decreasing contaminant concentration in *Temora longicornis* adult females (Seuront, 2010a). These results suggest that

i. *E. affinis* non-ovigerous females may have comparable sensory abilities irrespective of the concentrations of the contaminant, whereas males seem to identify more efficiently high-density contaminated patches than low-density ones, and
ii. the ability of copepods to detect patches contaminated with the soluble-fraction of diesel oil (hence with polycyclic aromatic and monoaromatic hydrocarbons, and their derived products), is likely to be both species- and sex-dependent.

The differences observed between male and female chemosensory abilities are consistent with the role played by chemoreception in males foraging ecology (e.g. Kiørboe et al., 2005; Bagøien & Kiørboe, 2005; Goetze & Kiørboe, 2008; Seuront, 2011a). Further work is, however, needed to generalize these results to a representative range of copepod species, and to assess the potential differences that may exist between the chemosensory abilities of different sexes and development stages.

3.3 Residence time in uncontaminated vs. hydrocarbon-contaminated patches

The residence time of both *E. affinis* males and non-ovigerous females in uncontaminated control patches significantly increased with the size of the patch ($P < 0.05$) and ranged from 8 to 47 s for patches 1 cm and 7 cm in diameter (Fig. 6). The residence time significantly decreased in contaminated patches for both males and females ($P < 0.05$), and does not exhibit any significant change with patch size ($P > 0.05$). The residence time observed for males significantly decrease with increasing concentration of the water-solube fraction of diesel oil, i.e. 0.83 s, 0.70 s and 0.60 s for patches contaminated at 0.01, 0.1 and 1% levels, respectively. In contrast, females residence time (0.34 s, 0.34 s and 0.33 s at WSF concentration of 0.01, 0.1 and 1%) did not significantly vary with the level of water contamination, but were consitently significantly shorter than male ones ($P < 0.05$).

These results are specified by the proportional residence time $PRT_{i,j}$ (Fig. 7). $PRT_{i,j}$ ranged from 1.6% to 8.8%, 1.3% to 7.2% and 1.2% to 6.3% for adult males, and from 0.6% to 3.1%, 0.5% to 3.1% and 0.6% to 3.6% for non-ovigerous females, at WSF concentration of 0.01, 0.1 and 1%. $PRT_{i,j}$ significantly decreased with the size of patches for both males and females for each WSF concentration ($P < 0.05$), and was significantly higher for males at each WSF concentration ($P < 0.05$).

These observations are consistent with the hypothesis that *E. affinis* non-ovigerous females may have comparable sensory abilities irrespective of the concentrations of the contaminant, hence exhibit a *'on-off'* behavioral response that leads them to escape a source of contamination, irrespective of the concentration of the contaminant, and ultimately lead them to minimize the exposure time to the contaminant. In contrast, males have a modulated behavioral response that is *sensu stricto* less efficient than female's behavior; it nevertheless also leads to minimize the exposure time to a contaminant through a density-dependent response. Ultimately, these behavioral changes are likely to avoid a stress-related reduction in individual fitness, which might in turn affect the whole zooplankton community.

Sex-specific response to hydrocarbon contamination and behavioral alterations for concentrations of the water-soluble fraction of diesel oil ranging from 0.01 to 1% have also been observed in another common calanoid copepod, *Temora longicornis* (Seuront, 2011a). This demonstrates the very acute chemosensory abilities of both *E. affinis* and *T. longicornis*, and generalizes previous work showing behavioral changes elicited by a variety of chemical cues (Katona, 1973; Doall et al. 1998; Weissburg et al., 1998; Yen et al., 1998; Woodson et al., 2007, 2008; Seuront, 2010a, b, 2011a; Cailleaud et al., 2011).

In this context, the next section investigate the behavioral stress that may be induced in the swimming behavior of *E. affinis* adult males and non-ovigerous females by the diffuse hydrocarbon contamination that is likely to follow any point-source contamination in estuarine waters.

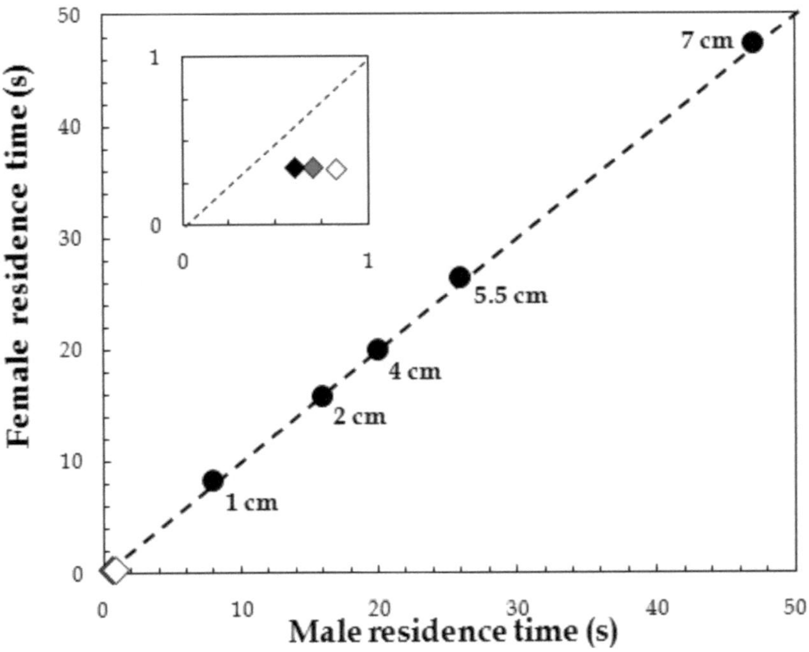

Fig. 6. Residence time of *E. affinis* adult males and non-ovigerous females in uncontaminated control patches of increasing diameter (black dots) and in patches contaminated with the water-soluble fraction of diesel oil (insert) at 0.01% (white diamond), 0.1% (grey diamond) and 1% (black diamond).

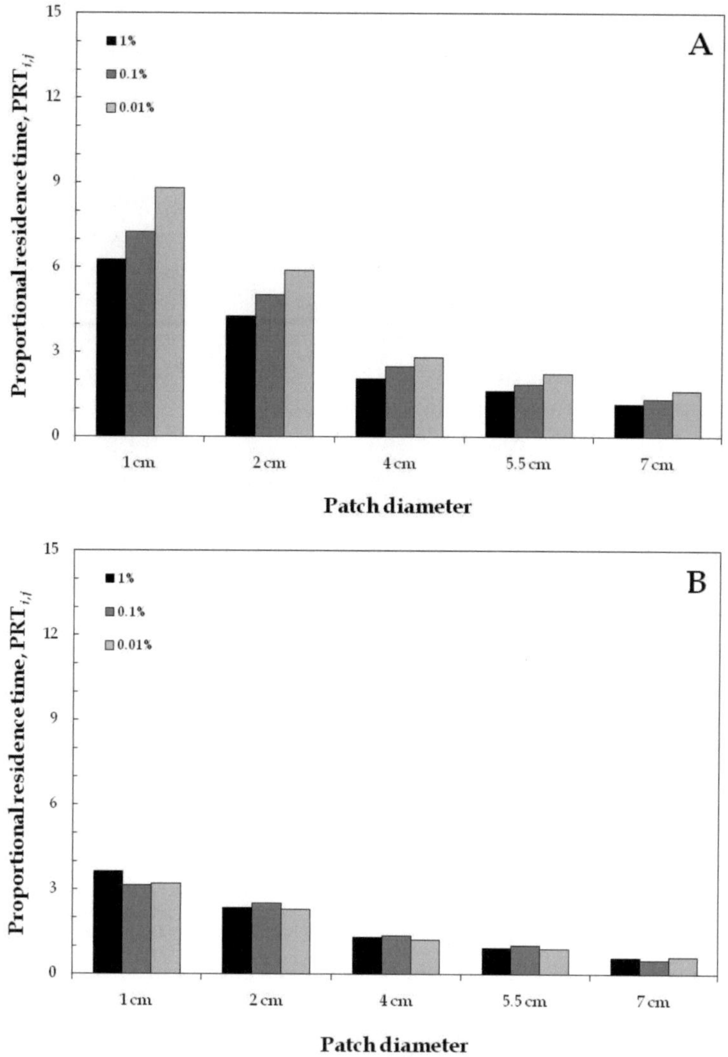

Fig. 7. Proportional residence time of *E. affinis* adult males (A) and non-ovigerous females (B) in patches of increasing diameter contaminated with the water-soluble fraction of diesel oil at 0.01% (black), 0.1% (dark grey) and 1% (light grey). The pro.3portional residence time ($PRT_{i,j}$) is the ratio of the time spent in a contaminated patch of diameter *i* and concentration *j* to the total time in spent in an uncontaminated patch of diameter *i*.

4. Behavioral response to a diffuse hydrocarbon contamination

4.1 Swimming speed and turning angle in hydrocarbon contaminated water

Four kind of swimming behaviors were further considered to quantify the swimming behaviour of males and non-ovigerous females (Seuront, 2011a): (i) cruising, in which their rostro-caudal body axes were aligned with the direction of motion, whether they were swimming up, down or horizontally, (ii) hovering, i.e. swimming upward at low speed, often with a horizontal component, with the rostro-caudal body axis oriented upward (Doall et al., 1998), (iii) passive sinking, i.e. downward vertical motion, with tail down, and (iv) breaking, in which they remain motionless, with their rostro-caudal axes oriented upward. Each behavioural activity was quantified in terms of time allocation percentage for each category of organisms.

In control, non-contaminated estuarine water, the swimming speeds of *E. affinis* adult males ($u_{m,c}$ = 2.3 ± 0.4 mm s^{-1}) and non-ovigerous females ($u_{f,c}$ = 1.8 ± 0.6 mm s^{-1}) were not significantly different (P > 0.05), and did not significantly differ from the swimming speed estimated in uncontaminated patches (i.e. $\overline{u}_{i,c} = 2.2 \pm 0.5$ mm s^{-1} for males and $\overline{u}_{i,c} = 1.7 \pm 0.3$ mm s^{-1} for females). Similarly, males and females did not exhibit any significant changes in their turning angle ($\alpha_{m,c} = 38.6 \pm 1.6$ ° and $\alpha_{f,c} = 39.2 \pm 1.4$ °), and were not significantly different from the turning angle $\overline{\alpha}_{i,c}$ estimated in uncontaminated patches, i.e. $\overline{\alpha}_{i,c} = 36.7 \pm 1.2$ ° a for males and $\overline{\alpha}_{i,c} = 37.3 \pm 1.2$ ° for females. The swimming speeds and turning angles of both males and females significantly differ between experimental conditions, were significantly smaller in contaminated water, and significantly decreased with the concentration of the water-soluble fraction of diesel oil considered (P < 0.05; Fig. 8). In contrast, sinking speed did not significantly differ between males (v_m = 0.5 ± 0.1 mm s^{-1}) and females (v_f = 0.6 ± 0.1 mm s^{-1}) in non-contaminated and contaminated estuarine waters (P > 0.05), and no significant differences were observed between control and WSF treatments (P > 0.05).

Two previous behavioural studies of *E. affinis* (Michalec et al., 2010; Cailleaud et al., 2011) considered sinking as "a swimming speed between 1 and 8 mm/sec and a direction straight towards the bottom, when the copepod is not swimming but sinks slowly due to the influence of gravity". This is highly questionable, as well as the results of the rather convoluted subsequent analyses, as the sinking speed reported for *E. affinis* falls in the range 0.4-0.8 mm s^{-1} (Seuront, 2006; present work). In addition, previous experiments conducted on various calanoid copepod species have reported sinking velocities typically ranging from 0.3 to 2.5 mm s^{-1}; see e.g. Tiselius & Jonsson (1990), and Weissman et al. (1993). Typical sinking speed obtained for anesthetised (hence likely to be the fastest ones in the absence of any appendage movement) *E. affinis* range from 0.4 to 0.7 mm s^{-1} for adult males (cephalothorax length, 0.81 to 0.83 mm) and 0.5 to 1.2 mm s^{-1} for non-ovigerous females (0.84 to 0.86 mm); Seuront, unpublished data. Those values are consistent with the passive sinking velocities reported for *E. affinis* (Seuront, 2006; present work) and other species (Tiselius & Jonsson, 1990; Weissman et al., 1993) where copepods were moving downwards, with their tail down, but also with copepod carcasses (Frangoulis et al., 2011). The slight differences observed between sinking velocities of non-anesthetised (Seuront, 2006; present work) and anesthetised *E. affinis* suggest that the influence of gravity was partially counterbalanced by the motion of feeding appendages, which is consistent with the intrinsic link existing between swimming and feeding behaviours in calanoid copepods (e.g. van Duren & Videler, 1995). However, the much higher velocities reported by Michalec et al. (2010) for *E. affinis* 'sinking' may rather suggest that the resolution of their camera, located 50 cm away from a 5 × 5 × 6 cm tank was not good enough to distinguish individuals actually sinking with their tail down from individuals actively swimming downwards.

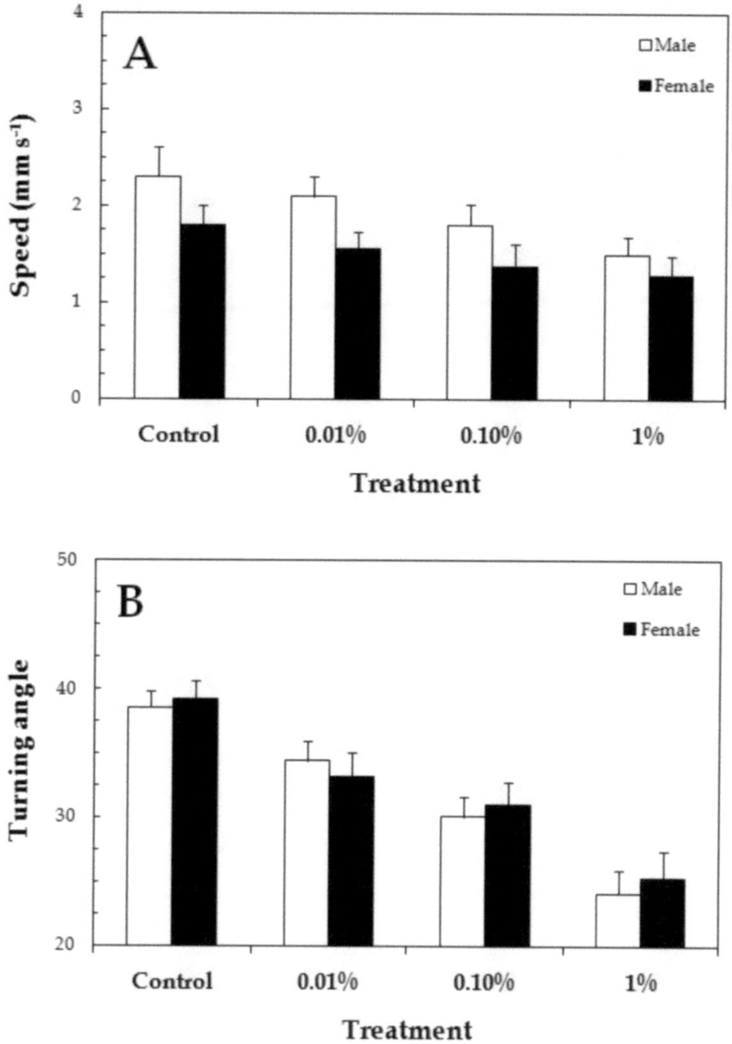

Fig. 8. Swimming speed (A) and turning angle (B) of *E. affinis* adult males and non-ovigerous females (B) in uncontaminated estuarine water and in estuarine water contaminated with the water-soluble fraction of diesel oil at 0.01%, 0.1% and 1%. The error bars are standard deviations.

Besides, both studies (Michalek et al., 2010; Cailleaud et al., 2011) report mean prosome lengths of respectively 0.85 mm and 0.95 mm for *E. affinis* males and females, which are consistent with the size of *E. affinis* individuals investigated previously (i.e. 0.81 to 0.83 mm for males and 0.84 to 0.87 for females; Seuront, 2006) and in the present work (0.83 to 0.86 mm for males and 0.87 to 0.92 for females). For sinking velocities of copepods of similar size to range between 1 and 8 mm s^{-1} as reported in Michalek et al. (2010) and Cailleaud et al. (2011), their density need to vary by a factor of 8 or to violate Stockes law which would both be unprecedented in the zooplankton literature.

Males and non-ovigerous females exhibited very comparable swimming paths (see Fig. 3a,b). The swimming activity of *E. affinis* is, however, clearly sex-dependent (Fig. 9). In uncontaminated estuarine water, males spend significantly more time cruising (76.3%) than hovering (16.7%), sinking (2.6%) and breaking (4.4%). In contrast, non-ovigerous females spend most of their time cruising (51%) and sinking (36.3%), with significantly less time spent sinking (7.4%) and breaking (5.4%). Under increasing condition of WSF contamination, both males and females decrease their swimming activity with a decrease in the time allocated to cruising (Fig. 9).

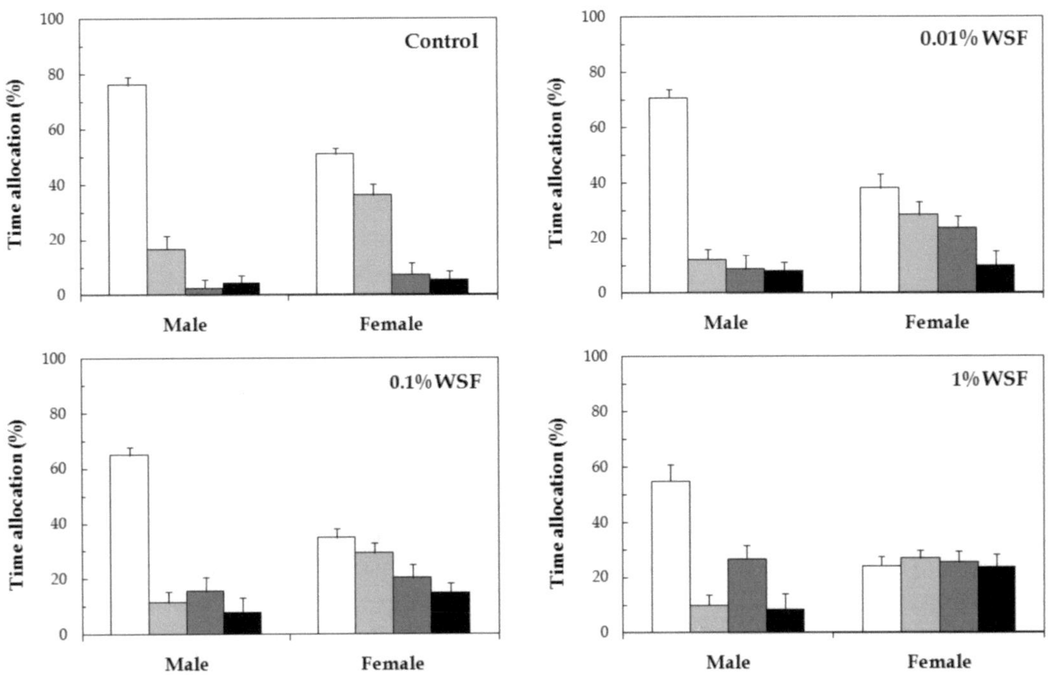

Fig. 9. Fraction of time allocated by *E. affinis* adult males and non-ovigerous females to cruising (motion with rostro-caudal body axis aligned with the direction of motion, whether they were swimming up, down or horizontally; white), hovering (upward motion at low speed, often with a horizontal component; light grey), sinking (downward vertical motion, with tail down; dark grey) and breaking (no motion, with rostro-caudal axis oriented upward; black) in control uncontaminated estuarine water and estuarine water contaminated with the water-soluble fraction of diesel oil at 0.01%, 0.1% and 1%. The error bars are standard deviations.

More specifically, while the time allocated to cruising significantly decreased from uncontaminated to contaminated waters, it remained the main behavioural activity for males in WSF contaminated water (Fig. 9). The time allocated to hovering slightly decreased with increasing WSF concentration, while sinking and breaking increased with increasing WSF concentration. In contrast, the time allocated by females to cruising and hovering significantly decreased from uncontaminated to contaminated water (Fig. 9); however, cruising and hovering respectively significantly decreased and did not significantly change with increasing WSF concentration. Sinking and breaking increased with increasing WSF concentration, leading to an even time allocation between cruising, hovering, sinking and breaking in water contaminated with WSF at 1% (Fig. 9).

The decrease in the swimming speed and swimming activity reported here of both adult males and non-ovigerous female swimming speed with increasing WSF concentration *a priori* contrasts with recent observations conducted on *E. affinis* adult males and females, which significantly increased their swimming speed following an exposure to 2 µg l^{-1} of 4-nonylphenol and nonylphenol-ethoxy-acetic-acid (Cailleaud et al., 2011). However, in this work (Cailleaud et al., 2011) the swimming behavior of *E. affinis* was recorded from the same individuals before and after the injection of 15 µl of test solution. As such, the behavioral observations conducted in contaminated water are more likely to result from the exposure to a gradient than a background concentration of nonyphenols, and/or to the stress response induced by the injection of the contaminant in uncontaminated water. This is consistent with *T. longicornis* and *E. affinis* males and females escaping at high velocities when reaching patches of WSF contaminated seawater (Seuront, 2010a; present work) and to the decrease in swimming speed observed in *T. longicornis* males and females under conditions of increasing WSF contamination (Seuront, 2011a), hence with distinct behavioral reactions following an exposure to a background concentration of contaminants and a gradient of contaminants.

To specify this, we compared the swimming speed of *E. affinis* males and females recorded over 3 successive 5-min intervals (i.e. 0-5 min, 5-10 min and 10-15 min) during the 15-min acclimation phase, with the swimming velocity recorded during the 30-min behavioral experiment *per se*. Note that no significant differences were found in the swimming speeds recorded over the 6 successive 5-min intervals available from the 30-min behavioral experiments ($P > 0.05$). This resulted in comparing swimming speeds between 4 time intervals, i.e. 0-5 min, 5-10 min, 10-15 min and 15-45 min (Fig. 10). *E. affinis* swimming speed did not significantly differ between the four temporal categories under conditions of uncontaminated water. In contrast, under conditions of WSF contamination, the swimming speed observed during the acclimation phase were significantly higher ($P < 0.05$) than those recorded during the behavioral experiment (Fig. 10). This effect increased with increasing WSF concentrations, and the the increase in swimming speed observed during the first 5-min interval (0-5 min) under WSF contamination leads to a decrease in swimming speed that ultimately converges towards the values observed during the 30-min behavioral experiment for both males and females; the higher the WSF concentration, the higher the swimming velocity during the acclimation phase (Fig. 10). This suggests that the claimed increase in *E. affinis* swimming speed in the presence of sub-lethal concentrations of nonylphenols (Cailleaud et al., 2011) may reflect a stress reaction related to the changes in water properties induced by the injection of nonylphenols rather than the actual effect of nonylphenols on behavior.

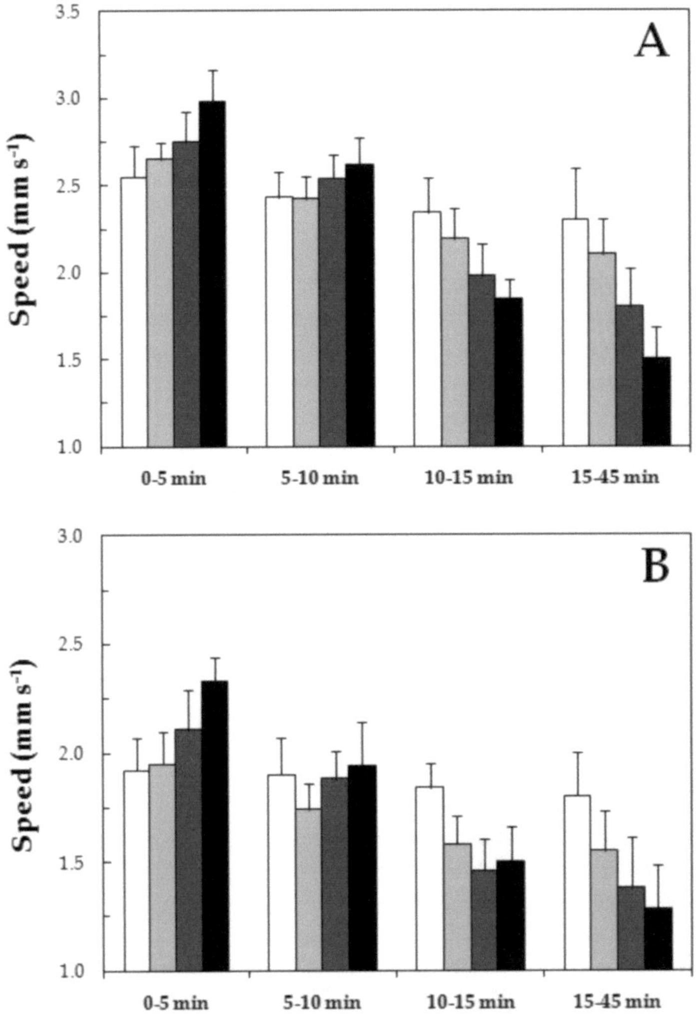

Fig. 10. Swimming speed of *E. affinis* adult males (A) and non-ovigerous females (B) in uncontaminated water (white) and in estuarine water contaminated with the water-soluble fraction of diesel oil at 0.01% (light grey), 0.1% (dark grey) and 1% (black), expressed as a function of the time elapsed since the start of behavioral experiments, that include acclimation (0-5 min, 5-10 min and 10-15 min) and experimental phase (15-45 min). The error bars are standard deviations.

Note, however, that the behavioral responses previously observed in zooplankton following water contamination range from hypoactivity to hyperactivity, depending on the species, sex, concentration and nature of the contaminant, and exposure time. For instance, *Daphnia magna* decreased their swimming speed after several days of exposure to cadmium (Baillieul & Blust, 1999) and a 9-h exposure to copper at 30 mg l^{-1} (Untersteiner et al., 2003). In contrast, no changes were recorded following a 3-h and 24-h exposures to copper at 0.01 mg ml^{-1} (Shimizu et al., 2002) and 5 mg l^{-1} (Untersteiner et al., 2003), and an increase in swimming speed occurred after a 24-h exposure to methyl-paraoxon at 0.7 mg l^{-1} (Duquesne & Küster, 2010). Similar results were obtained on the larvae of the cirriped *Balanus Amphitrite* (Faimali et al., 2006), the crustacean *Artemia* sp. and the rotifer *Brachionus plicatilis* (Garaventa et al., 2010) exposed to a range of chemical pollutants (i.e. antifouling biocides, neurotoxic pesticides, and heavy metals). In copepod ecotoxicology, *E. affinis* nauplii respond to sub-lethal copper concentration by successive phases of hyperactivity and hypoactivity, whose magnitude is season-dependent (Sullivan et al., 1983). In much shorter (30-min to 1-h) bioassays, exposure to sub-lethal hydrocarbon concentrations, leads to decrease the swimming speed of *T. longicornis* females (Seuront, 2011a) and both *E. affinis* males and females (present work). In contrast, no significant changes were observed in *Centropages hamatus* females (Seuront & Leterme, 2007; Seuront, 2010b) and *T. longicornis* males (Seuront, 2011a).

As stressed above from the patch avoidance response of *E. affinis* males and females, these observations imply that behavioral responses to pollution based on the analysis of swimming speed alteration are highly variable. As a consequence, and even if alterations in swimming speed have been detected at toxic compound concentrations well below LC50 values for a range of invertebrates (Avila et al., 2010; Ihara et al., 2010; Garaventa et al., 2010; Cailleaud et al., 2011; Seuront, 2011a, present work), its claimed use as a non-specifc behavioral end-point in marine ecotoxicology and environmental monitoring program (Faimali et al., 2006; Garaventa et al., 2010) cannot be warranted. A practical alternative based on the intrinsic fractal nature of behavioral properties and their subsequent modification under stressful conditions, applied in both invertebrate (Seuront & Leterme, 2007; Seuront, 2010b, 2011b) and vertebrate (Escós et al., 1995; Alados et al., 1996; Alados & Huffman, 2000; María et al., 2004; Seuront & Cribb, 2011), including humans (e.g. Togo & Yamamoto, 2000; Goldberger et al., 2002; West & Scaffeta, 2003), is provided in the next two sections.

4.2 Swimming path complexity in hydrocarbon contaminated water

The fractal dimensions $\bar{D}_b$ and $\bar{D}_m$ estimated from *E. affinis* males and females swimming paths were not significantly different ($P > 0.05$). This is in accordance with the theoretical formulation $D_b = D_m$ derived from Eqs. (2) and (4); see Seuront (2010b) for further details and theoretical developments. The fractal dimension D, i.e. $D = (D_b + D_m)/2$ was hence used hereafter to characterize the complexity of *E. affinis* swimming paths.

In uncontaminated estuarine water, male fractal dimensions ($D = 1.43 \pm 0.11$) were significantly higher ($P < 0.05$) than female ones ($D = 1.31 \pm 0.08$). This difference is consistent with the results obtained from *E. affinis* males ($D = 1.23 \pm 0.01$) and females ($D = 1.20 \pm 0.02$) in GF/C filtered estuarine water at 5 PSU (Seuront, 2006), and from *T. longicornis* swimming paths observed in GF/C filtered coastal waters (i.e. $D = 1.32 \pm 0.02$ for males and $D = 1.27 \pm 0.02$ for females; Seuront, 2011a). In contrast, the swimming paths of *Oncaea venusta* males and females were characterized by non-significantly different fractal dimensions, i.e. $D = 1.14 \pm 0.06$ for males and $D = 1.15 \pm 0.06$ for females, investigated in natural seawater (Seuront et al., 2004b). The differences reported above between *T. longicornis* and *E. affinis*, and *O. venusta* may, however, be due to the absence and presence of cues in their respective experimental set-ups. In the absence of cues, the different fractal dimensions estimated from male and female swimming paths in both *T. longicornis* and *E. affinis* may be related to intrinsic (i.e. innate) differences in male and female foraging strategies. Male fractal dimensions are significantly higher than female ones, hence males are engaged in more intensive foraging strategies than females, which is consistent with the reported behavior of males in the presence of female cues (e.g. Doall et al., 1998; Nihongi et al., 2004; Bagøien & Kiørboe, 2005; Goetze & Kiørboe, 2008; Yen et al., 2011). The similar complexity of the swimming paths of *O. venusta* males and females in natural seawater may be related to a common, hence sex-independent, adaptive behavioral strategy developed in response to the range of chemical cues (from e.g. preys, conspecifics and predators) that are likely to be present in their experimental containers. This is consistent with the fractal dimensions of *T. longicornis* adult females observed in natural seawater range, however, between $D = 1.18 \pm 0.04$ and $D = 1.82 \pm 0.05$ depending on the nature and abundance of the phytoplankton community and seawater viscosity (Seuront & Vincent, 2008). More generally, the fractal dimensions estimated from *E. affinis* female swimming paths observed in uncontaminated estuarine water are in the range of values observed for females of different calanoid species investigated using the same experimental protocol than the present work, i.e. $D = 1.25 \pm 0.02$ for *Acartia clausi*, $D = 1.37 \pm 0.03$ for *Centropages typicus*, $D = 1.27 \pm 0.02$ for *Pseudocalanus elongatus*, $D = 1.42 \pm 0.03$ for *Paracalanus parvus*, and $D = 1.25 \pm 0.03$ for *T. longicornis* (Seuront, 2011b).

Males and females fractal dimensions both significantly decrease ($P > 0.05$) with WSF concentration (Fig. 11a). The observed decrease is comparable to the decrease observed in *T. longicornis* males and female fractal dimensions under the same conditions of WSF contamination (Seuront, 2011a). In addition, the relative changes observed in *E. affinis* male and female fractal dimensions under control and WSF contamination range respectively from 3.5 to 11.9% and 3.8 to 11.5% in estuarine water contaminated at 0.01% and 1%. These rates are also similar to those observed in *T. longicornis* males (2.4 to 9.5%) and females (3.0 to 8.3%; Seuront, 2011a), suggesting a sex-independent response to WSF contamination. The swimming path complexity of adult females of 5 species of calanoid copepods did not change, however, under varying experimental light regimes (Seuront, 2011b). This divergence in the response of calanoid copepods to distinct sources of stress may indicate that the impact of stressful conditions on the fractal properties of movement behavior may be dependent on the nature of the stressor itself. The resolution of this specific issue is, however, far beyond the objectives of the present work.

In contrast, the fractal dimensions reported for *Daphnia* sp. swimming paths seem to consistently increase under stress. The fractal dimensions of *Daphnia magna* and *Daphnia pulicaria* swimming paths respectively increased in water contaminated by copper, organophosphorus (Dichlorvos) and carbamate (Propoxur) (Shimizu et al., 2002), and following a 1-min exposure to turbulence (Seuront et al., 2004c). Some of the fractal dimensions estimated for *Daphnia magna* swimming paths under conditions of chemical contaminations are, however, higher than the upper theoretical limit $D = 2$ (see Shimizu et al., (2002), their figures 4 and 5), which questions the relevance of their results. More generally, the application of fractals to cladocerans behavioral ecology and ecotoxicology is still far too limited (Shimizu et al., 2002; Seuront et al., 2004a,c,d; Uttieri et al., 2005; Ziarek et al., 2011) to allow reliable conclusions.

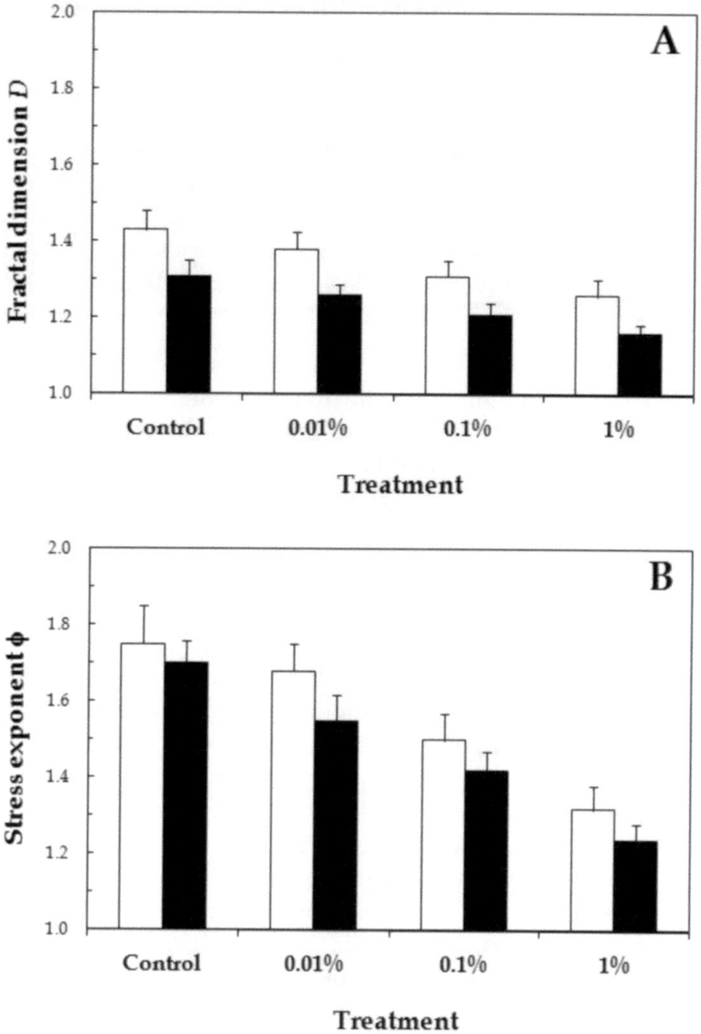

Fig. 11. The fractal dimension *D* (A) and the stress index ϕ (B) estimated from swimming paths of *E. affinis* adult males (white) and non-ovigerous females (black) in control uncontaminated estuarine water and in estuarine water contaminated with the water-soluble fraction of diesel oil at 0.01%, 0.1% and 1%. The error bars are standard deviations.

In contrast to the nature of alterations in swimming speed (i.e. hypoactivity vs. hyperactivity) which are likely to be sex- and species-specific as discussed above, the alterations observed in the fractal dimension of copepod swimming paths under conditions of hydrocarbon contamination seem to be both sex- and species-independent. Fractal dimensions are hence likely to provide a very sensitive and robust behavioral end-point in marine ecotoxicology and environmental monitoring program. Further work is nevertheless needed to confirm and generalize to other copepod species the congruent results obtained from *T. longicornis* (Seuront, 2011a) and *E. affinis* (present work) swimming behavior under conditions of hydrocarbon contamination.

4.3 Complexity of instantaneous displacements in hydrocarbon contaminated water

Log-log plots of $N(l \leq L)$ vs. l (not shown) were highly significantly linear ($P < 0.01$) and characterized by coefficients of determination r^2 consistently higher than 0.99 for both males and females. This shows that the cumulative probability distribution of move lengths L is compatible with a power-law behavior (see Eq. (5)), hence an underlying fractal structure. A major consequence of this fractal structure is that the statistical distribution of *E. affinis* swimming speed is far from Gaussian. As such, comparisons of experiments with different durations using mean values of standard behavioral metrics (e.g. swimming speed) that have a fractal structure are unlikely to be meaningful, because those mean values intrinsically depend on the duration of the experiment.

The stress exponent ϕ did not significantly differ between males ($\phi = 1.75 \pm 0.11$) and females ($\phi = 1.71 \pm 0.06$) in control experiments conducted in uncontaminated estuarine water (Fig. 11b). This suggests that the fractal structure of move lengths is similar for *E. affinis* males and females in the absence of any experimental stressors. In contrast, while the values of ϕ significantly decreased with increasing WSF concentrations for both males and females, they were consistently significantly ($P < 0.05$) lower for females than males at each WSF concentration (Fig. 11b). This indicates that WSF has a stronger disruptive effect on the swimming behavior of females, and specifies the conclusions obtained from the fractal dimension of male and female swimming paths (Fig. 11a).

The values found here for *E. affinis* in the absence of hydrocarbon stress are consistent with previous values reported for a range of calanoid copepods in the absence of any stressors, i.e. $\phi = 1.51 \pm 0.04$ in *Centropages hamatus* (Seuront & Leterme, 2007), and $\phi = 1.81 \pm 0.05$ in *Acartia clausi*, $\phi = 1.74 \pm 0.04$ in *Paracalanus parvus*, $\phi = 1.71 \pm 0.04$ in *Centropages typicus*, $\phi = 1.69 \pm 0.05$ in *Temora longicornis*, and $\phi = 1.61 \pm 0.04$ in *Pseudocalanus elongatus* (Seuront, 2011b). The observed inter-species variability in the values of ϕ nevertheless suggests that the value of the exponent ϕ, hence the fractal structure of move lengths, are species-specific. However, the 7 species of calanoid copepods investigated in the literature using Eq. (5) consistently show a decrease in the related stress exponent under stressful conditions, whatever the source of stress may be (Seuront & Leterme, 2007; Seuront, 2010b, 2011b; present work). More specifically, the relative decrease observed here in the exponent ϕ were 4.0%, 14.3% and 24.6% for males, and 8.8%, 16.5% and 27.1% for females at WSF concentrations of 0.01%, 0.1% and 1%. This suggests that the behavioral stress induced by WSF contamination of estuarine water at 0.1% and 1% is similar to the stress observed in other calanoid copepods during behavioral experiments conducted in the dark during daylight hours and in the light at night, with relative decrease in ϕ ranging from 10.9% in *P. elongatus* and 27.9% in *T. longicornis*; see Seuront (2010b) for further details. As a consequence, even if the magnitude of the changes observed in the exponent ϕ between control experiments and under stressful conditions is likely to be both sex- and species-specific, it is critical that ϕ consistently decreases under stressful conditions.

It is finally stressed, that the relative decrease observed in the exponent ϕ in Spanish ibex (*Capra pyrenaica*) parasited by the arthropod *Sarcoptes scabieis* (21.4%; Alados et al., 1996) and bottlenose dolphin (*Tursiops aduncus*) under various conditions of boat presence and traffic (8.7 to 31.5%; Seuront & Cribb, 2011) is also consistent with those reported for copepods (Seuront & Leterme, 2007; Seuront, 2010b, 2011b; present work). This may indicate that the differences between the values of the exponent ϕ observed for a given species or environment under stressful and non-stressful conditions might be more informative on the related behavioral changes rather than the absolute values of ϕ, as previously shown for several fractal and multifractal measures of behavioral and environmental complexity (Seuront, 2004, 2005, 2010b; Seuront et al., 2004a,b). Similar approaches, based on the structure of sequential behavior patterns (e.g. moving versus non-moving) have also been successfully applied to assess stress in a variety of terrestrial and aquatic vertebrates (Quenettes & Desportes, 1992; Carlstead et al., 1993; Escós et al., 1995; Alados et al., 1996; Alados & Weber, 1999; Alados & Huffman, 2000; María et al., 2004), and in the copepod *Centropages hamatus* under conditions of naphthalene contamination (Seuront & Leterme, 2007), and have consistently shown a decrease in the fractal complexity of behavioral display under stressful conditions.

5. Conclusion

This work investigated the ability of the estuarine copepod *Eurytemora affinis* to detect and subsequently avoid point-source contamination by non-lethal doses of the water-soluble fraction (WSF) of diesel oil, and their behavioral response to the diffuse contamination that is likely to follow any point-source contamination.

Both adult males and females have the ability to detect, consistently avoid and eventually escape localized patches of WSF contaminated estuarine water. They also exhibit a range of behavioral changes in contaminated estuarine water, i.e. a decrease in swimming speed and turning angle, a decrease in the fractal complexity of their swimming paths and a decrease in the fractal complexity of their successive displacements. The present work and published behavioral responses to pollution based on the analysis of swimming speed alteration indicate that the behavioral responses previously observed in zooplankton following water contamination range from hypoactivity to hyperactivity, depending on the species, sex, concentration and nature of the contaminant, and exposure time. As such, it is stressed that even if alterations in swimming speed have been detected at toxic compound concentrations well below LC50 values for a range of invertebrates (e.g. Avila et al., 2010; Ihara et al., 2010; Garaventa et al., 2010; Cailleaud et al., 2011; Seuront, 2011a), its claimed use as a non-specifc behavioral end-point in marine ecotoxicology and environmental monitoring program (Faimali et al., 2006; Garaventa et al., 2010) cannot be warranted.

A practical alternative is based on the intrinsic fractal nature of behavioral properties and their subsequent modification under stressful conditions. Both the fractal properties of swimming paths and instantaneous displacements exhibit a fractal complexity that, in sharp contrast to swimming speed alterations, is consistently decreasing under conditions of water contamination by the water-soluble fraction of diesel oil. A major consequence of the fractal properties observed in *E. affinis* swimming pattern in particular, but also in a range of invertebrates and vertebrates (e.g. Sims et al., 2008; Humphries et al., 2010), is their departure from Gaussianity. The use of the mean values of standard behavioral metrics such as swimming speed is hence unlikely to be meaningful, because they intrinsically depend on the duration of the experiment; see Seuront, 2010b for more discussion.

This generalizes and specifies previous claims that behavioral responses seem to be of similar sensitivity and efficiency as biochemical and physiological responses, thus allowing the field of behavioral ecotoxicology to expand (Dell, 2002). The observed changes in the fractal properties of swimming behavior indeed occur for very low contaminant concentration, and consistently converge towards a decrease in behavioral complexity under stressful conditions. As such, the use of fractal analysis is recommended in invertebrate ecotoxicology as a sensitive, non-invasive and robust behavioral sub-lethal end-point end-point with short-response times for toxicity bioassays, in particular as it is very sensitive to subtle behavioral changes that may be undetectable to other behavioral variables.

The application of fractals to crustaceans behavioral ecology in general (Coughlin et al., 1982; Bundy et al., 1983; Seuront, 2006; Seuront et al., 2004a-d; Uttieri et al., 2005, 2007, 2008; Seuront & Vincent, 2008; Dur et al., 2010, 2011a,b; Ziarek et al., 2011), and to crustacean ecotoxicology in particular (Shimizu et al., 2002; Seuront, 2010a,b, 2011a,b; present work) is, however, still in its infancy. Further work is needed to entangle the fractal complexity of behavioral properties and to generalize the use of fractal-based approaches to stress assessment in marine invertebrates.

6. Acknowledgment

This research was supported under Australian Research Council's Discovery Projects funding scheme (projects number DP0664681 and DP0988554). Professor Seuront is the recipient of an Australian Professorial Fellowship (project number DP0988554).

7. References

Alados, C.L., & Weber, D.N., (1999). Lead effects on the predictability of reproductive behavior in fathead minnows (*Pimephales promelas*): A mathematical model. *Environmental Toxicology and Chemistry*, 18, pp. 2392-2399

Alados, C.L., & Huffman, M.A., (2000). Fractal long-range correlations in behavioural sequences of wild chimpanzees: a non-invasive analytical tool for the evaluation of health. *Ethology*, 106, 105-116

Alados, C.L., Escós, J.M., & Emlen, J.M., (1996). Fractal structure of sequenctial behaviour patterns: an indicator of stress. *Animal Behavior*, 51, pp. 437-443

Amsler MO, Amsler CD, Rittschoff D, Becerro MA, Mc Clintock JB (2006) The use of computer-assisted motion analysis for quantitative studies of the behavior of barnacle (*Balanus amphitrite*) larvae. *Marine and Freshwater Behaviour and Physiology*, 39, pp. 259-268

Arias, A.H., Pereyra, M.T., & Marcovecchio, J.E. (2011). Multi-year monitoring of estuarine sediments as ultimate sink for DDT, HCH, and other organochlorinated pesticides in Argentina. *Environmental Monitoring and Assessment*, 172, pp. 17-32

Avila, T.R., Bersano, J.G.F., & Fillmann, G. (2010). Lethal and sub-lethal effects of the water-soluble fraction of a light crude oil on the planktonic copepod *Acartia tonsa*. *Journal of the Brazilian Society of Ecotoxicology*, 5, pp. 19-25

Bagøien, E., & Kiørboe, T., (2005). Blind dating-mate finding in planktonic copepods. I. Tracking the pheromone trail of *Centropages typicus*. *Marine Ecology Progress Series*, 300, pp. 105–115

Baillieul, M., & Blust, R., (1999). Analysis of the swimming velocity of cadmium-stressed *Daphnia magna*. *Aquatic Toxicology*, 44, pp. 245–254

Barata, C., Baird, D.J., Medina, M., Albalat, A., & Soares, A.M.V.M. (2002). Determining the ecotoxicological mode of action of toxic chemicals in meiobenthic marine organisms: stage-specific short tests with *Tisbe battagliai*. *Marine Ecology Progress Series*, 230, pp. 183-194

Barua, P., Mitra, A., Banerjee, K., & Chowdhury, S.N. (2011). Seasonal variation of heavy metals accumulation in water and oyster (*Saccostrea cucullata*) inhabiting Central and Western Sector of Indian Sundarbans. *Environmental Research Journal*, 5, pp. 121-130

Bartumeus, F., Peters, F., Pueyo, S., Marrasé, C., & Catalan, J., (2003). Helical Lévy walks: adjusting searching statistics to resource availability in microzooplankton, Proceedings of the National Academy of Science, 100, pp. 12771-12775

Bartumeus, F., da Luz, M.G.E., Viswanathan, G.M., & Catalan, J., (2005). Animal search strategies: a quantitative random-walk analysis. *Ecology*, 86, 3078-3087

Berdugo, V., Harris, R.P., & O'Hara, S.C., (1977). The effect of petroleum hydrocarbons on reproduction of an estuarine planktonic copepod in laboratory cultures. *Marine Pollution Bulletin*, 8, 138-143

Bundy, M.H., Gross, T.F., Coughlin, D.J., & Strickler, J.R., (1993). Quantifying copepod searching efficiency using swimming pattern and perceptive ability. Bulletin of Marine Science, 53, 15-28

Cailleaud, K., Michalec, F.G., Forget-Leray, J., Budzinski, H., Hwang, J.S., Schmitt, F.G., & Souissi, S. (2011). Changes in the swimming behavior of Eurytemora affinis (Copepoda, Calanoida) in response to a sub-lethal exposure to nonyphenolds. Aquat Toxicol 102, pp. 228–231

Calbet, A., Saiz, E., & Barata, C. (2007). Lethal and sublethal effects of naphathalene and 1,2-dimethylnaphthalene on the marine copepod *Paracartia grani*. *Marine Biology*, 151, pp. 195-204

Carls, M.G., & Rice, S.D., (1990). Abnormal development and growth reductions of pollock, *Theragra chalcogramma*, embryos exposed to water soluble fraction of oil. *Fisheries Bulletin*, 88, pp. 29-37

Carlstead, K., Brown, J.L., & Strawn, W., (1993). Behavioral and physiological correlates of stress in laboratory cats. *Applied Animal Behaviour Science*, 38, pp. 143-158

Charoy, C., & Janssen, C.R., (1999). The swimming behaviour of *Brachionus calyciflorus* (rotifer) under toxic stress. II. Comparative sensitivity of various behavioural criteria. *Chemosphere*, 38, pp. 3247–3260

Charoy, C., Janssen, C.R., Persoone, G., & Clément, P., (1995). The swimming behaviour of *Brachionus calyciflorus* (rotifer) under toxic stress. I. The use of automated trajectory for determining sublethal effects of chemicals. *Aquatic Toxicology*, 32, pp. 271–282

Chen, J. & Denison, M.S., (2011). The Deepwater Horizon oil spill: environmental fate of the oil and the toxicological effects on marine organisms. JYI, 21, pp. 84-95

Corner, E.D.S., Harris, R.P., Kilvington, C.C., & O'Hara, S.C.M., (1976). Petroleum compounds in the marine food web: short-term experiments on the fate of naphthalene in *Calanus*. Journal of the Marine Biological Association, 56, pp. 121–133

Coughlin DJ, Strickler JR, Sanderson B (1992) Swimming and search behavior in clownfish, *Amphiprion perideraion*, larvae. *Animal Behavior*, 44, pp. 427-440

Cowles, T.J. & Remillard, J.F., (1983a). Effects of exposure to sublethal concentrations of crude oil on copepod *Centropages hamatus*. 1. Feeding and egg-production. *Marine Biology*, 78, pp. 45-51

Cowles, T.J. & Remillard, J.F., (1983b). Effects of exposure to sub-lethal concentrations of crude oil on copepod *Centropages hamatus*. 2. Activity patterns. *Marine Biology*, 78, pp. 53-57

Dell, O.G., (2002). *Behavioral Ecotoxicology*, John Wiley & Sons, New York, NY, USA

Doall MH, Colin SP, Strickler JR, Yen J (1998) Locating a mate in 3D: the case of Temora longicornis. Phil Trans R Soc Lond B 353: 681-689

Dowling, N.A., Hall, S.J., & Mitchell, J.G., (2000). Foraging kinematics of barramundi during early stages of development. *Journal of Fish Biology*, 57, pp. 337-353

Duquesne, S., & Küster, E., (2010). Biochemical, metabolic, and behavioural responses and recovery of *Daphnia magna* after exposure to an organophosphate. *Ecotoxicology and Environmental Safety*, 73, pp. 353-359

Dur, G., Souissi, S., Schmitt, F.G., Cheng, S.H., & Hwang, J.S., (2010). The different aspects in motion of the three reproductive stages of *Pseudodiaptomus annandalei* (Copepoda, Calanoida). *Journal of Plankton Research*, 32, pp. 423-440

Dur, G., Souissi, S., Schmitt, F.G., Beyrend-Dur, D., & Hwang, J.S., (2011a). Mating and mate choice in *Pseudodiaptomus annandalei* (Copepoda: Calanoida). *Journal of Experimental Marine Biology and Ecology*, 402, pp. 1-11

Dur, G., Souissi, S., Schmitt, F.G., Michalec, F.G., Mahjoub, M.S., & Hwang, J.S., (2011b). Effects of animal density, volume, and the use of 2D/3D recording on behavioral studies of copepods. *Hydrobiologia*, 666, pp. 197-214

Elordui-Zapatarietxe, S., Albaigé, J., & Rosell-Melé, A., (2008). Fast preparation of the seawater accomodated fraction of heavy oil by sonication. *Chemosphere*, 73, pp. 1811-1816

Epstein, N., Bak, R.P.M., & Rinkevich, B., (2000). Toxicity of third generation dispersants and dispersed Egyptian crude oil on Red Sea coral larvae. *Marine Pollution Bulletin*, 40, pp. 497-503

Escós, J., Alados, C.L., & Emlen, J.M., (1995). Fractal structures and fractal functions as disease indicators. *Oikos*, 74, pp. 310–314

Faimali, M., Magillo, F., Piazza, V., Garaventa, F., & Geraci, S., (2002). A simple toxicological bioassay using phototactic behaviour of *Balanus amphitrite* (Darwin) nauplii: role of some cultural parameters and application with experimental biocides. *Periodicum Biologorum*, 104, pp. 225–232

Faimali, M., Garaventa, F., Piazza, V., Greco, G., Corra, C., Magillo, F., Pittore, M., Giacco, E., Gallus, L., Falugi, C., & Tagliafierro, G. (2006). Swimming speed alteration of larvae of *Balanus amphitrite* as a behavioural end-point for laboratory toxicological bioassays. *Marine Biology*, 149, pp. 87–96

Fielding, A., (1992). Applications of fractal geometry to biology. *Computer Applications in the Biosciences*, 8, pp. 359-366

Fisher, W.S. & Foss, S.S., (1993). A simple test for toxicity of number 2 fuel oil and oil dispersants to embryos of grass shrimp *Palaemonetes pugio*. *Marine Pollution Bulletin*, 26, pp. 385-391

Fockedey, N. & Mees, J., (1999). Feeding of the hyperbenthic mysid *Neomysis integer* in the maximum turbidity zone of the Elbe, Westerschelde and Gironde estuaries. *Journal of Marine Systems*, 22, pp. 207-228

Frangoulis, C., Skliris, N., Lepoint, G., Elkalay, K., Goffart, A., Pinnegar, J.K., & Hecq, J.H., (2011). Importance of copepod carcasses versus faecal pellets in the upper water column of an oligotrophic area. *Estuarine, Coastal and Shelf Science*, 92, pp. 456-463

Galante-Oliveira, S., Oliveira, I., Jonkers, N., Langston, W. J., Pacheco, M., & Barroso, C.M., (2009). Imposex levels and tributyltin pollution in Ria de Aveiro (NW Portugal) between 1997 and 2007: evaluation of legislation effectiveness. *Journal of Environmental Monitoring*, 11, pp. 1405-1411

Garaventa, F., Gambardella, C., Di Fino, A., Pittore, M., & Faimali, M., (2010). Swimming speed alteration of *Artemia* sp. and *Brachionus plicatilis* as a sub-lethal behavioural end-point for ecotoxicological surveys. *Ecotoxicology*, 19, pp. 512–519

Gerhadt, A., (1995). Monitoring behavioural responses to metals in *Gammarus pulex* (L.) (Crustacea) with impedance conversion. Environmental Science and Pollution Research, 2, pp. 15-23

Gerhadt, A., (2007). Aquatic behavioral ecotoxicology-prospects and limitations. *Human and Ecological Risk Assessment*, 13, pp. 481-491

Gerhadt, A., (2011). GamTox : a low-cost multimetric ecotoxicity test with Gammarus spp. For in and ex situ application. *International Journal of Zoology*, Article ID 574536, 7 pages

Gerhadt, A., & Janssens de Bisthoven, L., (1995). Behavioural, developmental and morphological responses of Chironomus gr. thummi larvae (Diptera, Nematocera) to aquatic pollution. *Journal of Aquatic Ecosystem Stress and Recovery*, 4, pp. 205-214

Gerhardt, A., Janssens de Bisthoven, L., & Soares, A.M.V., (2005). Evidence for the stepwise stress model: *Gambusia holbrooki* and *Daphnia magna* under acid mine drainage and acidified reference water stress. *Environmental Science & Technology*, 39, pp. 4150–4158

Gerhardt, A., Kienle, C., Allan, I.J., Greenwood, R., Guigues, N., Fouillac, A.M., Mills, G.A., & Gonzalez, C., (2007). Biomonitoring with *Gammarus pulex* at the Meuse (NL), Aller (GER) and Rhine (F) rivers with the online Multispecies Freshwater Biomonitor. *Journal of Environmental Monitoring*, 9, pp. 979-85

Goetze, E., & Kiørboe, T., (2008). Heterospecific mating and species recognition in the planktonic marine copepods *Temora stylifera* and *T. longicornis*. *Marine Ecology Progress Series*, 370, pp. 185-198

Goldberger, A.L., Amaral, L.A.N., Hausdorff, J.M., Ivanov, P.Ch., Peng, C.K., & Stanley, H.E., (2002). Fractal dynamics in physiology: alterations with desease and aging. *Proceedings of the National Academy of Science USA*, 99, pp. 2466-2472

Goto, T., & Hiromi, J., (2003). Toxiciy of 17a-ethynylestradiol and norethindrone, constituents of any oral contraceptive pill to the swimming and reproduction of cladoceran *Daphnia magna*, with special reference to their synergetic effect. *Marine Pollution Bulletin*, 47, 139–142

Guzmán del Próo, S.A., Chávez, E.A., Alatriste, F.M., de la Campa, S., Gómez, L., Guadarrama, R., Guerra, A., Mille, S., & Torruco, D., (1986). The impact of Ixtoc-1 oil spill on zooplankton. *Journal of Plankton Research*, 8, pp. 557-581

Harris, R.P., Berdugo, V., O'Hara, S.C.M., & Corner, E.D.S., (1977). Accumulation of ^{14}C-1-Naphthalene by an oceanic and an estuarine copepod during long-term exposure to low-level concentrations. *Marine Biology*, 42, pp. 187-195

Hashim, A.A., (2010). Effect of sublethal concentrations of fuel oil on the behavior and survival of larvae and adults of the barnacle *Balanus amphitrite* (Darwin). Turkish Journal of Fisheries and Aquatic Sciences, 10, pp. 499-503

Hastings, H.M., & Sugihara, G., (1993). *Fractals. A User's Guide for the Natural Sciences*, Oxford University Press, USA

Humphries, N.E., Queiroz, N., Dyer, J.R.M., Pade, N.G., Musyl, M.K., Schaefer, K.M., Fuller, D.W., Brunnscheweiler, J.M., Doyle, T.K., Houghton, J.D.R., Hays, G.C., Jones, C.S., Noble, L.R., Wearmouth, V.J., Southall, E.J., & Sims, D.W., (2010). Environmental context explains Lévy and Brownian movement patterns of marine predators. *Nature*, 465, pp. 1066-1069

Ihara, P.M., Pinho, G.L.L. & Fillmann, G., (2010). Appraisal of copepod *Acartia tonsa* (Dana, 1849) as a chronic toxicity test organism. *Journal of the Brazilian Society of Ecotoxicology*, 5, pp. 27-32

Janssen, C.R., Ferrando, M.D., & Persoone, G., (1994). Ecotoxicological studies with the freshwater rotifer *Brachionus calcyflorus*. 4. Rotifer behavior as a sensitive and rapid sublethal test criterion. *Ecotoxicology and Environmental Safety*, 28, pp. 244–255

Jose, J., Giridhar, R., Anas, A., Loka Bharati, P.A. & Nair, S., (2011). Heavy metal pollution exerts reduction/adaptation in the diversity and enzyme expression profile of heterotrophic bacteria in Cochin estuary, India. *Environmental Pollution*, 159, pp. 2775-2780

Janssens de Bisthoven, L., Gerhardt, A., & Soares, A.M.V.M., (2004). Effects of acid mine drainage on larval *Chironomus* (Diptera, Chironomidae) measured with the Multispecies Freshwater Biomonitor. *Environmental toxicology and chemistry*, 23, pp. 1123-1128

Jernelov, A., & Linden, O., (1981). Ixtoc I: a case study of the world's largest oil spill. *Ambio*, 10, pp. 299-306

Katona, S.K., (1973). Evidence for sex pheromones in planktonic copepods. *Limnology & Oceanography*, 18, pp. 574-583

Kendall, M., & Stuart, A., (1966). *The Advanced Theory of Statistics*, Hafner, New York

Kienle, C., & Gerhardt, A., (2008). Behavior of *Corophium volutator* (Crustacea, Amphipoda) exposed to the water-accommodated fraction of oil in water and sediment. *Environmental toxicology and chemistry*, 27, pp. 599-604

Kiørboe, T., (2008). Optimal swimming strategies in mate-searching pelagic copepods. *Oecologia*, 155, pp. 179-192

Kiørboe, T., Bagøien, E., & Thygesen, U.H., (2005). Blind dating-mate finding in planktonic copepods. II. The pheromone cloud of *Pseudocalanus elongatus*. *Marine Ecology Progress Series*, 300, pp. 117-128

Kirkpatrick, A.J., Gerhardt, A., Dick, J.T.A., McKenna, M., & Berges, J.A., (2006a). Use of the multispecies freshwater biomonitor to assess behavioral changes of *Corophium volutator* (Pallas, 1766) (Crustacea, Amphipoda) in response to toxicant exposure in sediment. *Ecotoxicology and Environmental Safety*, 64, pp. 298-303

Kirkpatrick, A.J., Gerhardt, A., Dick, J.T.A., Laming, P., & Berges, J.A., (2006b). Suitability of *Crangonyx pseudogracilis* (Crustacea: Amphipoda) as an early warning indicator in the multispecies freshwater biomonitor. *Environmental Science and Pollution Research International*, 13, pp. 242-250

Lee, C.I., Kim, M.C., & Kim, H.C., (2009). Temporal variation of chlorophyll a concentration in the coastal waters affected by the Hebei Spirit oil spill in the West Sea of Korea. *Marine Pollution Bulletin*, 58, pp. 496-502

Macedo-Sousa, J.A., Pestana, J.L.T., ., Gerhardt, A., Nogueira, A.J.A., & Soares, A.M.V.M., (2007). Behavioural and feeding responses of *Echinogammarus meridionalis* (Crustacea, Amphipoda) to acid mine drainage. *Chemosphere*, 67, pp. 1663-1670

Macedo-Sousa, J.A., Gerhardt, A., Brett, C.M.A., Nogueira, A.J.A., & Soares, A.M.V.M., (2008). Behavioural responses of indigenous benthic invertebrates (*Echinogammarus meridionalis, Hydropsyche pellucidula* and *Choroterpes picteti*) to a pulse of Acid Mine Drainage: a laboratorial study. *Environmental pollution*, 156, pp. 966-973

María, G.A., Escós, J., & Alados, C.L., (2004). Complexity of behavioural sequences and their relation to stress conditions in chickens (*Gallus gallus domesticus*): a noninvasive technique to evaluate animal welfare. *Applied Animal Behaviour Science*, 86, pp. 93-104

Matthews, B., Hausch, S., Winter, C., Suttle, C.A., & Shurin, J.B., (2011). Contrasting ecosystem-effects of morphologically similar copepods. *PloS One*, 6, e26700

McLusky, D.S., & Elliott, M. (2004). *The Estuarine Ecosystem : Ecology, Threats, and Management*. Oxford University Press, USA

Michalec, F.G., Souissi, S., Dur, G., Mahjoub, M.S., Schmitt, F.G., & Hwang, J.S., (2010). Differences in behavioral responses of *Eurytemora affinis* (Copepoda, Calanoida) reproductive stages to salinity variations. *Journal of Plankton Research*, 32, pp. 805-813

Morhange, C., Hamdan Taha, M., Humbert, J.-B., & Marriner, N., (2005). Heavy metal pollution exerts reduction/adaptation in the diversity and enzyme expression profile of heterotrophic bacteria in Cochin estuary, India. *Méditerranée*, 104, pp. 75-78

Munaron, D., Tapie, N., Budzinski, H., Andral, B., & Gonzalez, J.-L., (2011). Pharmaceuticals, alkylphenols and pesticides in Mediterranean coastal waters: Results from a pilot survey using passive samplers. *Estuarine, Coastal and Shelf Science*, in press

Nihongi, A., Lovern, S.B., & Strickler, J.R., (2004). Mate-searching behaviors in the freshwater calanoid copepod *Leptodiaptomus ashlandi*. *Journal of Marine Systems*, 49, pp. 65-74

Noaksson, E., Linderoth, M., Tjärnlund, U., & Balk, L., (2005). Toxicological effects and reproductive impairments in female perch (*Perca fluviatilis*) exposed to leachate from Swedish refuse dumps. *Aquatic Toxicology*, 75, pp. 162-177

Ohwada, K., Nishimura, M., Wada, M., Nomura, H., Shibata, A., Okamoto, K., Toyoda, K., Yoshida, A., Takada, H., & Yamada, M., (2003). Study of the effect of water-soluble fractions of heavy-oil on coastal marine organisms using enclosed ecosystems, mesocosms. *Marine Pollution Bulletin*, 47, pp. 78-84

Pavillon, J.-F., Oudot, J., Dlugon, A., Roger, E., & Juhel, G., (2002). Impact of the 'Erika' oil spill on the *Tigriopus brevicornis* ecosystem at the Le Croisic headland (France): preliminary observations. *Journal of the Marine Biological Association*, 82, pp. 409-413

Quenette, P.Y., & Desportes, J.P., (1992). Temporal and sequential structure of vigilance behavior of wild boars (Sus scrofa). *Journal of Mammalogy*, 73, pp. 535-540

Ren, Z., & Wang, Z., (2010). Differences in the behavior characteristics between *Daphnia magna* and *Japanese madaka* in an on-line biomonitoring system. *Journal of Environmental Sciences*, 22, pp. 703-708

Ren, Z., Li, Z., Ma, M., Wang, Z., & Fu, R., (2009) Behavioral responses of *Daphnia Magna* to stresses of chemicals with different toxic characteristics. *Bulletin of Environmental Contamination and Toxicology*, 82, pp. 310-316

Ren, Z., Zha, J., Ma, M., Wang, Z., & Gerhardt, A., (2007). The early warning of aquatic organophosphorus pesticide contamination by on-line monitoring behavioral changes of *Daphnia magna*. *Environmental Monitoring and Assessment*, 134, pp. 373-383

Reynolds, A.M., Smith, A.D., Menzel, R., Greggers, U., Reynolds, D.R., & Riley, J.R., (2007) Displaced honey bees perform optimal scale-free search flights. *Ecology*, 88, pp. 1955-1961

Rodrigues, R.V., Miranda-Filho, K.C., Gusmão, E.P., Moreira, C.B., Romano LA, & Sampaio, L.A., (2010). Deleterious effects of water-soluble fraction of petroleum, diesel and gasoline on marine pejerrey *Odontesthes argentinensis* larvae. *Science of the Total Environment*, 408, pp. 2054-2059

Rumney, H.S., Laruelle, F., Potter, K., Mellor, P.K., & Law, R.J., (2011). Polycyclic aromatic hydrocarbons in commercial fish and lobsters from the coastal waters of Madagascar following an oil spill in August 2009. *Marine Pollution Bulletin*, 62, pp. 2859-2862

Saeed, T., & Al-Mutairi, M., (1999). Chemical composition of the water-soluble fraction of the leaded gasolines in seawater. *Environment International*, 25, pp. 117-129

Samain, J. F., Moal, J., Coum, A., Le Coz, J. R., & Daniel, J.-Y., (1980). Effects of the "Amoco Cadiz" oil spill on zooplankton. A new possibility of ecophysiological survey. *Helgol. Meeresunters*, 33, pp. 225-235

Schnitzler, J.G., Thomé, J.P., Lepage, M., & Das, K., (2011). Organochlorine pesticides, polychlorinated biphenyls and trace elements in wild European sea bass (*Dicentrarchus labrax*) off European estuaries. *Science of the Total Environment*, 409, pp. 3680-3686

Seuront, L., (2004). Small-scale turbulence in the plankton: low-order deterministic chaos or high-order stochasticity? *Physica A*, 341, pp. 495-525

Seuront, L., (2005). Hydrodynamical and tidal controls of small-scale phytoplankton patchiness. *Marine Ecology Progress Series*, 302, pp. 93-101

Seuront, L., (2006). Effect of salinity on the swimming behaviour of the estuarine calanoid copepod *Eurytemora affinis*. *Journal of Plankton Research*, 28, pp. 805-813

Seuront, L., (2010a). Zooplankton avoidance behaviour as a response to point sources of hydrocarbon contaminated water. *Marine and Freshwater Research*, 61, 263-270

Seuront, L., (2010b). *Fractals and Multifractals in Ecology and Aquatic Science*, CRC Press, Boca Raton, FL, USA

Seuront, L., (2011a). Hydrocarbon contamination decreases mating success in a marine planktonic copepod. *PLoS ONE*, 6, e26283

Seuront, L., (2011b). Behavioral fractality in marine copepods: endogenous rhythms vs. exogenous stressors. *Physica A*, 309, pp. 250-256

Seuront, L., & Spilmont, N., (2002). Self-organized criticality in intertidal microphytobenthos patch patterns. *Physica A*, 313, pp. 513-539

Seuront, L., & Cribb, N., (2011). Fractal analysis reveals pernicious stress levels related to boat presence and type in the Indo-Pacific bottlenose dolphin, Tursiops aduncus. *Physica A*, 390, pp. 2333-2339

Seuront, L. & Leterme, S., (2007). Increased zooplankton behavioural stress in response to short-term exposure to hydrocarbon contamination. *The Open Oceanography Journal*, 1, pp. 1-7

Seuront, L., & Vincent, D., (2008). Impact of a *Phaeocystis globosa* spring bloom on *Temora longicornis* feeding and swimming behaviours. *Marine Ecology Progress Series*, 363, pp. 131-145

Seuront, L., Brewer, M., & Strickler, J.R., (2004a). Quantifying zooplankton swimming behavior: the question of scale. In: *Handbook of Scaling Methods in Aquatic Ecology: Measurement, Analysis, Simulation*, Seuront L. & Strutton P.G., CRC Press, Boca Raton, 333-359

Seuront, L., Hwang, J.S., Tseng, L.C., Schmitt, F.G., Souissi, S., Shih, C.T., & Wong, C.K., (2004b), Individual variability in the swimming behavior of the tropical copepod Oncaea venusta (Copepoda: Poecilostomatoida). Marine Ecology Progress Series, 283, pp. 199-217

Seuront, L., Yamazaki, H., & Souissi, S., (2004c). Hydrodynamic disturbance and zooplankton swimming behavior. *Zoological Studies*, 43, pp. 377-388

Seuront, L., Schmitt, F.G., Brewer, M.C., Strickler, JR.., & Souissi, S., (2004d). From random walk to multifractal random walk in zooplankton swimming behavior. *Zoological Studies*, 43, pp. 8-19

Seuront, L., Duponchel, A.C., & Chapperon, C., (2007). Heavy-tailed distributions in the intermittent motion behaviour of the intertidal gastropod *Littorina littorea*. *Physica A*, 385, pp. 573-582

Schmitz, O.J., (2008). Herbivory from individuals to ecosystems. *Annual Review of Ecology and Systematics*, 39, pp. 133–152

Shafir, S., Van Rijn, J., & Rinkevich, B., (2003). The use of coral nubbins in coral reef ecotoxicology testing. *Biomolecular Engineering*, 20, pp. 401-406

Shannon, K.L., Lawrence, R.S., & McDonald, D., (2011). Anthropogenic sources of water pollution: parts 1 and 2, In: *Water and Sanitation-Related Diseases and the Environment: Challenges, Interventions, and Preventive Measures*, Selendy J.M.H., John Wiley & Sons, Hoboken, NJ, USA

Shimizu, N., Ogino, C., Kawanishi, T., & Hayashi, Y., (2002). Fractal analysis of *Daphnia* motion for acute toxicity bioassay. *Environmental Toxicology*, 17, pp. 441–448

Sims, D.W., Southall, E.J., Humphries, N.E., Hyas, G.C., Bradshaw, C.J.A., Pitthchfor, J.W., James, A., Ahmed, M.Z., Brierley, A.S., Hindell, M.A., Morritt, D., Musyl, M.K., Righton, D., Shepard, E.L.C., Wearmouth, V.J., Wilson, R.P., Witt, M.J., & Metcalfe,

J.D., (2008). ling laws of marine predator search behaviour. *Nature,* 451, pp. 1098-1103

Sokal, R.R., & Rohlf, F.J., (1995) *Biometry*, Freeman, New York, USA

Soetaert, K., & Van Rijswijk, P. (1993). Spatial and temporal patterns of the zooplankton in the Westerchelde estuary. *Marine Ecology Progress Series,* 97, pp. 47-59

Sullivan, B.K., Buskey, E., Miller, D.C., & Ritacco, P.J., (1983). Effects of copper and cadmium on growth, swimming and predator avoidance in *Eurytemora affinis* (Copepoda). *Marine Biology*, 77, pp. 299–306

Tawfiq, N., & Olsen, D.A., (1993). Saudi Arabia's response to the 1991 Gulf oil spill Nizari. *Marine Pollution Bulletin*, 27, pp. 333-345

Thompson, B., Adelsbach, T., Brown, C., Hunt, J., Kuwabara, J., Neale, J., Ohlendorf, H., Scharzbach, S., Spies, R., & Taberski, K., (2007). Biological effects of anthropogenic contaminants in the San Francisco Estuary. *Environmental Research*, 105, pp. 156-174

Tiselius, P., Jonsson, P.R., (1990). Foraging behavior of six calanoid copepods: observations and hydrodynamic analysis. *Marine Ecology Progress Series*, 66, pp. 23-33

Togo, F., & Yamamoto, Y., (2000). Decreased fractal component of human heart rate variability during non-REM sleep. American Journal of Physiology-Heart and Circulatory Physiology, 280, pp. H17–H20

Untersteiner, H., Kahapka, J., & Kaiser, H., (2003). Behavioural response of the cladoceran *Daphnia magna* Straus to sublethal copper stressvalidation by image analysis. *Aquatic Toxicology*, 65, pp. 435–442

Untersteiner, H., Gretschel, G., Puchner, T., Napetschnig, S., & Kaiser, H., (2005). Monitoring behavioural responses to the heavy metal cadmium in the marine shrimp *Hippolyte inermis* leach (Crustacea: Decapoda) with video imaging. *Zoological Studies*, 44, pp. 71–80

Uttieri, M., Zambianchi, E., Strickler, J.R., & Mazzocchi, M.G., (2005). Fractal characterization of three-dimensional zooplankton swimming trajectories. *Ecological Modelling*, 185, pp. 51-63

Uttieri, M., Nihongi, A., Mazzocchi, M.G., Strickler, J.R., & Zambianchi, E., (2007). Precopulatory swimming behaviour of *Leptodiaptomus ashlandi* (Copepoda: Calanoida): a fractal approach. *Journal of Plankton Research*, 29, pp. i17–i26.

Uttieri, M., Paffenhöfer, G.A., & Mazzocchi, M.G., (2008). Prey capture in *Clausocalanus* (Copepoda: Calanoida). The role of swimming behaviour. *Marine Biology* 153, pp. 925-935

Van Duren, L.A., & Videler, J.J., (1995). Swimming behaviour of development stages of the calanoid copepod *Temora longicornis* at different food concentrations. *Marine Ecology Progress Series*, 126, pp. 153-161

Vane, C.H., Chenery, S.R., Harrison, I., Kim, A.W., Moss-Hayes, V., & Jones, D.G., (2011). Chemical signatures of the Anthropocene in the Clyde estuary, UK: sediment-hosted Pb, $^{207/206}$Pb, total petroleum hydrocarbon, polyaromatic hydrocarbon and polychlorinated biphenyl pollution records. Philosophical Transactions of the Royal Society A, 13, pp. 1085-1111.

Varela, M., Bode, A., Lorenzo, J., Alvarez-Ossorio, M.T., Miranda, A., Patrocinio, T., Anadon, R., Viesca, L., Rodriguez, N., Valdes, L., Cabal, J., Urrutia, A., Garcia-Soto, C., Rodriguez, M., Alvarez-Sarez-Salgado, X.A., & Groom, S., (2006). The effect of the "Prestige" oil spill on the plankton of the N-W Spanish coast. *Marine Pollution Bulletin*, 53, pp. 272–286

Venkateswara Rao, J., Kavitha, P., Jakka, N.M., Sridhar, V., & Usman, P.K., (2007). Toxicity of organophsphates on morphology and locomotor behavior in brine shrimp, Artemia salina. *Archives of Environmental Contamination and Toxicology*, 53, pp. 227–232

Villa, M., Manjón, G., Hurtado, S. & García-Tenorio, R., (2011). Uranium pollution in an estuary affected by pyrite acid mine drainage and releases of naturally occurring radioactive materials. *Marine Pollution Bulletin*, 62, pp. 1521-1529

Visser, A.W., (2007). Motility of zooplankton: fitness, foraging and predation. *Journal of Plankton Research*, 29, pp. 447-461

Weissburg, M.J., Doall, M.H., & Yen, J., (1998). Following the invisible trail: kinematic analysis of mate-tracking in the copepod *Temora longicornis*. *Philosophical Transactions of the Royal Society of London A*, 353, pp. 701-712

Weissman, P., Lonsdale, D.J., & Yen, J., (1993) The effect of peritrich ciliates on the production of *Acartia hudsonica* in Long Island Sound. Limnology & Oceanography, 38, pp. 613-622

West, B.J., & Scafetta, N., (2003). Nonlinear dynamical model of human gait. *Physics Review E*, 67, pp. 051917

Whaltham, N.J., Teasdale, P.R., & Connolly, R.M., (2011). Contaminants in water, sediment and fish biomonitor species from natural and artificial estuarine habitats along the urbanized Gold Coast, Queensland. *Journal of Environmental Monitoring*, 13, pp. 3409-3419

Woodson, C.B., Webster, D.R., Weissburg, M.J., & Yen, J., (2007). Cue hierarchy and foraging in calanoid copepods: ecological implications of oceanographic structure. *Marine Ecology Progress Series*, 330, pp. 163-177

Woodson, C.B., Webster, D.R., Weissburg, M.J., & Yen, J., (2008). The prevalence and implications of copepod behavioral responses to oceanographic gradients and biological patchiness. *Integrative and Comparative Biology*, 47, pp. 831-846

Xuereb, B., Chaumot, A., Mons, R., Garric, J., & Geffard, O., (2009a). Acetylcholinesterase activity in Gammarus fossarum (Crustacea Amphipoda). intrinsic variability, reference levels, and a reliable tool for field surveys. *Aquatic Toxicology*, 93, pp. 225–233

Xuereb, B., Lef`evre, E., Garric, J., & Geffard, O. (2009b). Acetylcholinesterase activity in Gammarus fossarum (Crustacea Amphipoda): linking AChE inhibition and behavioural alteration. *Aquatic Toxicology*, 94, pp. 114-122

Yang, F., Zhang, Q., Guo, H., & Zhang, S., (2010). Evaluation of cytotoxicity, genotoxicity and teratogenicity of marine sediments from Qingdao coastal areas using *in vitro* fish cell assay, comet assay and zebrafish embryo test. *Toxicology in vitro*, 24, pp. 2003-2011

Yen, J., Weissburg, M.J., & Doall, M.H., (1998). The fluid physics of signal perception by mate-tracking copepods. Philosophical Transactions of the Royal Society A, 353, pp. 787-804

Yen, J., Sehn, J.K., Catton, K., Kramer, A. & Sarnelle, O., (2011). Pheromone trail following in three dimensions by the freshwater copepod *Hesperodiaptomus shoshone*. *Journal of Plankton Research*, 33, pp. 907-916

Zar, J.H., (2010). *Biostatistical Analysis*, Prentice Hall, USA

Ziarek, J.J., Nihongi, A., Nagai, T., Uttieri, M., & Strickler, J.R., (2011). Seasonal adaptations of *Daphnia pulicaria* swimming behaviour: the effect of water temperature. *Hydrobiologia*, 661, pp. 317-327

11

Interactions Between Marine Ecosystems and Tourism on the Adriatic and Mediterranean

Stanko Geić, Jakša Geić and Sanja Rašetina
University of Split,
Croatia

1. Introduction

Tourism today within its quantitative proportions in domestic and international travels around the world has more than 3.5 billion participants, which makes a half of world ´s population. International tourism as a vital economic factor of a large number of developed and transition countries is approaching already one billion participants, with the economic effects that exceed 800 billion USD, and makes one of the most distinct economic sectors in world economy (UNWTO 2010).

The largest part of these migrations in the last two centuries is in the Mediterranean, being the most important world tourism mega region that accepts a third of tourists in the international tourism in more than 100 million accommodation units, which makes more than 300 million tourists from abroad and the same number of domestic visitors (Geić S., 2010).

This enormous tourist traffic, supported by the appropriate receptive and municipal transport infrastructure, takes place mostly on the narrow coastal, insular and sea areas of the Mediterranean, using its balneological benefits of the marine and recreational areas of the Euro-Afro-Asian Mediterranean as the most enjoyable climate for tourist recreation, which was recognized already on the UN World Conference on the Human Environment (1972) when this area was declared the **World recreation zone**.

Considering the fact of enormous processes of littoralization that attracted many tourist facilities to the Mediterranean coast, as well as other economic activities that are directly or indirectly related to the sea or naval communications, but also considering galloping urbanization of coastal cities and towns, as well as weekend houses (residencies), one gets to a logical conclusion of the growth of all kinds of pollution that legally or illegally use the Mediterranean waters as an „unlimited" natural recipient. If one adds frightening pollutions by numerous river flows of the industrially developed European, Asian or African countries that also end in the Mediterranean, situation really becomes alarming.

Namely, the Mediterranean as a relatively closed sea with purifying options by changing water masses that are limited with the narrow Gibraltar defile and the Suez Canal, has been called symbolically for some time in the professional literature „ a black hole" of Europe. It effects naturally and in the long run the sustainability of the Mediterranean itself as a quality, economic and fishing resource and its function in the process of health and quality of life preservation for the inhabitants of the coast and the islands, therefore logically tourists as well.

When analyzing the geographical position, distinctive tightness and fewer possibilities of selfpurification by water masses change, as well as ever more intense, traffic, economic and urban progresses, the stated above can also be applied to the Adriatic, although in a much more complex version, being potentially along with the Black Sea the most endangered part of a wider Mediterranean area.

The European and World science have referred to the stated problems on numerous international congresses, on the topic of environment protection and sustainable development, mostly under the auspices of the UN and its specialized organizations, which is very important for sustainable management of marine, economic and tourist resources of the Mediterranean (Stockholm 1972, Barcelona 1975, 1976, 2005, Rio de Janeiro 1992, Dakar 1993, Lanzarote 1995, Sevilla 1995, Istanbul 1996, Kyoto 1997, Poznan 2008, Cancun 2010).

In the context of protection and sustainable management, as well as control of marine and coastal resources of the Mediterranean, one needs to stress in particular the regulations of the Barcelona Convention, i.e. (since 1995) the Convention for the Protection of the Marine Environment and the Coastal Region of the Mediterranean as well as the Mediterranean Action Plan (MAP 1975 and 2005) that all foresee:

- Supervision and effective control of pollution in the sea,
- Sustainable management of marine and coastal resources,
- Effective integration of environment in social and economic development
- The protection of sea and coast with the measures of prevention and pollution elimination,
- Protection of natural and cultural monumental heritage,
- A solid cooperation between the Mediterranean coastal countries on the level of ecological protection,
- A joint effort to improve the quality of life (UNEP, 2005).

In terms of the above, presented work is operationalized through the following matrix:

1. Introduction
2. Approach to the problem and the concepts of coastal and marine resources management in the Adriatic and Mediterranean tourism
3. Selected concepts and models for ecosystems management
4. Conclusion

2. Approach to the problem and the concepts of coastal and marine resources management in the adriatic and mediterranean tourism

There was a relatively quick reaction in the Adriatic to the ever more frequent scientific and professional as well as publicist warnings about the danger that threatens with the pollution of environment and of natural areas in this precious and ecologically very fragile area. Crucial task in that sense was made by a group of extraordinary international and local scientists within two epochal projects under the auspices of the UN, which defined the concept of coastal management and potentials of the Adriatic. These are:

- UN Development program „Physical Development Plan for Yugoslav Adriatic Region", Split 1970
- UN Development program - Protection of the Human Environment Project for the Adriatic Region, Rijeka 1978

All of these documents on track of the most excellent achievements of former science and profession in the field of spatial planning and natural and anthropogenic resource protection were later, even today, a foundation of analysis in the sphere of the sea and coastal protection in the Adriatic as well as the basic resources in spatial planning documents of the Adriatic local and regional administration, especially in cultural and natural heritage protection plan, but also areas and zones that are intended for public recreation or tourism construction as a sustainable tourism development.

Cognition and activities of these as well as of the operative projects in progress, connected to protection of the Adriatic, are going to be reasserted in this article, and they are of enormous importance for intensive development of the Adriatic tourism and protection of marine ecosystem. These are the Integral Project of the Kaštela Bay Management by the Split University, UNEP and HAZU and the WHO. There is also a mega-project of the pollution management „Project Adriatic", which operationalizes the potentials of 20 Adriatic cities and municipalities, with funds of the CBRD, the World Bank and the EBRD.

It is followed by very successful and ever more frequent micro-projects for the particular sites (holiday resorts participants revitalized by tourist function etc.) (Geić S., Voloder F., 2009.).

Projects mentioned above in their ecological, technical and financial component and meaning for the regional tourism and economical development will be compared in this paper to analogue programmes throughout the Mediterranean (Cote d'Azur, Toulon, Monaco, Rhodes, Athens, Istanbul, Rimini, Barcelona, the Balearic Islands, the Canary Islands, Akaba - Israel, Cyprus and Al Hoceima and Agadir - Morocco, etc.).

All of these mega-projects will be analyzed and actualized critically, according to contemporary multinational models of the sea and coastal area management (PREPARE-2001 in authorship of British and Spanish scientists, META by the English experts and famous American LAC PLANNING SYSTEM - Limits to Acceptable Change), or will be put in a concept of a strategic management planning of the coastal area according to the PEST model that looks at problems through **P**olitical, **E**conomical, **S**ocial and cultural and **T**echnological research component, pointing out to a contemporary macro-concept "Tripple Helix" which assumes coordinated economy activity, local administration and the State, in context of the sustainable development and especially in the development planning of the sustainable contemporary tourism at sea and on the coast. The author's research team of this paper was also set on this concept-project in terms of the multidisciplinary experts and scientists of the economic, geographic and sociological provenance (Geić S. et. al. 2010.).

2.1 Approach to the problem of water resources in the Mediterranean

Tourism as one of the most distinct modern socio-economical phenomena of the world in its quantitative indicators is approaching gradually in the framework of international relations to billions of participants, with the economic effects that exceed the 800 billion USD. These sizes are almost triplicate by inclusion of domestic tourist migration, which gives more impressive social, economic and ecological dimension to a whole figure.

The biggest part of European travel trends, which makes over half of total world tourism in the area of tourist offer and demand, is realized in the Mediterranean coastal region, intensively estimating its water resources, especially the sea as the most important component of attractiveness of the content of the tourist offer.

There are of course needs for enormous amount of fresh water used in the frame of facilities of tourist supra-structure or in complementary sectors, especially agriculture as an important manufacturer of a wide range of raw materials and final goods needed for meet the catering tourist demand.

Given the characteristics of Mediterranean climate with a extremely summer arid period in time of **greatest demand for water** to meet the needs of multiciplinary number of consumers following intense sequences of tourist season in many Mediterranean tourist destinations, **rational management of these resources** represents an important component of the economic policies of all Mediterranean countries and regions.

The fact that the **largest part of pollution** of local population, tourists and economic activities through the river flows and coastal discharges, directly or indirectly, traditionally **oriented towards the Mediterranean Sea** as the most important recipient, is causing increasing concern.

Namely, the Mediterranean is a relatively closed sea whose options of purifying by changing water masses are limited by narrow Gibraltar defile and the Suez Canal, so the professional environmental literature have been calling it symbolically the largest "black hole" of Europe for a long time, expressing **extreme concern for the long-term sustainability of the Mediterranean Sea** as a quality tourism and fishing resource, and resource to the maintenance of health and quality of life for tourists and coastal residents, which by all intensive processes of urban and economic littoralisation is becoming more endangered.

Giving consideration to geographical position, the relative tightness and increased traffic, economic and urban developments, brought in much complex variants can be applied to the **Adriatic, which is with the Black Sea, the most endangered part of the Mediterranean** and its surroundings.

Bearing in mind the fact that the UN experts have declared yet in seventies of last century, how even than, and especially in the foreseeable future that evident are three key problems of humanity in a **deficiency of energy, food and drinking water**, which is particularly actual in the relatively arid areas including the Mediterranean, the rational management of water resources has becoming one of the fundamental conditions of survival of civilization.

Conclusions that have been carried out are precisely elaborated in a number of global conferences beginning with the world conference on sustainable development (Rio 1972...), and the UN Conference on Climate Change (Kyoto 1997, Poznań 2008...), and from the tourist aspect, the **tourist experts speak about it anxiously**, especially since the World Conference on sustainable tourism development (Lanzarote 1995), International Conference MAB (Seville 1995), and many prominent world statesmen and scientists gathered at the World Summit Habitat II (Istanbul 1996...) which was dedicated to these problems.

All this resulted in the special **hiring of numerous international bodies and organizations** under the auspices of the United Nations (UNEP, UNESCO, FAO, IUCN...) or other regional international organizations (OECD, EU, etc...) including the increasing number of environmental organizations and associations throughout the world, and subsequently rerouting of new ways of international and national policies and legislative measures towards preservation and rational use of water.

This resulted in **sensibilization of businesses subjects** especially broadest part of the population, and by making appropriate more rigorous measures and technical procedures which allude protection of water resources through more efficient technical procedure of conditioning of waste water including mechanical, chemical and biological processes that protect land and water surface, and certainly the Sea as the largest global recipient.

The **modern technologies are appearing and are enabling recycling processes** with which the subdued volumes of potable water are saved in conditioning treatment, transforming waste water in the so-called Industrial water suitable for use in industry, agriculture, tourism, and utilities sectors.

Through this procedure, for example Israel, turns its desert regions into a fertile agricultural area with enormous yields, where organic fertilizers are making the contribution as a by-product in the process of recycling waste water. Naturally, these processes and technologies are much cheaper and more efficient than **procedures of desalting of sea water** as a common practice in the rich Arabic coastal regions, especially relying on the possible effects which are made by Conditioning processes to protect the sea and its fishing, recreational and health components.

These experiences across Europe and the world, especially in the **function of sustainable tourism development** that seeks to successfully develop over 190 countries around the world with the support of UNWTO and the number of regional associations, can be of exceptional benefit for rerouting of specific activities and projects at national and regional level in Croatia. This is particularly important for its coastal part oriented towards intensive tourist economy that is, based on extremely valuable potential provided by marine resources and a Mediterranean climate, already declared as the **global zone for recreation** at the international conference on the environment (Stockholm 1972).

Therefore, while Croatia is approaching toward Euro-Atlantic integrations, a **sustainable concept of rational management of environment**, with the effective **protection of most valuable resources**, especially the **sea and the water**, for the needs of the present generations and for the next generations should be created.

2.1.1 Tourist significance of water supply and pollution management projects in the Croatian Adriatic area

Several key world problems were identified at the UN Conference on the Human Environment in Stockholm (1972), which urgently need to be solved with joint forces by the

IC, and in order to sustain life on our Planet. All of these problems are directly or indirectly closely related to water resources on Earth. It is by all means logical when you look at the fact that water covers 70% of the Planet in all its physical states and therefore represents the most significant resource; the very fundamental source of life indeed.

The Conference accentuated in terms of the facts presented even then as today, and almost for the eternity the ever present issues and questions:

- issues of ever more frequent pollution of international waters and air,
- issues of the rapid population growth, followed by the threats of water shortage, needed for different activities,
- lack of enforceable research programme of protection of the ever more endangered ecosystem including the international waters´ system,
- world's incapability to limit the usage of persistent toxic pollutants that mostly settle into the waterways above and under surface, and derogate the quality of drinking water, as well as water for agriculture and food industry, but also water resources for the health and recreational purposes,
- lack of efficient waste recycling systems which results in its discharging on the surface or into waterways, mostly into sea, being the biggest but not the most infinite recipient,
- absence of a convenient development plan for usage of the new energy sources, which includes insufficient exploitation of waterways for new hydroelectric power plants, lack of pumped-storage hydroelectric power plants, lack of usage of the enormous wave power, oscillation tides and similar water potentials,
- insufficient investment of public and private sector in pollution prevention and environmental improvement, including the most threatened water systems and which is still considered to be an unnecessary cost, and not a condition for a local community's and even the global civilization's survival,
- incompetence of states and the international community to develop an enforceable system of ecological surveillance and cooperation, especially when it comes to water resources that are mutually used as recipients, or as waterways in production of drinking water and food, tourist recreation, health etc. (Geić, S., 2002).

All of the problems mentioned above obviously fit into a framework of 3 crucial questions that are an existential burden for the humanity, **i.e. shortage of drinking water, food and energy** and which is directly or indirectly connected to water resources. The UN and its specialized organizations (UNEP, FAO, UNICEF, UNESCO and others) as well as the regional integrations (OECD, EU) specifically emphasize these problems, while intensively trying to remediate current state, with help of numerous non-governmental organizations.

It is urgent to establish a **sustainable water resources management**, as soon as possible, as a significant contribution to the global efforts in this field, in the context of this paper and the correlative terms of water management and tourism development in the Croatian Adriatic that was recognized as an international recreation zone, rich in natural and anthropogenic sources.

In this context should the regulations of the **Barcelona Convention and the Mediterranean Action Plan (MAP) for the Protection of the Mediterranean (1975)** be emphasized and which, according to the Supplement in 2005, predict in the Programme for the Protection of the Marine environment and the Coastal Region of the Mediterranean:

- Supervision and effective control of pollution in the sea,
- Sustainable management of marine and coastal resources,
- Effective integration of environment in social and economic development
- The protection of sea and coast with the measures of prevention and pollution elimination,
- Protection of natural and cultural monumental heritage,
- A solid cooperation between the Mediterranean coastal countries on the level of ecological protection,
- A joint effort to improve the quality of life (UNEP, 2005) (www.unep.org).

There were relatively prompt reactions to the more and more frequent scientific, professional and journalistic warnings in the Adriatic about the danger that threatens with the pollution degradation of the natural environment. A complex project supported by the UN, called „Protection and improvement of environment in the region of the Yugoslav Adriatic" (IET 1978) was realized after this initiative in the seventies and which resulted in an important action and preservation of the coastal water resources for numerous other purposes of our civilization.

It is important to state following important actions in this area of operalization of local and regional administration, supported by the State and the international community, which had a direct impact on the quality of tourist services:

- Activities for improvement and **construction of necessary utilities, infrastructure and especially water supply facilities**, where numerous projects of local and regional water supply constructions were realized in the seventies and eighties, as well as the actions for the water supply recovery sources and pollution protection, but also actions for the establishment of the sanitary control on the water supply facilities. A specific realization of the regional water supply systems, supported by the international financial institutions began in that sense in the seventies, throughout Dalmatia-in the Šibenik region, for the regions of Omiš, Brač and Hvar, for the Makarska Riviera, Ploče region, Pelješac, Korčula and Mljet, area of Dubrovnik and also joint water supply for Sinj municipality and Split hinterland, that are for several years in function. Unfortunately, the intensity of these activities after the nineties did not follow the objective needs, therefore are the water supply problems still evident on the Dalmatian islands and in the hinterland, but also insufficient supplies on the coast followed by the price that practically obstructs rational agricultural production, especially in the periods of drought. New projects are therefore very significant, where regional water supplies are connected with particular systems on the Jadro, Cetina and Krka river sources, forming a unique ring of a much bigger capacity, that should ultimately solve in the long run the problems of water supply in the coastal area, hinterland and on the islands, being a basis of a socioeconomic development of Central Dalmatia. It should be mentioned that water for agricultural production purpose is free of charge in many developed countries, as an act of support for this activity. These projects were preconditions for an intensive tourism development on the coast and islands.
- Intensive activity for **realization of more regional ecoprojects** with purpose of introducing contemporary disposal of total waste material, meaning contemporary solutions to the city sewage systems and construction of secondary network from the housing, catering and industrial establishments, by water waste conditioning and

permanent sea pollution check-ups as well as bathing and fishing prohibition in certain polluted areas.

The Adriatic and Croatian coastal area are both distinctive and unique because of the richness of life, purity and transparency of the sea, and well-indented and dynamic landscape. In that sense is also Croatia's basic strategic orientation the sustainable management of the Adriatic Sea, coast and islands, and conservation of invaluable treasures and natural diversity of marine ecosystems and coastal areas, allowing balanced development of economic activities in the coastal and island areas. There are the most valuable, but also the most sensitive natural ecosystems in areas of the Adriatic Sea, coast and islands. There are processes carried out that depend on a mutual activity of the sea and mainland. All these peculiarities require very careful and thoughtful management and control. Developing pressures and negative impacts on natural ecosystems are getting more and more evident and come from intensifying processes of the universal littoralisation which endangers coastal and island landscape as well as water surfaces, and especially the sea where pollution of all kinds end up directly or indirectly.

In this sense almost 20 years ago began the **operative program of the sea bathing water quality testing** on beaches, and went through many modifications, adjustments and changes. Legal basis for implementation of this program are the Regulation on the quality of bathing water **(NN 110/2007)** and the Environmental Protection **(NN 82/1994).** Program of the sea bathing water quality testing on beaches is financed by the budget of the coastal boroughs, and the testing itself is conducted by the authorized laboratories. Sea bathing water quality monitoring means sampling and analyzing sea water for bathing and recreation, and constant public information. The Ministry of Environmental Protection, Physical Planning and Construction is responsible for coordination of the program, consolidation and valorization of data as well as for informing the public.

Aims of the program are:

- Protecting the health of bathers and public health education,
- Sustainable beach management, due to preserving their natural features,
- Pollution source detecting, functioning of the existing waste disposal system and monitoring of the sewage system construction
- Public information (Geić S., 2007).

For example, the sea in Split-Dalmatian County is sampled at 140 points. Observing the individually evaluated samples in the season 2008, 59,70% of samples were evaluated as being the sea of high quality, 39,37% as suitable for bathing and 0,93% as being moderately polluted, which are relatively favourable indicators if compared to competitive regions.

Results of the sea bathing water quality testing on beaches during 2008 indicate that the sea bathing water in Croatia is of high quality because even 99.01% of samples meet strict criteria stipulated by the Regulation on the standard bathing water quality on beaches. These results can be used in tourism promotion which is particularly effective in example of their publication by ADAC organization, intended for motorized tourists throughout Germany and neighbouring countries (www.mzopu.hr).

In order to preserve high quality of sea on beaches and their natural features, one should approach the issue of beach management and sea bathing water and recreation completely, bearing in mind economic and management importance on one hand and sea and marine environment protection on the other.

Along with all positive economic effects, tourism brings also a great burden for marine and coastal environment and represents a major risk factor in terms of possible sea quality deterioration, therefore is in that context an especially important for Croatia development of the sustainable tourism.

Bearing in mind the statements already mentioned in this paper, "Eco Kaštela Bay" Project and „the Adriatic Project" supported by international financial institutions throughout Croatian Adriatic will be particularly analyzed, and that will despite the elements of imperfection of the system have extraordinary positive repercussions for the tourism development and level of health education in the Adriatic sea.

- **"ECO Kaštela Bay" project**

Having considered the objective needs that Adriatic tourist destinations, in accordance with the requirements of the tourist demand and growth ecological awareness and health culture of the local population, **effectively manage all water resources** of land and sea as the fundamental elements of quality of life, economy and tourism, is logical that they led in the implementation of **modern system of responsible water management** and in complex weather conditions tend to provide enough water for the population, tourism and supporting economy.

This process, although with the time distance, follows a successful **policy of many EU member states** that by its legislation and practical measures and incentives in the coastal regions are achieving exceptional results in the segment of water supply as well as recycling and conditioning of pollutions.

On the relative **backwardness of Croatian regions** for these examples in the near and distant history, says the fact that also today, a significant part of the Dalmatian hinterland and islands is outside of water supply systems, and systems of pollutions conditioning are still in the process of realization.

This is a **paradox**, bearing in mind the fact that in environment of only one hundred km in Central Dalmatia, **there are abundant water flows** (Krka, Jadro, Cetina) sufficient for multiple users, which were possible to be link into a unique system and to ensure appropriate protection, and treat them as an economic resource in the terms of the future when drinking water lack almost half of humanity.

With declared, it should accent the fact that in these regions during the 20th century, economic system was developed by the principle of "it is worth-it damage the environment", which has **completely devastated some of the best spatial coastal resources** (Bays of Kaštela, Šibenik, Bakarska) known for a century traditional farming economy and fisheries, and then tourism.

This fact has contributed to the tragic **process of depopulation** of hinterland and significant negative processes of excessive **coastal littoralisation** with the devastation consequences of coastal zones and arable area and water resources.

In an effort to radically change the status, in the mid-eighties of 20th century the University of Split, in cooperation with HAZU and with the help of international specialized UN organizations, the World Bank and the European Bank, and the World Health Organization, a multidisciplinary project "**Economy of the broad area of Kaštela Bay**" began.

The basic elements of the project are based on the following conclusions:

- Kaštela Bay as the epicentre of region is contaminated, and its tremendous natural resources does not give positive effects in economic development, and quality of life has been seriously violated.
- Scientific and practical basis of the project is on the modern paradigm of care and improvement of the environment and the strong role of science in the study of eco-systems, in the study of the consequences of development on the environment and measures for preventing pollution and ways of improving the environment.
- Scientific approach is based on the method of integrated approach by the principles of sustainable development and precaution as measures of modern caution to the obsolete but also new technologies.

The project consists of three sub projects: "Methods and Models," The natural environment and eco-systems" and socio-economic and spatial systems". Developer scenarios showed the necessity of changing the strategy of the coastal economy by turning toward tertiarisation of economy and European trends of sustainable development, with maximum protection of environment.

Valuable basis for further work was provided by **Ecological sub project** with a detailed analysis of the situation in all spheres of environment, pedological map, the state of water resources, pollution of air, land and sea etc. As the result, the sub project "Integral ecological project of the city of Split with the environment", later called "ECO Kaštela Bay" was conceived (UNIST & UNEP 1993).

ECO-Kaštela Bay is the largest project of this kind in the Mediterranean. In order to achieve a common goal - the crystal clear sea and plenty of drinking water - four cities: Split, Solin, Kaštela and Trogir with Split-Dalmatia County, the Croatian Water and Split Water and Sewage, founded in 1998, the Agency ECO-Kaštela Bay. Price is projected at 143.2 million euro. Sources are the European Bank for Reconstruction and Development (30.7 ml euro), the World Bank (33.2 ml), Republic of Croatia (63.3 ml), Croatian Waters (4.5 ml), and the citizens from the basis of increased prices (11.5 ml). The project includes three sub projects: the sewage system Split-Solin, sewage system Kaštela-Trogir and upgrading of water supply system Split-Solin-Kaštela-Trogir, where the primary task is finding solutions of sewage systems and waste water (www.ekz.hr).

Long-term oceanographic research gave result that the sea of Kaštela Bay, which is a **semi-closed aquarium**, can not be the recipient of consolidated waste waters, so the solutions were found in the drainage of waste water into the sea of large capacity for receipt in this case, the Brač and Split Channel, which according to preliminary research may receive a larger amount of pollutions without a fear of consequence to the ecosystem of the sea.

This has enabled the phased construction of the **facilities for purification of waste water** where the mechanical purification should be started, and then with a chemical biologically systems. While approaching technically to the area of coverage as an optimal solution, two sewer systems were accepted, with two devices for treating and two submarine outfalls in the Split and Brač channel which have defined sewage systems for cities of Kaštela and

Trogir and sewerage system Split - Solin. This region has a population of 300,000 inhabitants, whose number almost doubles in the tourist season (Andročec V. et al., 2003).

It is logical that this mega project is appropriate treated in the **spatial plan** of Split Dalmatia County, in which a significant place is occupied by water supply and drainage along with statements that supply with drinking water from rivers, and the appropriate use of the potential sources has priority in relation to the use of water for other purposes.

Drainage systems in accordance with the plan should be lead to steady relationship with the **water supply systems**, and their development or construction is necessary to adjust to protected areas and established criteria of protection and to primarily protect the drinking water and protection of the sea (PP ŽSD 2002).

In the part of the Plan that relates to the **protection of the sea** is particularly underlined the need to reduce the amount of the consumption of technological and industrial water through the models of use of consolidated waste water, as is the trend in developed parts of the world. This imposes the statement that the entire project is only the first step in the management of waters, and what follows are the **recycling systems** and methods of possible use of so-called "Gray water" in agriculture, tourism and other economic and social sectors.

With the solution to the sewerage system Split-Solin in the flow area of Solin, adopted was a concept of connection of concentrated pollution flow, which is the reason that waste water of municipalities Dugopolje and Klis are taken outside of flow area of Jadro River, and where complete protection of this vital source is achieved.

For the rest of the settlement of Splitska Zagora Located in the zone of sanitary protection of the Jadro source and Žrnovnica, as a drainage solution, proposed is to build more local systems with a common device for purification. After that, the consolidated pollutions would be drainage in soil, and could be potentially used for irrigation of nearby debris fields.

Regarding the protection of the sea, the Plan emphasizes measures for preventing and reducing pollution from land through the **construction of modern sewage system**, which is the basic sanitary-health standard, of course, with the appropriate construction of the central device for the purification of the submarine outfalls.

The plan accepted the basic elements of the **study of monitoring the environment status** after the construction of the system as of: the sea as a base recipient with 9 normal parameters, and the atmosphere as recipient of potential odour and noise. (Ivanović M. et al., 2002).

In this sense, it specifies the necessity to complete the mechanical (primary) degree of purification before the under-sea disposition, according to "Directive Council of Europe" about the treatment of waste water of settlements (EEC 271/91)", which specifies how it is necessary to build facilities for consolidation of second degree for the cities greater than 15000 EP (Equivalent population), and on the utility systems with the greatest load of pollution.

Unfortunately, the current status of Conditioning in the system of ECO Kaštela Bay predicts this level only when needed and in the second stage in the foreseeable future when they reach the appropriate percentage of secondary network connectivity. Moreover, it is not a subject of presented project, which is a serious drawback, which can cause unwanted effects of reducing the quality of the sea in the attractive seaside localities in the vicinity of submarine outfalls, and with the work of sea currents and the wider area.

This, in accordance with the **Study on the environmental impacts** (Ivanović M. et al, 2002) for the Split channel, will yet be the subject of oceanographic research (monitoring) by the system incorporation.

In this context we find it necessary to announce that many analyzed similar systems in the world and especially in the **Mediterranean** ensure **implementation of all three systems** of treatment to preserve the sea as a vital resource in tourist recreation (author's comment).

The plan specifically emphasizes how to **prevent the contamination of underground** with a permanent rehabilitation and prevention measures, and commits industrial facilities in the coastal region on the pollution pre-treatment and connection to public systems, and gradual transition to the application of modern environmentally acceptable technologies, with the recycling and reuse of water.

To prevent pollution of coastal sea caused by marine traffic and port activities, the necessary protection **measures are set out in accordance with international conventions and national regulations**. All these solutions require significant resources as well as effective supervision, which, in the present moment are almost unreachable and represent a serious ecological problem.

Considering that the project does not include the construction of secondary network, which will represent an additional cost of population and economic entities already strained with dedicated separation of the price of water, **the doubt of the functioning system** can be concluded, whose facilities are designed to specific quantities of pollution flow. The next technical problem may be in the dispose of solid waste considering uncompleted systems use of the same in agriculture, and the current problem of the location of Regional Centre for the disposal of waste which in public provoke a number of controversies (author's comment).

The Adriatic project

This long-term megaproject to protect Croatia's coastal area from water pollution was launched in 2000 by the government of the Republic of Croatia and the **Croatian Waters Company. The Croatian Waters Company - the Adriatic project** carries out the project based on the precalculation of the value of the project in the amount of 280 million euro. The realization of the project has been supported also by the International Bank for Reconstruction and Development (IBRD) which in 2004 granted a loan in the amount of 100 million euro over the term of 15 years to be used for financing the improvement of communal wastewaters collection, transport and purification systems for the towns and villages along the Adriatic coast and on the islands.

The project is going to be financed from the **available funds of the Republic of Croatia** (state budget, the fund for the development of islands, water protection funds of the Croatian Waters company, etc.), while the loan is going to be repaid by **the end users** of the loan - **utility companies** - through raised water price, as well as by the Croatian Waters company. The financial participation of IBRD in the planned costs of the project is 50%, the Republic of Croatia 22%, local self-government units 19% and the Croatian Waters company

12%. **The basic premise for this project** is the fact that the management of wastewaters is currently at an unsatisfactory level which implies the inadequate protection of the sea and a possible threat to ground waters. This in its turn poses a serious problem in light of development of Croatian tourism and with regards to the public health. The alarming facts are that only 40 percent of households and 40 percent of industrial facilities at the level of the Republic of Croatia are hooked up to sewerage systems, and less than 12 percent of all collected wastewaters, including those in coastal areas, is being treated. This is especially disturbing in light of necessity to use the Adriatic Sea as a main resource for tourism and recreation. For example, merely 7% of the Sava river basin pollution is purified while in the case of the Danube river basin this percentage is significantly higher – 37% (HRT, 2009).**The deterioration of sea water quality** in some Croatian regions as a consequence of inappropriate discharge of untreated wastewaters has already caused visible problems, including eutrophication and phytoplankton bloom at individual locations, as well as less evident pollution of sea life by organic and inorganic micro-pollutants imposing a continuous threat to tourism, recreational and fishing activities. In this regard, the basic objectives of the project are protection and preservation of water quality to secure unobstructed growth of tourism and economy in accordance with environmental protection requirements which is to be achieved through implementing an internal and external monitoring and management of pollution and sea quality.

For the needs of the preparation of the Adriatic project 1030 cities and villages in 7 counties on the Adriatic coast were analyzed, i.e. 1,149,130 inhabitants, while the selected systems encompass 308,891 inhabitants with the total load of 1,224,800 equivalent inhabitants. The Adriatic project is currently being implemented in 15 towns and municipalities on the Adriatic coast.

The basic objectives of the project are:

1. Protection and preservation of water quality;
2. Establishment of conditions for safe economic development in accordance with environmental protection requirements
3. Preservation and improvement of the achieved degree of environmental protection;

The construction of the suggested projects is going to solve pollution and sewerage problems for several upcoming decades, which is going to secure unobstructed growth of tourism and economy in general in the coastal region. The Project is expected to produce **positive effects on the environment,** including the **improvement of public health** on account of **improved water quality** in the swimming areas and areas inhabited by bivalves, **better ecological conditions** on account of more reliable wastewater treatment, and greater viability of fisheries and aquaculture.

All works encompassed by the Adriatic Project are technically rational, socially and economically justified and acceptable from the standpoint of environmental protection based on the undertaken environmental impact study. **Sea monitoring** which is regularly performed since the 1970s is carried out within the Adriatic Project as well through:

- **a. Internal monitoring** – at the level of program implementation with the objective of determining the status of an area, wastewater characteristics and the characteristics of treated wastewaters;
- **b. External monitoring** – as a control of the implementation of the program as a whole

The Adriatic Project is of **regional character** and implemented **in the coastal region,** whose tourism potential and importance for the development of Croatian economy are great. 'Coastal region' has been defined for the needs of implementation of this Project as the area whose final wastewater disposition is discharged directly into the sea. Several regions that have also been taken into consideration owing to their exceptional natural value (national parks) or their pronounced indirect influence on certain sensitive areas are an exception.

The subprojects whose documentation was better elaborated and which had to be included mandatorily in order for the planned needs of utility companies to be considered more integrally have been selected for implementation in the first stage. The classification of systems by order of priority was carried out by applying **the difficulty classification method** based on **seven criteria**:

1. Equivalent inhabitants (the criterion that reflects the extent of pollution)
2. Recipient's sensitivity to wastewater discharge (the criterion of the sensitivity of the region)
3. Impact on the development of tourism (the influence on the tourist capacities of the degree to which the system has been constructed)
4. The degree to which the system has been constructed (preference is given to projects enabling the completion of the system in order for it to achieve full functionality)
5. Preparedness of project documentation (preference is given to projects at a higher level of elaboration)
6. The degree to which the region is developed (the criterion according to which preference is given to regions that are less developed)
7. The height of the investment in relation to the number of inhabitants

The Adriatic project is currently being implemented in 15 towns and municipalities including 6 (Lovran, Matulji, Novigrad, Opatija, Pula, Rijeka) under **Rijeka water management** and 9 under **Split water management** (Biograd, Dugi rat, Makarska, Omiš, Opuzen, Pakoštane, Rogoznica, Sv. Filip i Jakov and Zadar).

The analysis of individual subprojects has shown that unfortunately wastewaters are only **mechanically treated** and that they are discharged into the sea in this condition. This is the main characteristic but also one of the major shortcomings of both megaprojects considering the experiences in other parts of the world where higher level of purification is obtained by adopting chemical and biological procedures with an auxiliary recycling method which enables the reusage of greywater but also of hard substances in agriculture and municipal use (www.jadranski.projekt.hr).

2.1.2 Experience of tourist regions of the Euro-Afro-Asian Mediterranean in marine and land water resources management

The **common denominator** of all tourist regions of the **Euro-Afro-Asian Mediterranean** save for the extraordinary climate with the plurality of natural and social attractions, is the problem of **quality of water supply**. Connected with this is also the **problem of wastewater management**, in other words protection of the **sea quality** as the primary tourism resource which is the basis of tourism as the main economic activity of the region.

In the context of resolution of such problems in the Croatian part of the Adriatic sea, it is useful to analyse the applied **modalities and concepts in water supply management and**

waste water management, in other words protection of the sea from pollution, in the relevant Mediterranean countries and destinations which may serve as role models, but which are also our competition in the tourism market. The following is an analysis of national and regional policies, and concrete projects in the sphere of management of these pertinent segments of the tourism offer infrastructure in various developed tourist countries and regions of eastern and western Mediterranean.

The **Mediterranean** as the most notable tourist mega region in the world will remain for a long time, if judging by future prognosis, **the most important receptive macro space**. But for this reason, and given the accompanying ever growing urban development and economic activities with notable wastewater present in the watercourses, this is **an area of particular environmental endangerment**. This problem is further emphasized by the fact that the Mediterranean is the busiest world waterway with enormous amount of traffic related to the transport of oil and other toxic matters, and enormous quantities of weapons, including chemical and nuclear weapons, being transported, which raises possibilities of maritime incidents with grave repercussions.

Many international conventions have been negotiated under the auspices of the UN and regional organisations and associations which regulate this matter from a legal and practical point of view, in an attempt to preserve the Mediterranean and its valuable tourism and fishing potentials. In addition, contribution has also been made by introducing optimisation of systems for monitoring and intervention in the cases of incidents, which need to be continuously improved through international cooperation. All of this remains of particular interest to the tourism organisation.

In this sense, one must emphasize the relevance of provisions from the **Barcelona convention for the protection of the Mediterranean** (1976) and the **Mediterranean Action Plan** (MAP) which operates within the framework of the very Convention.

- **Barcelona Convention and the Mediterranean Action Plan (MAP)**

The **Mediterranean Action Plan** (MAP) was adopted under the umbrella of the United Nations Environment Programme (UNEP) in 1975 as the first of the Regional Seas Programme. MAP was adopted with an aim to secure the quality of life in the Mediterranean, as well as to establish and strengthen cooperation and harmonise strategies of common natural coastal resources management. The goals of MAP are the protection of the environment, promotion of the sustainable development model, as well as harmonizing of relations between Mediterranean countries in relation to the said problems.

The 16 Mediterranean member states Barcelona convention from 1976 was transformed in 1995 in the Convention for the Protection of the Marine Environment and the Coastal Region of the Mediterranean which today has 22 contracting parties, and represents the legal framework for the operation of MAP.

Protocols which regulate the **activity of the Barcelona Convention** with regard to water protection planning are:

- ***Dumping Protocol*** - Protocol for the Prevention of Pollution in the Mediterranean Sea by Dumping from Ships and Aircraft or incineration at sea (1976., 1995.)
- ***Emergency Protocol*** - Protocol Concerning Cooperation in Combating Pollution of the Mediterranean Sea by Oil and other Harmful Substances in Cases of Emergency (2002.)

- ***LBS Protocol*** - Protocol for the Protection of the Mediterranean Sea against Pollution from Land-Based Sources and Activities (1980, 1996)
- ***SPA and Biodiversity Protocol*** - Protocol concerning Specially Protected Areas and Biological Diversity in the Mediterranean (1995.)
- ***Offshore Protocol*** - Protocol for the Protection of the Mediterranean Sea against Pollution Resulting from Exploration and Exploitation of the Continental Shelf and the Seabed and its Subsoil (1994.)
- ***Hazardous Wastes Protocol*** - Protocol on the Prevention of Pollution of the Mediterranean Sea by Transboundary Movements of Hazardous Wastes and their Disposal (1996.)
- ***ICZM Protocol*** - Protocol on Integrated Coastal Zone Management in the Mediterranean (2008.)

6 Regional Activity Centres (RAC) operate under the framework of MAP, which are all located in the Mediterranean countries

- **REMPEC** (Regional Marine Pollution Emergency Response Centre for the Mediterranean Sea)
- **BP/RAC** (Blue Plan Regional Activity Centre) – in charge of protection of the environment in the context of sustainable development in certain regions of the Mediterranean.
- **PAP/RAC** (Priority Actions Programme Regional Activity Centre) – established with the goal of integrated coastal area management to alleviate and prevent negative effects on the environment resulting from development in built-up coastal areas.
- **SPA/RAC** (Specially Protected Areas Regional Activity Centre) – focuses on the protection biodiversity in the sense of protection of Mediterranean species, their habitats and ecosystems.
- **INFO/RAC** - Centre which provides communication services and technical support to the MAP Secretariat and other MAP regional components. The Centre also focuses on enhancing public awareness concerning environment while establishing working partnerships that enable sustainable development across the Mediterranean region.
- **CP/RAC** (Cleaner Production Regional Activity Centre) - promotes the reduction of industrial waste in the Mediterranean.

Mediterranean Commission on Sustainable Development (MCSD) was established in 1996 as the counselling body of MAP, which also produced the Mediterranean strategy for sustainable development. The role of the **Program for marine pollution assessment and control in the Mediterranean (MED POL)** is particularly important as it represents the scientific and technical component of MAP (www.unep.org).

The Adriatic macro region which is particularly analysed in this paper is actively involved in numerous programs under the umbrella of the Barcelona Convention and MAP, which is realised through the activities of the Split Institute of Oceanography, the University Centre for the Study of the Sea, the UNEP PAP-RAC Centre in Split, as well as the University in Dubrovnik.

In accordance with outlined programmes and global activities, attitudes towards water resources across the globe which make up 70% of the surface out of which 97% is the sea and only 0,01% drinking water, is rapidly changing. This is particularly the case in Europe and the Mediterranean, with many water supply projects and conditioning and recycling of wastewater projects being in place for the protection of fresh water and seas areas as recipients, and which programs involve international and regional financial institutions. Relevant research in the EU indicated that 75% of the Europeans consider water quality in their country a major problem. Some 27% of EU population or 130 million inhabitants are facing this problem. The Greeks (90%), in particular Cypriots (97%) are especially affected given that they lack their own sources of drinking water (Geić S. et al. 2009).

Climate change has particular influence on the drinking water supplies which is evident given the smaller capacity of certain European rivers especially in the Mediterranean region with a projected fall of 10%. With global warming and reduction in rainfall together with the rising of the temperature and often droughts in the areas of southern and middle Europe which have since the 1970-ties been more intensively affecting Italy, Portugal, Spain, France, Malta, Cyprus, Greece, even Croatia, enormous damage is done to agriculture and tourism, with fires of more and more catastrophic proportions destroying forest areas and creating a new microclimate with even less amount of rainfall.

In line with the new trends from the beginning of the new millennium, the problems of droughts and water shortage in the EU countries have often been debated. Accordingly, the European Commission has drafted an all encompassing assessment of the state in certain member states which envisages more rational handling of drinking water resources, which also presumes more intensive recycling of wastewater for repeated use of water in agriculture, tourism and related industries.

- **Program Blue Flag**

Blue Flag Idea was born in 1985 in France when criteria of environment protection of beaches and water quality were defined. French concept was accepted by the European commission with inclusion of other environment management segments. In 1987 year of environment only 244 beaches and 208 marines from 10 countries were carriers of blue flags.

In 2001 European foundation for education on environment, in the capacity of project coordinator (FEEE) became global organization and was transformed to Foundation for Environmental Education (FEE), it's the most important partners became **UNEP and UNWTO**. This resulted with the fact that in 2008 Blue Flags was fluttering on 2633 beaches and in 620 marines around the world. In program which extends over the boundaries of the European Union 38 countries are already involved.

On the area of the regions that this work relates to, we record in Croatia 125 beaches and 21 marina, Spain 442 i.e. 72, France 225 and 67, Italy 210 and 56, Greece 416 and 8, Turkey 258 and 13, Montenegro 18 and 13 in Morocco (www.blueflag.org, 2008.).

2.1.3 Problems of water management program in the eastern Mediterranean

- **Regions of Afro-Asian Mediterranean**

Greece and the entire south-east of Europe are in a complex situation in relation to water supply, in particular **Cyprus and Malta** that have no watercourses. Cyprus must import water from Greece and Turkey that, among other, exports water even to Israel on the basis of the "water for arms" agreement. Construction of a water supply system from Turkey to the Turkish part of Cyprus is also envisaged, as well as the construction of a major waste water treatment and recycling facility in the locality of **Mia Milia**. It has been estimated that this project will cost 30 million Euros, and it will be finalised by both Cypriot communities.

The weight of the present day situation in Greece is proven by the example of the island of **Rhodes** as one of most developed tourist regions in Greece with 15% accommodation capacity, 27% of foreign visitors per night, and 20% of foreign currency influx of Greece, which has based its development on the mass tourism development with all of its negative consequences for the environment, water supply and problems of wastewater. This island without continuous watercourses and with only 800 mm of rain a year, with intensive tourism development and enormous water consumption and wastewater problems, required a new concept for tourism and economic development. This necessity was comically announced in 1993 by Greek authors in a meaningful title "His Rodos hic Saltus". According to them, Rhodes has, with its excessive tourism development become a synonym for millions of tourists, but also millions of showers, and enormous amount of wastewater, piles of garbage, frequent catastrophic fires, thermal energy, with all of its pollutants. ..."Rhodes is a constantly milked cow which is inadequately fed or not fed at all" (Tamorri M. et al., 1993).

In an attempt to exert changes, "**Action Plan - Rhodes**" will be operational until 2010 within the framework of UNEP's programmes for priority actions in the Mediterranean which estimated the possibility of sustainable carrying capacity, and ordered territorial relocation and modernisation of tourism superstructure aiming to avoid the formula of mass tourism, and accompanied with implementation of the projects of conditioning and recycling of wastewaters. The result of conditioning and recycling of wastewater activities is that 48 beaches had become part of the Blue Flag which is a 50% in comparison to the 1990-ties.

Turkey has in recent years, with its fast growing tourism development and impressive tourism superstructure, become one of the top 10 most developed countries of the world with around 20 million of foreign visitors and 19 billion USD of foreign currency influx (UNWTO 2008). The majority of the tourism activity is realised in the **wider region of Istanbul** (Sea of Marmara, Istanbul Bosporus) with over 30% of overall tourist aggregates of this country. However, the unregulated organisation, wastewater and maritime incidents in this busy region endanger the natural anthropogenic features with potentially catastrophic consequences given the intensive traffic of ships carrying oil and weapons, including atomic weapons (Geić S., 2007).

This considered, an urgent plan of activities was drafted by the Ministry of culture and tourism titled "**SOS for Istanbul**" and presented at the international conference Habitat II in 1996. This plan popularises the GAIA doctrine from 1970 as the basis of a new environmental conscience with the message "all of us are in the same boat and it's called the Earth, and there is no nation without responsibility for pollution of the sea, continent and air, concluding how it is a moral and legal obligation of the civilised world to preserve Bosporus and Istanbul. (Binan C 1996).

This programme in its environmental component seeks reduction of the transport of dangerous cargo via these waterways, and seeks implementation of a major project for conditioning of wastewater of Istanbul and its wider area. The programme's impressive technical version of mechanical, biological and chemical treatment of wastewater has been presented at the fair which was followed by an international congress, with a high level of recycling of wastewater and a variety of ways to use the so-called "grey water" in agriculture, industry and tourism.

The **Turkish region of Antalya** with its hundreds of thousands of foreign visitors, is an example of extraordinary activity in the field of tourism development due to foreign investments, which having developed detailed Physical and Master plans until 2015, successfully realised its projects related to heritage protection, as well as water resources protection projects following the concept of sustainable development.

Also being a part of this, is the large **integrated environmental project of water supply** and wastewater conditioning worth 235 million USD which has been realised with the help of the World and European banks (Deverci A. et.al., 1996).

As a result, in this region which is nowadays considered the most developed Turkish Riviera, one can find 143 beaches and 2 marinas falling under the EBF system. All of this enabled the expansion of tourism investments of multinational companies turning Alanya into a modern tourist town and an organised tourist destination of the highest category. This town had the strength and funds to resolve its traffic problems which resulted in major pollution with a system of public transport involving the construction of an underground with stations linking the airport, the city centre and all other tourist localities with rich cultural, sports, recreational and health features necessary for modern all year round tourism traffic.

Egypt - Project Alexandria Integrated Coastal Zone Management (MAP): aims to organise sustainable wastewater management in the region of El Mex Bay and Alexandria, as well as to support the efforts for sustainable management of Egypt's coastal resources.

As for water supply as the basis for tourism and agricultural development of Egypt, projects of construction of grand dams and hydropower plants on the Nile are particularly important. They have, with the help of the international community, secured electrical energy and irrigation of large areas of desert, as well as helped tourism development on the Nile and artificial lakes of upper Nile.

2.1.4 International programs for protection of water resources in Adriatic

- **Programs in the Italian regions**

Italy as one of the leading industrial and tourist countries of the world, faced among the first counties with the dangerous problems of environmental pollution and in particular on its Adriatic coast, which is the recipient of the biggest part of pollution, and also a significant tourist area that exactly for this reason, is losing on the attractiveness and competitiveness. Especially well-known problems of pollution are at the coastal of tourist region of **Veneto** (Venice) and **Emilia Romagna** (Riviera Romagnola).

This is why in the seventies of the 20th century, the extensive research has been carried out which showed that in 80% of cases the sea along with Italian Adriatic coast was polluted and dangerous to the health of bathers. Therefore, to solve the pollution problem, Italy has extracted the amount of 6000 billion lira, and invested this amount in the next 15 years, which has greatly improved situation on the main tourist sites, especially on the affirmed Riviera at the regions of Veneto and Emilia Romagna.

As a result of these and similar actions, today Italy within the frame of international project **Blue flag** with 38 participating countries, has even 210 beaches and 56 marinas in the European blue flags system, which is almost double from Croatia or fourfold more quality beaches compared to the end of the nineties. Emilia Romagna with the famous Riviera Romagola which, from the former "Golden beach of Europe" was ecologically devastated and with no blue flags, today has 25 beaches and 5 marinas in the European blue flag system, the region of Veneto 9, i.e. 4, and Friuli Venezia Giulia 12 beaches i.e. 4 marinas, as a result of intensive fight for pollution conditioning from the mainland and the prevention and systematic maintenance of the sea surface in difficult conditions following a still large pollutions in the Po river basin as the epicentre of Italian industry and urbanization, and by the frequent navigational route towards large northern Adriatic ports and shipyards.

In this context, in subsequent of global climate change and increasing of sea level as a consequence of inappropriate relation of a man regarding the every kind pollution in the waters, land and air, there is a great project SOS for Venice with participation of UNESCO and UNEP, the Italian government and a number of world foundations. Except the restoration of cultural goods, project provides also the rehabilitation of the watercourses and in particular the construction of dams and the seawall, which will prevent the frequent floods in the city of lagoons, and stop the tragic process of Venice depopulation following the decrease in the level of life quality (Geić S., 2007).

- **Program BIOLFISH - Adriatic**

In the context of more and more sophisticated requests of tourist demand and the environmental awareness of tourists and also of domestic population, very important are more frequently combined inter-regional eco projects in Adriatic, which systematically solve problems of pollution disposal on the land and in the sea, which is most closely related to organic production in the tourism and agriculture.

Same is the inter-regional **project Biolfish**, i.e. the program of ecological agriculture and sustainable Adriatic fishing from the coasts, with the goal of improving the quality and evaluation of olive oil, fish and seafood.

Program applicant is the Apulia region, Italy and the project partners ICEA - Institute for Ethical and Environmental certification from Udine, Padua and Rimini; AIAB Veneto - Italian association of ecological agriculture for the region of Veneto, CiBi - Italian Consortium for Organic Agriculture, Bari; ECO LIBURNIA, Rijeka, Albanian Association of Organic Agriculture, Tirana and Dubrovnik-Neretva County (www.seadriatic.net/aii).

- **Program I.W.M.A. - Korčula**

This is an integrated approach of pollution management (Adriatic New Neighbourhood Program) for the island of Korčula. The applicant is the Dubrovnik-Neretva County, and the project partners are the Province of Ravenna and interested county entities.

The aim of the project is the preservation and revitalization of the natural and cultural heritage of the island of Korčula, through the implementation of the project that will strengthen the existing capacities of local municipal institutions and improve the quality of island life and tourism in accordance with standards of EU for the protection of the environment.

- **Global Environment Facility project (GEF)**

GEF represents a program of strategic cooperation in the Mediterranean, made by UNEP and the World Bank, which relates to programs of protection against pollution of this area and conservation of biodiversity on the shores and in the waters of the Mediterranean. Besides Croatia, participants come from 14 Mediterranean countries, and project partners: World Bank, UNEP, FAO, UNESCO and UNIDO, and numerous associated partners, and donors are Spain, France and Italy.

For the purposes of this work, especially important are management problems in the area surrounding the Trebišnjica river delta, and the delta of the river Neretva, which concerns to all regions of the **southern Adriatic**.

a. **Trebišnjica Management Project** supports intervention for improving integrated management of surface and ground cross-border water resources which is of extreme importance for the water supply region of southern Dalmatia, Herzegovina and Montenegro's coast.
b. **Neretva River delta** project is occupied with sustainable management marsh ecosystem and the priority investments to reduce water pollution in the Neretva basin.
c. **Tourism development project in Montenegro** is the concept of development of alternative forms of tourism, including effective methods of disposal of waste water in Bar and Ulcinj, the management and preservation of wetland areas of the Bojana-Buja delta, and the expansion of the protected area from the coastal section to the inland.

- **World Wildlife Fund Programs on South Adriatic**

Although Mediterranean Sea represents only 1% of the world's sea surface, it contains around 6% of sea species including endangered species and specially protected Mediterranean monk seal, sea turtles, dolphins and blue tuna. Former civilization cradle today is faced with many threats like pollution, pollution from the coast and sea traffic, fish catching beyond measure and excess construction of the coast which is influenced by one third of total world tourism which is accepted by Mediterranean regions.

In that sense organization **WWF** conducts training and protection programs in order to help local organizations in area of nature protection and sustainable forests management, freshwater and sea surroundings and other natural resources.

On wider area of regions this work relates to, WWF helps project „Living Neretva" and programs of protections of "Livanjsko polje" and "Skadarsko jezero", program of protections of Mediterranean monk seal and sustainable development.

Program Living Neretva deals with river basin Trebišnjica which is through karstic terrain connected to Neretva river, on the surface of 10,100 km^2. On this area of special attractiveness urban and industrial and tourist development, agriculture, construction of hydroelectric power station and hydro-technical objects influenced greatly on area causing numerous problems specially in a part of pollution management which seriously endanger this unique natural phenomena.

Project "Living Neretva" began in 2006 in both entities in Bosnia and Herzegovina, in order to manage with natural resources of basin Neretva across the border in accordance with basic principles and criteria of General European Union Directives about waters and habitations.

Complete vision of the project is to secure protection of nature and sustainable development of the area that is known as Neretva basin, which will be the basis for long term prosperity and development in the region and direct aim is to develop model of integral management, connecting water management and protection of biodiversity. Project is financed by the Government of Norway and is performed by Mediterranean program of WWF in cooperation with WWF Norway (Geić, S., Voloder, F., 2009.).

- **Protection project of the Montenegro-Albania Green Belt – WWF**

From west Montenegro to north Albania attractive Durmitor massif and the mountain Prokletije spreads, it is area formed by ice blocks and karstic phenomena and river straits deeply carved in landscape which includes longest and deepest, European formed Canyon of Tara River. Some of the most protected forests of southeast Europe are situated here, which are habitation of numerous herbal species and wildlife predators, which was the main reason that this area was included in **UNESCO natural heritage register**.

Despite this values and global protection this area is endangered by clearance of forests and especially by planned dumps and hydroelectric power station constructions along the Drina river coast. The goal of the project is creation of favourable conditions for long term protection of biological diversity and sustainable usage of natural resources in Durmitor, Tara, Prokletije landscape and to establish grounds for long term engagement in Dinaric massif and Dalmatian coast, as well as in south part of Mediterranean eco-region.

2.1.5 Water management programs in west Mediterranean region

Due to the fact that on the area of west Mediterranean the most intensive tourist valorization with almost two-hundred years tradition is recorded, it is logically that we measure here the first activities on protection of water resources, specially the sea as the fundamental tourist resource. In this part of the work we shall present chosen projects which were carried out on French and Spain Mediterranean destinations within the new tourist politics towards environmental protection within sustainable development of tourism development conception initiated in seventies in last century.

- **Programs in tourist regions of Cote D'Azur, Toulon and Corse**

France as one of the leading tourist countries with centuries long tradition of tourism development and with realization of 83 million foreign visits and 54 billion USD in foreign currency influx (UNWTO 2008), is at the forefront of environmental protection activities, in particular in view of the fresh water and sea as important tourism resources.

In this sense, it is particularly important to analyze many mega projects of conditioning of wastewater which have been realized firstly in the destinations of Cote d'Azur, and later throughout the French Mediterranean. The immediate trigger for these projects is the Blue Flag system which assesses the quality of the sea and the beaches and that was started in France in the eighties, marrying activities of the ministry of environment, tourism infrastructure, and health and protection of the coastal area. Activities such as education in environmental protection, wastewater treatment, environmental protection, organization of waste management, quality of tourism services on the beaches and beaches management system, are all under direct jurisdiction of an evaluation team which implements the politics of sustainable tourism.

In the middle of the nineties, **tourism destination of Antibes** was declared a world champion due to a modern system for mechanical, biological and chemical treatment of pollutants and other environmental protection activities. This system is implemented for 172 000 users at the expense of over 300 million Euros, which is double the value of the investment for half the amount of pollutants and users then the ECO-Kaštela Bay project. This speaks of the level of excellence of this very project and vice versa.

A similar project under the name of AMFITRA was realized in the French Riviera area of **Toulon.** This project had the capacity for 550.000 users which is double the number of than inhabitants in that area. Realization of this project that includes all three levels of purification, save for the protection of bathing zones in the Bay of Toulon, has notably contributed to the saving of a rare natural phenomenon of the national part Prot Cros in the wider area of the island Iles d Hyeres.

Monaco joined in these activities and initiated in relation to the Monaco Riviera an eco project of complex conditioning of pollutants with a capacity for users (100,000) three times greater than the number of inhabitants, bearing in mind the intensive tourism development (OTV 1996, 1997, 1998.).

The Sovereign Prince of Monaco Albert II is particularly active when it comes to environmental protection, in particular in relation to the sea, with its ecological trust fund dedicated to projects in the Mediterranean drawing on the **EU's program for the Mediterranean** headed by France, also including Monaco.

All of these activities for the protection of physical space and the bay area follow the critical development of the mass tourism of today. It was back in the 1960s that French scientist B. Kayser (1960) wrote: "*....once the most beautiful natural Riviera, after a hundred years of chaotic construction today resembles a harlequin...*". In the context of mentioned above has a concept of tourism penetration from the coast to hinterland of Provence been elaborated, which creates a syntagm "provencelization" recognized all over the world.

Comparison of the technical systems management pollution Antibes, Toulon, Monaco (France), Antalya (Turkey), Rhodes (Greece) and Eco-Kastela Bay (Adriatic Project) is given in Table 1.

Tourist region of the Corsica island (Corse) covers area of the French mountainous island in the Mediterranean sea of 8722 km^2 with about 300 000 inhabitants. Woods occupy 20%, underbrush 45%, while the rest is under pastures and arable land.

From the eighties became Political initiatives frequent, supported by France and Italy, but also the EU, in order to realize the economic, cultural and ecological international **interregional cooperation** of two neighbouring islands of Corsica (France) and Sardinia (Italy). There is an area of the Strait of Bonifacio with the Archipelago of neighbouring islands, which is ecologically particularly interesting. In this spirit was the Paris Convention adopted in 1986, with the project of cooperation and marine delimitation in the Strait of Bonifacio, followed by the resolution of the European Parliament in 1989.

Protecting natural and historic values program was developed and institutionalized on both islands. All the way back in 1972 Corsica declared the nature park of an area of one third of the island with 250,000 acres, that covers mountainous area but also coastal area on

Suspension systems / sites	Antibes	Monaco	Toulon	Antalya (Turkey)	Kastela Bay	Rodos (Greece)	Project Adriatic (7 counties)
Number of rezidents	172.000	100.000	550.000	600.000	279.000	128.000	308.100
Per day range pollution	40.500	31.000	103.000	120.000	130.000	35.000	140.000
Capacity m3/h	3.375	1.800	9.000	15.000	13.000	2.800	15.000
lit/sec		500	2.500	3.700	3.650	2.600	4.000
Price in mil.euros	300	150	500	350	143,2	200	280

Reference: - OTV, Mediterrane amphitra, L'usine de Depolluation des laux du Cap Sicie, Toullon, 1997.,3.
- OTV, Wastewater treatment plant, Monaco, 1998.,and Antibes 1996
- Eko Kaštelanski zaljev, Inegralni ekološki projekt Kaštelanski zaljev, Split 1996.,2.
- UNEP, Mediterranean Action Plan-Rodos-PAP-RAC, Split, 1993.
- Deveci, A. et. al., Antalya – world city, Turkey, 1996.
- www.jadranski.projekt.hr

Table 1. Technical characteristics of the eco-system of Antibes, Monaco, Toulon, Eco-Kastela Bay, Antalya, Rhodes, and the project Adriatic during the mid nineties

the northwest of the island, including the coastal sea with significant restrictions in economic exploitation of land and marine resources. The southern part of the Corsica coastline is especially important, where several smaller nature reserves were formed with purpose of protecting the geological forms as well as native animal and plant species of the area. They include the coast and the islands in the Strait of Bonifacio, where the Cerbicale islands represent a specific nature reserve on 5 small islands that close Porto Vecchio.

This extraordinary valuable area in the surrounding of Cape Girolata and Cape Porto, including the nature reserve Scandola, is already in the World Natural Heritage register by the UNESCO organization.

An insular marine reserve around Lavezzi islands was founded in 1982 on 5000 acres of sea surface and 80 acres of Cavallo island. Every possibility of urbanization is here excluded, in order to protect geomorphological forms of flora and fauna.

The INTER-Reg Programme was developed from 1991 to 1993 under the auspices of the EU, and it predicts, among other things, formation of an international reserve in the Bay of Bonifacio. Regional governments of Corsica and Sardinia are working on the programme, as well as the Universities of Corsica and Sassari (Sardinia) and also local and regional authorities (Merler, A., 1993, 39).

The biggest help in the project was prohibition of oil tankers transportation and other ships with dangerous cargo through the Strait of Bonifacio, being a precondition of nature reserve existence on the islands between Corsica and Sardinia.

- **Programs in the Balearic and Canary islands, as well as in Catalonia**

Spain has for decades been one of the world tourism leaders with 60 million foreign visits and 58 billion USD income (UNWTO 2008). This was possible based on an impressive tourism superstructure which in many regions superseded the carrying capacity of the area, even the social surroundings, and had a negative effect on what use to be extraordinary natural and social values.

Spain has for the past two decades been the leader in activities for a new balanced tourism development pursuant to the sustainable development concept as part of the so-called "**Politics of third generation**" accentuating the concept of "**Total Quality of Management**". Many programs for improvement of the quality of the environment titled "**Planes de execelenica**" (plans for improvement) with full participation of regional and local authorities were initiated, and many international conferences were organized such as the **Man and the Biosphere World Conference** MAB – Seville 1995, as well as the **International Conference on sustainable tourism -** Lanzarote 1996. These events drew attention to the need to maximize activities for the overall protection of natural and social resources on the continent as well as the sea.

The **Balearic islands** as one of Spain's top tourist regions with over 20% of all tourism aggregates of Spain, are experiencing an extraordinary environmental endangerment as a result of overly intensive development of mass tourism in the 1970-ties, so much so that the term "balearisation" was coined to describe this negative occurrence. This was an incentive for the drafting of the 1970-ties Tourism Management Plan for the Balearic islands ("**POOP**") envisaging strict conditions for new tourism construction and reduction of capacities, with realization of many environmental projects for conditioning of pollutants. A direct result of this initiative and programs is the fact that many beaches already have the Blue Flag status since 1995.

The project **EKOMOST** organized by the German Tourism Institute and the University of Munich enabled the possible sustainable development of Balearic islands and the island of Rodos which have been, according to J. Batle and M. Robledo (2000) under attack from the virus "Eristion" - a Greek god eating its own flesh - meaning tourism which is destroying its own essence. The goals of this project are sustainable environmental tourism in line with objective tourism capacities of the area and with maximum protection and revitalization of natural resources on the continent and the sea, as well as diagnosing and resolving problems of water supply management and pollution. (Faige M., 1995).

In order to protect a very attractive area of Cabrera islands, south of Mallorca, it was declared a national park, which includes 10 000 acres of land surface and the waters. Numerous species of Mediterranean flora and fauna on 1863 acres of land and marine surface were protected by the strictest regime according to situation in other 8 national parks in Spain-while 5 of them are on islands-4 on the Canary islands and one on the Balearic islands (ICONA, 1995).

A similar project **PREPARE** was realized by scientists from the international organization WTFC for towns **Alcudia (Balearic islands**) and numerous tourism centres in the United Kingdom and the Netherlands, also being supported by the EU Program Archway. This project established a model promoting sustainable development policy through responsibility and consciousness concerning the environment, and with necessary activities including a developed procedure for reporting and review. This also includes sustainable management of water supply and wastewater in line with RIO Agenda 21 ((Bruce M.D. et.al. 2000).

The **MENORCA** island (705 km, 62,000 inhabitants) along with its natural attractions, is characterized by numerous **prehistoric sites and monuments from the later periods of turbulent history**, scattered around the island also known as an open-air museum (MIT 1976).

Because of its exceptional **natural and historical heritage,** Menorque d'Astudis Institute (IMA) and the Spanish MAB Committee applied to UNESCO in 1989 that the islands becomes one of 13 reserves of Spanish biosphere, so that this status does not disturb the economical development of the island. UNESCO accepted it in 1994 by declaring Menorca a Biosphere Reserve.

The geological diversity of islands represents also its **picturesque landscape** with an interesting **Mediterranean ecosystem** preserved also in the inhabited areas (1,000 types of vegetation, with 7% endemic plants; 25 types of land birds and many sea birds as well - several out of them are endemic.

Extensive tourism growth in the eighties threatened the Menorca ecosystem seriously. Large public works and projects connected to tourism often neglected the landscape and rural architecture values, even the native way of living of the inhabitants, causing painful migrations and depopulation of the villages and also causing flora and fauna reduction. Beginning of the nineties was a critical turning point of the trends that took some different courses after the Declaration of Biosphere Reserve.

Backbone of the **Biosphere Reserve of Menorca** are 3 Natural Park zones, declared by the regional government's decision. First zone on the northeast is S'ALBUFERA DES GRAU with lagoons that are long up to 10 km (70 acres), numerous endemic vegetation and abundance of wetland birds threatened with extinction. Second zone is LA VALL, a northwest sparsely populated part of the island with rich holly oak and pine forests and also underbrush, cliffs 100 m high, attractive system of „dunes" in a closed bay with numerous bird communities. There is also, opposite the coast, the best preserved marine ecosystem of Menorca, in the buffer zone.

The third zone of reserves is the south of the island with three canyons about 7 km long, 100 m wide and 50 m high with the pertaining coast. Flora and fauna are subtropical. Among the zones of reserve, which represent Menorca's ecosystem, there are forests and agricultural surfaces with farms, as the counterpoint of natural and anthropological system.

Thanks to these projects Menorca has, with over a million visits per year, become a part of 7,000 declared protected zones in the world and of a special **UNESCO system of islands biosphere reserves**, such as for example: Aleutians and Virgin Archipelago - USA, the Channel and Juan Ferdinandez Archipelago - Chile, Galápagos Archipelago - Ecuador, Yakushima Archipelago - Japan, Rhum and St. Kilda Archipelago – Great Britain, Molène Archipelago - France, Hiiumaa Archipelago - Estonia, Zembra and Zembretta Archipelago - Tunis, and other (Geić S., 2002).

Menorca is a unique example of an island with relatively developed economy that has accepted the program of further economic and tourist development, but with systematical protection of natural and historical heritage that are guaranteed by UNESCO, Spanish and county government, MAB committee and NGO organizations (Rita, J., 1993).

Could these examples be an incentive to associations of protection and relevant structures of Split-Dalmatia County to start a similar initiative for a beautiful and still preserved archipelago that closes the Bay of Trogir and Split Channel from the islet Arkanđel via Drvenik to Šolta, which in addition to natural attractions and rarities has an exceptional cultural and historical component, especially the one closing the Bay of Kaštela with UNESCO sites of Split and Trogir? (author's comment)

Canary Islands, with almost 20% tourist dimensions of Spain, have recorded an intensive tourist development threatening to natural and anthropological resources. In that sense the **GUMP program** has been developed at the initiative of the Multidisciplinary Centre of University of Canary and numerous world scientific institutions, with a famous Institute of New Technologies and Sustainable Development. The project has been developing a pilot program of alternative energy use, seawater desalinization, purification and economic use systems, waste waters, along with revitalization of agriculture, cattle breeding, fishery and mariculture as traditional activities on islands together with modern selective forms of tourism (CIT 1995).

This is where the **ECO-ISLAS project**, sponsored by the EU, has been realized, which covers smaller islands of **La Palma and Fuertoventura** (Canaries) and islands of **Elba** (Italy), **Alonnisos** (Greece) and **Pelworm** (Germany). The project offers modern solutions of resources registration, identification of acceptance capacity and water management, pollutions, energy and traditional economy, all of this in the attempt to revitalize autochthonous economy and tourism for the purpose of preventing islands depopulation (Cavana, M., 1995).

LANZAROTE is famous as a Canary island with 300 extinct volcanoes that as late as at the beginning of 19th century created and modified its landscape, flora and fauna with its eruptions, creating the world's **unique ambience** resembling the lunar surface.

Organized tourism in this island was declared as the UNESCO world heritage and a part of MAB program (1993); it seems especially interesting in the context of development orientations of tourism on the insular part of Croatian Adriatic.

Timanfaya National Park is the most attractive among the world nature rarities of volcanic origin with surprising relief forms, geological forms and manifestations in the environment of specific vegetation and human creation in the fight for survival on a waterless inhospitable area of the island.

Since the 1960s, **tourism has expanded** thanks to **intensive investments** and **revitalization program** of these areas with a special treatment of Spanish government and the European Community, with the concept of classic forms of credit arrangements and joint investments of Spanish and foreign investors (Marin, C., 1993).

It makes sense that such **tourist development** has completely **changed the quality** and **style of life** on the island. However, thanks to a very serious program of local and regional governments with the cooperation of science, numerous accompanying negative effects have been avoided, which are much present and fatal on many other destinations around the world. A great contribution was given to this by C. Manrique, a local artist with the world reputation. Thanks to him **modern postulates in urbanism and protection of heritage** have been accepted, so it is difficult today to distinguish where autochthonous rural architecture of the island villages ends and where a tourist settlement begins.

Exceptional **care of natural environment** is also proven by the Canary Island Natural Areas Protection Act from 1987, by which 12 localities, or **70% of the island's surface area**, obtained a special treatment of protection (Echenique, J. 1993).

*It is evident that in these areas the reasoning of the great Michelangelo has been fully accepted: "...**The true work of art is but a shadow of the divine (natural) perfection**" (author's comment).*

In the context of such, for Croatian conditions almost unbelievable information and the fact of years-long wanderings of the current Croatian economic, spatial and tourist policy, the following question makes sense: Are we situated in the same European and Mediterranean environment, do we belong to the same civilization circle???? (author's comment)

The mega project of the coastal area revitalization of the Catalonian capital **Barcelona**, within the scope of preparing the city for the Olympic Games in 2002, is also worthy of attention. By an exceptional project solution, once an unsightly port area has been turned into a modern attractive tourist and business zone that has enriched this Mediterranean city, the leading cruiser port on the Mediterranean. The project has created 18 km of beach with numerous tourist, catering and recreational facilities on the land and sea with the state-of-the-art pollution purification systems, by which Barcelona has become an exclusive holiday tourist destination (Geić, S., 2007).

- **Projects on the island Sardinia (Italy)**

Tourist region of Sardinia (Sardegna) represents an Italian island in the west Mediterranean with the surface area of 23,812 km^2 with 1,601,586 inhabitants (1979). It is separated by 12 km from Corsica by the Strait of Bonifacio and by the Tyrrhenian sea from the Apennine Peninsula.

Its relief is the combination of a hilly area and gorges with steep shores intertwined by wetlands and lagoons. The island has lately recorded intensive **tourist development of elite and various selective forms of tourism**, especially on the north-west coast (Costa Smeralda) and represents Italian "hit" destination.

Natural and historical places of interest, due to the development **various civilizations** during three thousand years, form a part of tourist attractions of the island that at the end of the 1990s managed 4.1% tourist capacities of Italy.

Within the framework of **protecting and valorizating the island's natural attractions**, a significant role is played by the **National Park Orosei**, organized at the area of 100,000 ha in the bay with the same name that also covers the mountain massif Gennargentu and the island Asinara. This is the only nature park of Sardinia that at the same time protects the land and the marine area. In compliance with the foundation deed from June 1992, special protection measures are foreseen on the island Asinara at the territory of 51.9 km^2, which covers 100 km of marine coastal line and expands up to 17 km into the island's interior.

In the same year Italian Ministry of the Environment established the **marine reserve** around the island **Budelli** in the **La Maddalena archipelago**. **Protection measures** were supposed to save natural specificities of advancing expansive tourism and sudden urbanization. Almost entire archipelago is now monitored and protected as the nature park; one also organized the provision of normal activity of local population within the areas between protected parts and zones of controlled economical development. A special regional legislation has been made for this, which is supplemented by the one on the national level and within the framework of European Union. This is exceptionally important since the development of the Aqua Marine Park and its integration with the planned **international nature park** in the Strait of **Bonifacio,** enables a special integration of the entire area into the social and economical fabric of the Mediterranean.

In addition, further specific initiatives on the northern part of Sardinia are underway that need to be economically and protectively integrated to the trans-border protected area with the south of Corsica (France). Also interesting is the **Tavolara-Capocoda-Cavallo Nature Park** that contains a combination of a coastal and marine Aqua Park.

Events within the scope of natural and cultural resources have resulted at the beginning of the 1990s in the new significant development processes with both positive and accompanying negative connotations. **The positive ones** are primarily a successful fight of local population and regional authorities for the prohibition of hazardous cargo transport in the strait based on the agreement between the two authorities from 1993. At the same time an interstate and interregional deed was signed for the foundation of the **international nature park in the Strait of Bonifacio**.

In the same year a feasibility study was finished on the **protected marine area** Tavolara - Capocoda Cavallo between Capo Ceraso (Olbia) and Cala Finocchio (San Teodoro) by which the Italian Government aimed at activating the national park in this area. An international workshop sponsored by the UN was also supposed to contribute by models of mutual cooperation, which was organized by the University of Sassari and the research centre of Sardinia Co.Ri.Sa. with the topic of aqua marine parks.

Dubious factors in these cases are too sudden decisions with huge restrictions without sufficient appreciation of local attitudes and undue communication with local population.

Negative factors of these processes are specific contradictions, such as the case of keeping the function of a correctional institution on the islet of Asinara, which has impeded the realization of the national park project. There is also a sporadic military use of the area that was as early as in 1993 defined for the foundation of the **Gennargentu National Park** in the central part of Sardinia (Merler, A., 1993, 39 - 42).

Therefore, the problems due to awkward attempts of considering physical and geographical elements of the area as separated from social and cultural components of urbanized zones were evident in this environment as well. Evidently, the population, its culture and tradition and the relation to the area always need to form integral part of protected zones and be an active factor on which a successful protection and revitalization depend a great deal.

These valuable international projects necessitate great efforts of both regions, countries, international community and multidisciplinary science. This implies active inclusion of local population, but also adequate relation of numerous visitors that visit this area via forms of excursion tourism.

- **Programs of water supply and pollution in Morocco**

Morocco, a developing country with strong tourism ambitions in the Mediterranean and the Atlantic, is also developing tourism policies of maximizing protection of the natural resources, in particular in its coastal area. In this sense – **the Project of sustainable development for Nador Lagoon and Al-Hoceima (MAP)** envisages protection of the coastal eco-system of Morocco and global development of Moroccan coastal area pursuant to sustainable development principle which is particularly relevant for tourism ambitions of this region.

This is practically continuance of a major project for **revitalizing of the tourist region of Agadir** devastated by a catastrophic earthquake in the 1960s, which with the help of major projects initiated by the international community, became a modern tourist destination with organized water supply and wastewater management systems enabling further projects of tourism superstructure in accordance with the models "concession" and "joint venture" (Mohamed VI. 2004).

2.2 Tourism in programs of "Adriatic-Ionian Initiative" organization

Within the scope of new development concepts of "soft" tourism, which wants guests to see and experience more in the shortest possible period, international and interregional cooperation of neighbouring Mediterranean regions and the joint appearance on the market have become ever more frequent, which is also especially emphasized by the EU, via the program AII.

On a global scale, exceptional natural and social tourist resources of a relatively small area of the Adriatic and the Dinarides are based on exceptional attractiveness of the coastal area and hinterland with attractive climatic, relief and vegetational characteristics.

They are nowadays protected and tourist valorized via eight marine and land national parks and numerous other relevant institutions of heritage protection in compliance with the international legislation, which has turned them in recognizable tourist attractions at the international plan.

Within the scope of this natural matrix, by human action throughout millennia within the Dinara and Mediterranean cultural circle, and under the influence of various civilizations and cultures, an impressive cultural heritage has been created that is nowadays established internationally via 8 historical sites protected as world cultural assets included in the UNESCO register, which are first-class tourist attractions. Together with natural localities, they form over 80% of total attractive resources treated by the neighbouring South Adriatic countries, even the basis of their tourist and economical development.

The Adriatic-Ionian initiative as a regional coordinative body, founded at the Conference on the Security and Development of the Adriatic and Ionian Sea, held in Ancona on 20 March 2000, obligates all the member countries (Greece, Italy, Croatia, Slovenia, Albania, Monte Negro, Serbia and Bosnia and Herzegovina) to co-operate according to the documents of the UN, OESS, COE and EU.

Parliamentary cooperation provides basis and guidelines on these levels: economy, maritime affairs, transport, tourism, environmental protection, science, culture, fighting all forms of organized crime and cross-border co-operation which brings to social and economical development and strengthens the peace, regional security and stability.

The AII bodies are the Adriatic-Ionian Council and the High Officials' Committee as the main executive body. The administrative body is the permanent administrative secretariat that operates through Round Tables in six thematic units: Small- and Middle-Sized Entrepreneurship; Transport and Maritime Affairs; Tourism, culture and Interuniversity and Education Cooperation; Environmental Protection and Sustainable Development. The Forum of the Adriatic and Ionian Chambers of Commerce Forum is especially active, and it operationalizes initiatives on the regional, economical, and especially tourist collaboration, ever since the year 2001. (www.uniadrion.net).

The representatives of the Chambers of Commerce and the other institutions that participate in the Forums of the AII try to conciliate and mutually promote the communication policy in order to successfully represent the Adriatic-Ionian region as an attractive macro region and a unique tourist entity with a possibility of creating a competitive tourist product in a way that individual, mutual and specific resources combine through an innovative and effective transborder cooperation, creating a unique image of a region within the boundaries of a homogeneous tourist system.

Its purpose is to unite the regional, natural, cultural and service resources as well as to enable the competitiveness of the region on an international level. In that way you emphasize traditions and various natural, historical and ecologically preserved entities and patterns the total tourist offer that is then recognizable on the international tourist market.

Along with the monitoring and analysis of the tourist movements, and especially within the selective means of tourism, follows the concept of harmonization and coordination of the tourist marketing policy and the coordination of the tourist legislature. There is also a continuous collaboration and communication with the university and scientific institutions with a purpose of stimulating young people to get involved with the tourist activities and improvement of tourism. Special emphasis is on promotion of nautical tourism by connecting the nautical ports with the attractive cruising destinations in the Adriatic and Ionian Seas. In the context of protection of tourist resources, there is a concept of a joint approach to preventing and improving, in case of natural disasters or possible eco incidents at sea or on coastland.

Taking in consideration that there are no separate funds dedicated solely to economical and social subjects, the **AII stimulates more projects and measures, as well as activities** within the Initiative as following:

- Cooperation with PADMA LAB project (Pan Adriatic Destination Management Learning Laboratory),
- Project MARINAS (Modelling Adriatic routes integrating networks and areas in the Adriatic sea)-Interreg IIIA – with a purpose of integrating of the Adriatic tourist offer,
- Project ITAC (Innovation technology for Adriatic competitiveness) - Interreg IIIA NPPA – with a purpose of stimulating small- and middle- sized companies as well as the public and private research centres, development of new products and services (www.seadriatic.net/aii).

The role of AII is particularly important in exchanging experiences connected with the environmental protection and preservation and in particular of the sea ecosystem by the intersectional approach. In these terms the following activities and documents occur:

- Protocol on integrated coastal zone management in the Mediterranean that needs an urgent ratification and implementation.

- The Mediterranean strategies of Sustainable Development and the directives of the EU connected with the maritime strategies, as important frames for the maritime activities in the region of the AII countries.
- Studies based on ecological research and monitoring for the purpose of objective and comprehensive picture of ecological condition of the region.
- Intensifying of use and exchange of results in the regional research studies, in particular in fishery monitoring, debalasting, eco incident prevention plan and similar.
- Declaration of the Adriatic Sea as a Particularly Sensitive Sea Area (PSSA).
- Improving of port capacity for the acceptance of pollution in order to insure a more efficient sea protection (www.seadriatic.net/aii).

2.2.1 Selective tourism in protected nature parts of the South Adriatic

A wider area of the South Adriatic is rich in protected natural resources starting from national parks and other forms of protected areas as oases of untouched nature with exceptional possibilities for the development of alternative aspects of tourism. Exactly hundred-years long bad traffic connectivity and isolation, as well as a weak economical development of some parts of this area, have contributed to the preservation and numerousness of natural attractions, which will be faced with ever more intensive tourist valorization with the possibility of creating numerous tourist itineraries throughout the entire year.

In that sense the UNESCO locality of Durmitor National Park with the surroundings need to be pointed out, which has already been intensively touristically valorized, and then Bay of Kotor as an exceptional natural phenomenon, the nearby Skadar Lake and many other natural attractions in Montenegro.

Natural attractions that are in the protection regime on the territory of south Dalmatia are the localities of Mljet National Park, archipelago and Neretva Delta. In Herzegovina these are the Middle Neretva, where the World Wildlife Found project of river preservation was developed, as well as the Hutovo Blato Natural Park.

What especially stands out is the attractive area of **Lower Herzegovina**, recognizable in terms of rich natural heritage. This area represents a developed karst relief with Popovo Polje in the centre, the site of Zavala and preserved Vjetrenica cave, then the locality of Grepca with tens of caves, rivers Bregava and Radimlja on the north, river Neretva and Hutovo blato on the west. On the right side of Neretva the attractive karst relief has developed in tectonic and morphological sense (Ravlićka cave) with water islands and famous travertine waterfalls (www.rb-donjahercegovina.ba).

Nearby are also recognizable traces of distant and recent past: Illyrian sites of Badanj and Daorson, medieval archaeological site of Mogorjelo, numerous findings of standing tomb-stones (*stećci*), medieval towns - fortifications, water supply cistern on Hrasno etc. All of this makes it the unique treasury of cultural and natural heritage, which is supported by the growth of interest of experts and scientists, but also of tourists, lovers of nature and history.

In geological, tectonic, hydrological, hydrogeological, speleological, biological and ethnological sense, the area of the Lower Herzegovina, with specific karst processes and underground world of water architecture finally connected with the sea, represents one of the most interesting areas in Europe, where 500 different karst formations have been recorded significant for scientific, research, as well as for tourist development. This special natural underground world with 74 endemic species, along with other natural and cultural heritage, needs an organized care of today's people and systematic documentation with a stronger support of international scientists and associations. This is also supported by a long history of research, and even a partial tourist valorization that goes back to the beginning of the 19th century.

This also refers to Trebišnjica river basin (98 km) with underground flows 187 km long, as the biggest sinking river in the Balkans and one of the biggest in the world. Trebišnjica river basin also includes sub-basins of the river Mušnica - Sušica, majority part of Ombla sub-basin in Dalmatia with the adjoining underground flows and more than a hundred water sources with the end in the sea near Dubrovnik. This area in Herzegovina and southern Dalmatia is protected in various ways and it represents an exceptional tourist value for both regions (www.trebinje.rs.ba).

In karst ambience on the left bend of the river of Neretva, Hutovo Blato Nature Park is a unique sub-Mediterranean wetland in Europe, important from ornithological, ichthyological, scientific, ecological and tourist aspects as one of the biggest winter quarters of birds within the area of Europe.

Skadar Lake, at the border of Montenegro and Albania, which was declared a national park in 1983, located in Zetsko-Skadarska Valley, keeps the contact with the Adriatic Sea via Bojana river. With the total surface area of 530 km^2, it represents the biggest lake in the Balkans and the most interesting biotope of the region with exceptional limnological characteristics, exceptional abundance of ornithofauna and ichthyofauna as well as abundant wetland-like vegetation. As a significant habitat of water birds, Skadar Lake was registered in the World List of Wetlands of International Importance at the Ramsar Convention in 1996 (www.nparkovi.cg.yu).

All the afore stated offers exceptional possibilities of developing selective aspects of a wider region's tourism, which has also been increasingly recognized by the modern tourist demand, so it combines numerous attractions and activities within the scope of a top quality active holiday.

Coastal and island natural attractions of the **southern Dalmatia** are led by the **Mljet National Park** declared as one in 1960. It is located on the north-western part of the southern Dalmatia, with 5,375 hectares of protected land and surrounding sea, and it represents the first institutionalized attempt of protecting the original ecosystem on the Adriatic. Its specificity is the combination of natural and historical heritage due to the values of a Romanesque Benedictine monastery on the islet in Great Salt Lake.

The neighbouring **Lastovo archipelago** is a nature park with the surface area of 195.83 km^2 that covers Lastovo with 44 surrounding islands, small islands, cliffs and reefs, and which is one of the most preserved marine areas on the Adriatic. It is protected as a "spacious natural and partially cultivated area of land and sea with ecological features of international and national value, with emphasized landscape, educational, cultural and historical as well as tourist and recreational values" (Strčić et al. 1983).

Neretva Delta covers the area of about 20,000 ha, of which 12,000 ha is in the Republic of Croatia, and the rest in the Republic of Bosnia and Herzegovina, and it has been increasingly frequently in the tourist valorization in selective tourism. In the area of the Delta there some of rare remaining European wetland areas enlisted as Ramsar areas and protected by the Barcelona convention as specially protected Mediterranean areas. The valley of Neretva river includes five protected areas with the total surface area of 1,620 ha. These are ornithological reserves Pod gredom, Prud and Orepak, ornithological and ichthyological reserve of Neretva Valley and protected areas Modro oko and Desne lake. Kuti and Parila lakes are in the procedure of being declared ornithological and ichthyological reserves (www.panda.org).

- **Selective forms of nautical tourism on the South Adriatic**

The Adriatic Sea is a relatively closed no part of the Mediterranean Sea. The border of the Adriatic Sea and Ionian Sea are the Strait of Otranto 75 km wide between Italy and Albania. The characteristics of the eastern Adriatic coast is high coast with a mountain range above the sea and exceptional indentedness as the consequence of the sunken mountain relief forms due to melting after the Ice Age, when the sea level elevated for 100 m, so that the peaks of the former mountains became the islands and the valleys bays and straits. This area is considered to be the most attractive and clearest sea surface in Europe, so it is, along with the preserved historic nucleuses, the reason of increasingly intensive development of the nautical tourism, including the round trips of large cruisers (Šamanović, 2002).

A special jewel is Dubrovnik as a magnet with the port in front of which numerous "floating hotels" and anchored. Around Dubrovnik there is a series of natural and historical pearls attracting the navigators, as well as the cruisers. Nearby is Korčula, an attractive island of famous wines, beautiful beaches, specific urbanity radiating the history, as well as the peninsula Pelješac, the home of skilful seamen, wine producers and vineyards, rural settlements and peculiar flora and fauna. Starting from Mljet to Dubrovnik towards Cavtat (Epidaurum), there is a collar of attractive Elaphiti Islands.

Nautical tourism and round trips have a long tradition in this area. Development of this aspect of tourism has been intensifying due to important improvement projects as well as the AII projects. However, the lack of nautical berths is still evident, as well as the lack of modern ports for big cruisers, although Dubrovnik has become the second cruiser port on the Mediterranean and the town of Korčula has become an important cruiser point. Two ACI marinas in this area (Korčula and Dubrovnik) contribute to the advancement of nautical tourism of the South Adriatic, which still lacks the high quality supporting facilities.

Also in accordance with this are the activities in Montenegrin Littoral that has realized intensive investments in the last couple of years as well as the creation of recognizable tourist destinations. These are two probably greatest investments on the Mediterranean in the realization phase that also include nautical facilities.

This is a tourist and business settlement **Astra Montenegro in Budva** and Tivt marina in Boka Kotor Bay next to which a real tourist town **Porto Montenegro** is going to be built. Thee value of the project is estimated at about EUR 600 million and it covers the yacht marina with 800 berths (150 for mega yachts), series of luxury residential, business and tourist buildings, two high class hotels that are constructed by the multinational tourist giant Four Seasons. Here are also the attractive public areas, galleries, squares, nautical museum and numerous other facilities that will give a dimension of an exclusive tourist site to this area (www.nacional.hr).

Huge foreign investment is also provided on the cape of Zavala in Budva with the Hyatt Hotel and luxurious apartments and business facilities. Business and tourist settlement **Astra Montenegro** includes 2 luxury hotels with 600 rooms and 40 elite villas covering 66 thousand square meters.

These investments of mainly foreign capital will raise the level of tourism of Montenegrin destinations that were until today mostly known for the massive bathing tourism. This will also affect the neighbouring observed regions. However, when analyzing these facilities, greater attention should be given to the estimation of accepting capacity and the concept of sustainable development, which the foreign capital is often not ready to accept and which then causes the long-term damage to natural and anthropogenic resources. These problems are also emphasized by numerous round tables within the framework of the AII.

2.3 Tourist policy of the EU regarding sustainable tourism

The European Union is a community of countries in which **strictly specified rules are applied** as well as standards on tourism so that it could improve its quality level and develop in compliance with the wishes of the modern tourist demand. It increasingly prefers new whole-year selective forms of offer and challenges new processes in the management of total tourist destination oriented to the satisfaction of consumer requests, with maximum ecological protection, especially of the coastal area.

Tourist policy of the European Union in the upcoming years will be - according to the forecasts in relation to the past ones - more quality, imaginative and efficient. It will adjust to the current opportunities, which will contribute to the increase of alternative aspects of tourism. The EU in the role of a moderator approximates various interests of the southern host countries (primary growth of tourism) with general interests (avoiding harmful consequences for the environment, respecting cultural heritage, quality of services) by means of elaborated measures of the joint tourist policy.

The EU especially advocates the assurance of a quality and standardized service in tourism, protection of tourists and above all protection of natural and cultural heritage, by setting high ecological standards that can ensure sustainable tourist development and satisfaction of consumers.

Individual features of the tourist policy of relevant EU tourist countries, as well as Croatia and Turkey as soon-to-be members, regarding environmental protection can be seen in the following table 2.

Namely, the measures for the increase of EU potentials in tourism put a special emphasis on the promotion of environmental protection and sustainable development of tourism, more specifically via the policies covering natural and cultural heritage as well as transport and energy.

Strategic goals	Croatia	France	Italy	Greece	Spain	Turkey	Cyprus
Protection of human environment	Eco project and Blue Flag, educati-on, eco services	Special attention paid to the ecologic al aspect of tourism	Care of enviro-nmental protection and prevention of possible ecological incidents	Systematic enviro-nmental protection, stimulation of eco projects	Develo-pment of sustainable tourism, elimination of damages in the enviro-nment caused by tourism	Elimination of pollutants for the region Istanbul-Bosporus	Eco projects in the sphere of pol-lutions, eco tourism

Reference: OECD, Tourism Policy and International Tourism in OECD Member Countries, Paris 1992/96 according to Petrić L. 1998, 152. - Geić, S., 2007.

Table 2. Tourist and ecological strategic goals of Mediterranean countries, current and soon-to-be EU members

In the year of 2000 a new directive was adopted with the following goals in environmental protection, and especially of waters (Water Framework Directive (2000/60/EC):

- protection of all waters, lakes and rivers, underground and coastal waters and ensurance of quality of these waters,
- water management based on water capacity and not on administrative borders by engaging the government of citizens and interest groups,
- prevention of pollution by controlling emissions and setting the quality standards for all waters.

The implementation of the Natura 2000 network is s a part of a nature protection activity. The European Commission especially points out the possibilities for sustainable tourism in the areas covered by projects of Natura 2000. LIFE-Nature projects refer to activities regarding the resources preservation in the areas covered by Natura 2000 projects. The goal of this activity is to prevent uncontrolled tourist activities and to raise awareness regarding natural values of the area.

As early as in 1999 a **document of spatial development ESDP** (European Spatial Development Perspective) was adopted and the ESPON (European Spatial Planning Observatory Network) was established, with the aim of ensuring tools for the implementation of spatial management measures, which directly and indirectly affects tourism by following developmental trends in especially sensitive areas (islands, mountains, sparsely populated areas, coastal zones and similar) and regional management of natural disasters as well as potential technological incidents in general and regarding climatic changes. The document is especially important for the area of the Mediterranean as a most attractive European and world mega region for tourist, which is threatened by different human activities, and therefore by tourism as well.

3. Selected concepts and models for ecosystems management

After several decades of extensive development in tourist travels, with neglecting of accompanying problems and limiting factors in the environment itself, the world **UN Conference on Sustainable Development** (Rio, 1992), by Agenda 21, finally at the highest level promotes the need of taking care of **ecologically rational as well as socially and culturally sustainable tourism**.

It also identifies **three basic means** that must be used for the realization of this goal in all activities that use nature as a resource or are allocated in such a way that they may have impact on the environment. These are:

- Introduction of a **new regulation** in the assurance of environment protection and human health as a whole,
- Introduction of such **market mechanism** that assures that the price of goods and services also includes the price of **environment protection**, taking into consideration the sensitivity and limitation of spatial and water resources and energy around the globe,
- Introduction of **ethical management** in all activities, with attention to **health and ecology**, by integrating them into elements of planning and making developmental decisions, along with enabling the widest dialogue with employees and international public (UN Agenda 21, 1992).

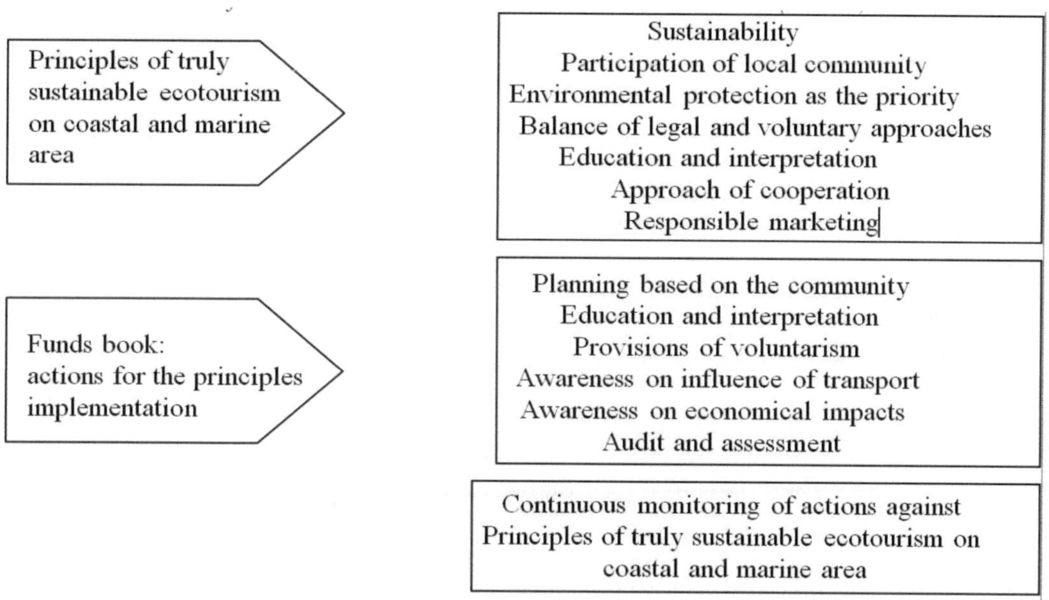

Reference: Group of authors, Planning for Marine Ecotourism in the EU Atlantic Area, UWE, Bristol, 2001.

Scheme 1. Model of ecotourism in coastal and marine environment (META)

Along these lines, the scientists of the Bristol University West of England have been elaborating the **system of META** (Transnational Model for Marine Ecotourism) for the coastal and marine area of the Atlantic tourist mega region Europe, which specifies the necessary actions of all participants in the segment of developed forms of ecotourism on this coastal and marine area that may also be applied on the Adriatic area.

When developing a program on a limited Adriatic area according to the formula of sustainable tourist development on ethical, ecological and economical principle, it is necessary to take into strict consideration the notion of "**carrying capacity**", which talks about the optimum possibility of tourist construction or optimum valorization of a specific area from the aspect of its physical, emotional, legal or economical limits that depend on the sensitivity of an individual locality. On this level, Hendee, Stankey and Lucas (1990) have elaborated the **LAC PLANNING SYSTEM**, i.e. LAC (Limits to Acceptable Change) model that defines the scope of acceptable changes on a specific locality via four basic principles:

- Specification of acceptable and achievable elements of an individual resource and its social conditions defined by a series of measurable parameters,
- Analysis of the relationship between the acceptable and unacceptable conditions,
- Identification and judgement of management actions necessary to achieve wanted acceptable goals,
- Program of evolution monitoring and control of management effectiveness. These principles are elaborated by the authors in **nine phases** of the LAC model:

1. **Identification of the scope of the problem and subject matter** represents the foundation of each decision-making in the area, especially from the aspect of its tourist valorization.
2. **Definition and description** of all segments of the area that is the subject matter of investors, which includes its immediate and wider environment due to polyfunctional tourist phenomenon that forms a unique area with important interactive functions. In this context the existence of physical plans and plans of social and economical development, including global and national strategy, especially important due to timely elimination of possible incompatible activities on the same areas and their immediate environment. This is especially relevant for the precious insular areas of the Adriatic.
3. **Selection** of numerous quality indicators of **tourist resources**, as well as of very complex **social conditions** on appropriate area, is necessary to identify and quantify forms of tourist development. Namely, they need to be in complete compliance with natural and social environment, which is the prerequisite of a successful long-term option of tourist and general economic development of a specific region.
4. In compliance with the afore stated, it is necessary to **inventorize and classify** all natural and historical attractions that form the resource basis of tourism, as well as to analyze social conditions in which these resources exist. Namely, it is exceptionally important for tourist development to have a number and quality of cultural and natural monuments that are usually categorized as monuments of the world, national, regional and ambiental value and are protected via various forms, in accordance with the legislative of individual countries (national parks, nature reserves, natural monuments...). This also includes provisions of relevant international organizations (UNESCO, UNEP, IUCN, MAB and similar). In the context of **social conditions**, which need to be analyzed and adequately valued, numerous examples of the world attractive localities are known that exactly due to specific social conditions (religious, political and similar motives) that are not at all and cannot be the subject of tourist valorization.

5. The level of tourist valorization of a specific area depends on the set **standards** that a specific area defines in the context of resources protection and valorization. If standardization and appropriate social indicators relevant for tourist development of a specific area do not exist, it is necessary to, in compliance to the LAC model, specify them within the scope of the project task itself and to incorporate them in a concrete investment report, i.e. development program, as an important indicator in making tourist development decisions.
6. With respect to the importance of opinion and attitudes of the local community, which implies the need to timely consult appropriate institutions, the public and the entire population (public insight, referendum and similar), it is necessary to **identify** and offer a few alternative solutions out of which the optimum one will be selected with general acceptance.
7. Within the scope of this procedure it is necessary to specify and foresee all elements and necessary **actions** (financial, technical and similar) by precisely stating environmental impact as well as the plan for each of the offered alternatives.
8. After the **identification and selection of a preferred alternative**, there is a procedure of selecting the most favourable alternative in given conditions, which also implies the appropriate process of decision-making in proper state and local authorities.
9. According to the adopted program (project), the **implementation of appropriate actions** takes place along with constant **monitoring** of the project realization elements.

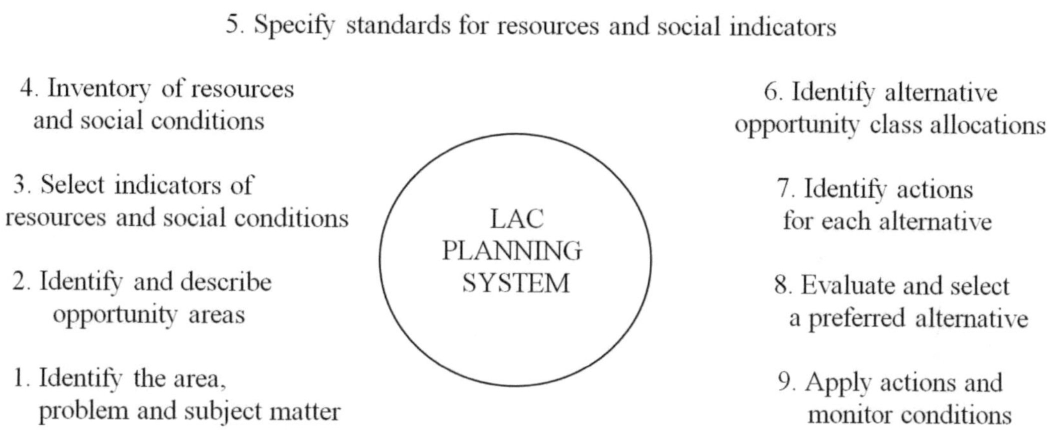

Reference: Hendee, Stankey and Lucas, Wilderness Management, International Wilderness Leadership Foundation, North American Press, Colorado, 1990

Scheme 2. Limits to Acceptable Change (LAC model)

Contemporary British tourist theoreticians that have analyzed the problem matter of sustainable tourism, with a special sensibility to protection of natural, as well as historical and cultural tourist resources, have elaborated an interesting model called **PREPARE** (Policy, Responsibility, Eco-awareness, Programme Audit and Review). This model implies the combination of eighteen activities grouped in six phases and notions implying the segments of tourist policy, responsibility, ecological sensibility, program of activities with needed indicators, registration and constant audit of plans and procedures as well as specification of the reporting procedure for the segment of sustainable tourism. The concept was elaborated on the example of four European cultural and tourist localities: Conwy, Chapstowe (Wales), Naarden (Netherland) and Alcudia (Spain) included in the international association WTFC that has, a later also included Slovenian Piran, obtained the support of the program and fund EU ARCHWAY. By analyzing and quantifying the results in the period from 1993 to 1996, a significant progress of all five localities has been indicated regarding the protection of tourist resources and the complete application of the sustainable tourism concept based on the application of tourist policy in compliance with Rio - Agenda 21.

Application of this model in the Balearic destination Alcudia has shown excellent results regarding the protection of marine and terrestrial ecosystems, resulting in a number of international environmental awards, and the same can be applicable to other Mediterranean regions.

		Portion (%)
P	Prepare: prepare **policies** for sustainability Initial audit of the city tourist offer Identification of sustainable tourism system Policy for sustainable tourism	20
R	pRepare: prepare **responsibility** and resources Responsibilities within local authorities or other public body Resources for measurement of key environment indicators Executive official responsible for policy development	15
E	prEpare: prepare **ecological** awareness/communication and education Communication procedures within and outside local authorities Ecological and ambiental awareness within and outside local authorities Register of relevant documents for operationalization and development of tourism	15
P	prePare: prepare action **programs** with objectives and indicators Register of ambiental indicators for tourism Ambiental objectives Program of improving environmental protection for tourism Local manual of sustainable tourism of the local tourist industry	30
A	prepAre: prepare the procedure of reports controlling **(audit)** Procedure of controlling tourism sustainability Plan of controlling tourism sustainability Reports for control of tourism sustainability	15
Re	prepaRe: prepare policy **revision** Procedure for audit of sustainable tourism policy Audit of sustainable tourism policy	5

Reference: Bruce, M.D. et al., A Model to Aid the Development of Tourism Policies, THR-UNIS, Great Britain, 2001

Scheme 3. PREPARE – Analytic network with phases and elements

In the same context one may also use the modern model of **PEST analysis** that implies the method of analysing business environment and represents the basis for strategic planning. Its name derives from the acronym for political and legal, economic, social and cultural as well as technical and technological environment, which applied to Adriatic conditions looks as follows:

- *Political and legal environment* – The problem of inconsistent and inefficient county organization of the Republic of Croatia, which is outside historical tradition and breaks the entirety of the region of Dalmatia and the unity of Istria and Kvarner as two traditional historical regions to artificial political formations – counties that are most often the purpose to themselves for the formation of political personnel of the governing structures and obsolete local officials who have lost the elective legitimacy due to their incompetence. In that sense, there is a lack of educated destination managers in tourism, even the conjoint personnel and institutions that are competent for monitoring of marine and coastal ecosystems. For example, no quality institutions with a long tradition, such as Hydrographic Institute in Split and Institute for Health Protection, are prominent here, and maritime guard has not yet been established, while inspection services for monitoring are inefficient and unequipped.
- *Economic environment* – in our conditions it is observable via the problem of monocultural development of the coast towards tourism and speculative trade, while agriculture and fishery have experienced the collapse due to decades of harmful import policy, high exchange rate of kuna and prevailing trade lobby in politics. Shipbuilding with a centuries-old tradition and exceptional references is nowadays also almost collapsing, which leads to even bigger monocultural orientation of economy. Small entrepreneurship is the holder of economy and employment due to tourist development, while huge systems have been destroyed in criminal transformation, which refers even to huge tourist systems and systems in tourist entrepreneurship that could have generated new investments in growth and sustainable development.
- *Social and cultural environment* – shows a shattered cohesion in small tourist places that are nowadays turned into tragicomic political stage with excesses of incompetent and corrupted politicians. Tourist ambitions of populations are reduced to survival from season to season, and instead of own product guests are offered imported goods of foreign retailers. The same can be said for cultural and entertaining attractions that are more numerous, but without a firm program scheme and appropriate quality. Preparation of the season is most often late due to comprehension of tourism as an exclusively seasonal activity, which is wrong with respect to year-round nautical, culturological and natural resources. Rustic areas of the coast should be an attribute and something new in tourist offer, and they are nowadays turned into so called "news-stand culture" and resemble more to flea markets borrowed from eastern cultures, devastating the beauty of the area and traditional culture.
- *Technical and technological environment* – Competitive regions have advantage since they have had exceptional organization for decades as well as the system in utilities, roads and all other technical and technological programs that are requested by modern tourism on land and sea, which is still missing on the Croatian coast, especially including the systems of sea protection and pollution management, which have only started to be solved.

Within the scope of tourist destination management, the concept of "**triple helix**" needs to be noted, which implies harmonized activities of **science, state** and **entrepreneurship** for the purpose of long-term sustainable management of valuable natural and cultural resources, especially in the outspread sphere of modern selective forms of tourism. All of this is exceptionally actualized by the latest global economic and financial crisis that has shaken the foundations of past (prevailingly liberal) economic development. The stated may result in realization of optimum economic and social impacts enabled by tourism in the framework of its polyfunctional social and economic functions and characteristics with minimizing supporting negative manifestation on the spatial and social and economical matrix, especially on the sensitive world's valuable areas of the Adriatic and Mediterranean.

4. Conclusion

The Mediterranean and the Adriatic Sea in their marine and coastal area, based on outstanding natural and cultural attractions, are the most important tourist world mega region that realizes over 30% of global international tourist arrivals and foreign exchange inflows (UNWTO 2010), taking in addition the flow of domestic tourist traffic.

For the purpose of endulging a huge number of over half a billion tourists, a grandiose tourist receptive and utility infrastructure and superstructure on the coast and sea was built, that in conjunction with tourists and local population affects with different types of pollution the degradation of marine ecosystems. They are a valuable resource of current and future tourism development, whose contents and the so-called selective forms of these activities are increasingly oriented toward the sea. These are of course ever more intensive sport and recreational activities followed by numerous complex water-transport systems in the open sea, ports and coastal areas, as well as the ever more intense development of fisheries and mariculture with purpose of doubling the number of consumers during the summer season with its attendant pollution and many other degradational effects.

When added to current huge amounts of pollution from land that follow drainage infrastructure directly or through rivers, often untreated flow into the Mediterranean, the situation is becoming alarming and requires urgent intervention by the appropriate national and international institutions within the UN, EU, Council of Europe, OECD and its specialized institutions and associations.

In this context, since the 70-ies of the 20th century a particularly active programme of the UN Environment Programme (UNEP) which, in the framework of the Barcelona Convention and the Mediterranean Action Plan (MAP) to its many projects and protocols, more successfully regulates the issue of protection and sustainable management of marine ecosystems and the Adriatic. In the framework of these international, then national and regional programs for the protection of marine and terrestrial ecosystems we can point out that they were realized on the developed Mediterranean tourist areas, which in the recent past have been leading to programs of sustainable tourism development, especially in the field of water management and land attractive resources, which is the basis for tourism development.

On that note, the paper critically analyzed and compared projects in several regions of the Mediterranean, which were implemented with the support of international financial institutions (EBRD, IBRD) and organizations (UNESCO, UNEP, IUCN, EU-ESDP) and others. These projects logically use the latest models for the management of coastal and marine systems based on the achievements of contemporary science and practice in the sustainable management of these resources. These are the model of eco-tourism in the coastal and marine area - META, developed by a group of English scientists, then the model of American scientists, which defines the width of the permitted changes in a particular locality through four basic principles - LAC PLANNING SYSTEM, then the model dedicated to the issue of sustainable tourism, with special emphasis on the protection of natural and historical and cultural tourism resources - PREPARE, which was developed by a group of scientists in the WTFC, supported by the European Union, and finally the PEST analysis and the "triple helix" activities that involve complex environment of tourist destinations in the affirmation of the role of science, government structures and of entrepreneurship. This is the case with programs in the implementation phase of the Croatian Adriatic (Eco-Kastela Bay, Adriatic Project) regarding pollution control programs for the protection of marine systems with a comparison to several complementary systems in the Mediterranean region (Côte d'Azur, Rhodes, Antalya).

5. References

Andročec V. et al., Projekt Eko- kaštelanski zaljev, Građevinar 55/7,377-381,Zg, 2003

Batle J, Robeldo M.A., Rethinking tourism in the Balearic Islands, Annals of Tourism Research, Vol. 27, No. 4, Elsevier, 2000.

Binan, C., Sea Accidents in the Bosphourus, Loss of Cultural space, Habitat II City Summit, Istanbul, 1996.

Bruce, M.D. i dr., A model to aid the development of tourism policies, THR-UNIS, Velika Britanija, 2001.

Cavana, M., Projecto ecoloislas, Conferencia Mundial de Turismo Sostenible, Lanzarote, 1995.

CIT—Centres Insularies du Tourisme, Canaries Date Base, Tenerife, 1995.

Deveci, A. i dr., Antalya a World City, Antalya Beyeksehir Belediyesi, Turkey, 1996.

Echenique, J., Holidays in Lanzarote, Ediciones AM, Madrid, 1993.

Eko Kaštelanski zaljev, Inegralni ekološki projekt Kaštelanski zaljev, Split 1996.,2.

Faige M., Sustainable tourism in Mediterranean, CMTS, Lanzarote, Spain, 1995

Geić, S., Turizam i kulturno civilizacijsko naslijeđe, Veleučilište u Splitu, Split, 2002.

Geić, S., Organizacije ekoturizma Kostarike,, Acta turistica, EF, Zagreb, 16/2004.

Geić, S., Organizacija i politika turizma, Sveučilište u Splitu, Split, 2007.

Geić, S., Maršić, N., Geić, J.: Problems of water management in Croatian and regions of the Mediterranean, 8. International Conference Challlenges of Europe, University of Split , Bol, 2009.

Geić, S., Voloder, F.: Turist cooperation of South Adriatic border Regions as a part of AII, 7. International Conference, University of Rijeka, Ljubljana, Antwerp - Opatija 2009.

Geić, S. et al.: Upravljanje vodama u hrvatskim turističkim regijama, Naše more, 57/2010, Sveučilište u Dubrovniku, 2010.

Grupa autora, Planning for marine ecotoutrism in the Atlantic area, UWE, Bristol, 2001.

Hendee, Stankey and Lucas, Wilderness Management, International Wilderness Leadership Foundation, North American Press, Colorado, 1990.

Ivanović M. et al., Studija o utjecaju na okoliš sustava Eko kaštelanski zaljev (Podsustav Trogir, Kaštela), Rijeka, 2002.

Kayser, B., Campagnes et Villes de la Côte d' Azure, essai sur les consëquences du dëveloppement urbain, Monaco, 1960.
Marin, C., Lanzarote Island of Volcanoes, Insula No. 2., Malta, 1993.
Merler, A., Transboundary Cooperation betwen Corsica and Sardinia, Insula br. 2, Paris, 1993.
Mediterranean Action Plan/MAP, Carrying Capacity Assesment for Tourism of Central-Eastern Part of Rhodes, PAP - RAC, Split, 1993.
MIT—Ministerio de Information y Turismo, Balerishe Inseln, ARO Madrid, 1976.
Mohamed VI., Message for participants of FIJET Congress, Marraquech, 2004.
OTV, Waste Water Treatment Plant, Antibes, Monaco, Toulon, 1996.,1997.,1998.
OTV, Mediterrane amphitra, L'usine de Depolluation des laux du Cap Sicie, Toullon, 1997.
Rita, J. i dr., Menorca Biosphere Reserve, Insula No. 2, Malta, 1993.
Strčić P. et al., Jadranski otoci Jugoslavije, Poslovna politika, Beograd 1983.
Šamanović, J., Nautički turizam i management marina. Split: Visoka pomorska škola, 2002. UNEP
Tamorri M. et al, Hic Rodus, hic Saltus, Insula br. 2 , Malta 1993.
UNEP, Mediterranean Action Plan-Rodos-PAP-RAC, Split, 1993.
UNIST & UNEP, Integral ecological project of the region of Split, Split., 1993.

5.1 Internet pages

http://mkt.unwto.org/sites/all/files/docpdf/unwtohighlights11enhr_0.pdf
www.blueflag.org (službene stranice Blue flag asocijacije)
www.donjahercegovina
www.ekz.hr (Eko kaštelanski zaljev , 2009 god)
www.hvjp.hr (Hrvatske vode - Jadranski projekt)
www.jadranski.projekt.hr
www.nparkovi.cg.gv
www.seadriatic.net/aii (službene stranice Jadransko-jonske inicijative)
www.unep.org (United Nations Ennvironment Programme)
www.unep.org/pdf/un-list-protected-areas.pdf, UNEP 2003.
www.UNWTO Tourism higlights 2010., službene stranice UNWTO
www.uniadrion.net, Virtual University oft he Adriatic